Hydrology and Water Resources Management in a Changing World

In Focus – Special Book Series

Hydrology and Water Resources Management in a Changing World

Editors
Kolbjørn Engeland and Knut Alfredsen

IWA PUBLISHING

Published by **IWA Publishing**
Republic – Export Building, 1st Floor
2 Clove Crescent
London E14 2BE, UK
Telephone: +44 (0)20 7654 5500
Fax: +44 (0)20 7654 5555
Email: publications@iwap.co.uk
Web: www.iwapublishing.com

First published 2020

British Library Cataloguing in Publication Data
A CIP catalogue record for this book is available from the British Library

ISBN: 9781789062168 (paperback)
ISBN: 9781789062175 (eBook)

Contents

Editorial: Hydrology and water resources management in a changing world

The XXX Nordic Hydrological Conference (NHC) was held in Bergen with 145 participants. The conference was organized by the Norwegian Hydrological Council in cooperation with the Nordic Hydrological Association (NHF). A total of 114 contributions were submitted from 24 countries, of which 30 were given by master or PhD students. In addition, the five invited speakers gave presentations at the conference. Most of the contributions came from Nordic and Baltic countries which constitute the NHF's area of activity. The main theme of the conference was 'Hydrology and water resources management in a changing world'. This theme reflects important challenges for both researchers and practitioners in the public and private sectors. Sub-topics were: water in urban areas; groundwater; floods; climate services; hydrological processes and models; hydropower; water consumption; environmental impact and water quality.

As part of the conference, we had a tour of 'Bryggen', which gave us insight into how blue-green solutions were used to save this UNESCO world heritage site. Rory Dunlop from NIKU, Guri Venvik from NGU and Floris Boogaard from Tauw have worked for several years preserving 'Bryggen', and gave insightful scientific guidance to the challenges and solutions.

After the conference, authors were given the opportunity to submit their original contributions for publication in a special issue of *Hydrology Research*. After a rigorous peer-review process, 15 papers were accepted for publication in this special issue covering a range of the conference themes outlined above.

Six of the papers address advanced methods and technology in hydrological modelling. Hydrological modelling is instrumental both for scientific application and for providing public services, and the papers present algorithms, tools, platforms and systems used within modelling different aspects of the hydrological cycle. To better simulate peak flows with short time resolution, Tsegaw *et al.* (2020) extend the DDD-model (Skaugen & Onof 2014) by introducing a dynamic river network that depends on the catchment saturation.

Correct classification of precipitation phase is important for hydrological modelling in Nordic and high altitude areas. Both Grigg *et al.* (2020) and Feiccabrino (2020) investigate approaches for this classification. Grigg *et al.* (2020) present three methods for utilising physiographic catchment descriptors in precipitation phase determination for 169 meteorological stations in Norway and Sweden. Feiccabrino (2020) estimates uncertainty in precipitation phase models using thresholds of air temperature, dew-point temperature and wet-bulb temperature. The analysis was done for time of observations and averaged data for readings of different time intervals. It is shown that the best option is to use wet-bulb temperature or reduce the time-step from 24 hours to 3 or 1 hours.

The availability of input data is often a limitation for precipitation-runoff modelling. Both the density and representativity of the gauging stations is challenging in mountainous areas with large precipitation gradients and most of the gauges located at low altitudes. Sivasubramaniam *et al.* (2020) compare the use of temperature and precipitation from observations and from meteorological models for several catchments located in mid-Norway. It is shown that using input data from meteorological models is almost as good as, and in some cases better than, using gauge data. Bhattarai *et al.* (2020) compare different global forcing datasets for hydrological modelling of the Narayani catchment in Nepal. The paper evaluates the variability and uncertainty of discharge in the catchment for the different forcing datasets. The results also show that the global datasets could be highly useful as input for modelling in areas with sparse observations.

Snow is an important component in hydrology in cold climate areas, and observations of snow can be used for improving hydrological forecasts by data assimilation. In Magnusson *et al.* (2020) snow observations are used to update the snow water equivalent, an important state

doi: 10.2166/nh.2020.000

variable for correct prediction of runoff during the snowmelt period. The method is tested in a streamflow forecasting framework.

Hydrologic indicators are often used to describe changes to the hydrological regime due to changes in anthropogenic or climatic factors, and in the case of a lack of data indicators are estimated based on hydrological models. Massmann (2020) investigates how good models will simulate changes in hydrological indicators based on calibration strategies, catchment properties and properties of the hydrological indicator itself. The studies show that indicators of magnitude are estimated reasonably well, but indicators related to dynamics of the flow regime are problematic.

Climate change is impacting the high latitudes more rapidly and significantly than any other region of the Earth, and two of the papers address climate change impact on streamflow. Akstinas *et al.* (2020b) combine three climate models with different emission scenarios with a statistical downscaling approach to estimate uncertainty in runoff projections in Lithuanian rivers. It is shown that the global climate models (GCMs) are the dominant uncertainty source in the runoff projections.

Zhang *et al.* (2020) use the VIC-glacier model to study the contribution of melt water and rain water to streamflow of two basins on the Tibetan Plateau. The model is calibrated using streamflow, glacier mass balance data and MODIS snow, and used to study changes in streamflow between the baseline period (1964–1990) and the 1990s and 2000s. An increasing trend in runoff during the past 50 years is shown, mainly explained by increasing rainfall, but also some increase in glacier melt.

Floods are a major issue in hydrological analysis, and are the main topic of two of the papers. The papers present contrasting flood types, covering both a rainfall driven flood in a small catchment and flood in a large basin driven by rainfall and snow melt. Extreme localized precipitation with short duration is expected to become more frequent in the future, and Bruland (2020) estimates extreme unit discharge in a small catchment after such a flood event in Storelva in Utvik, Norway. A common problem for many analyses like this is lack of data, and the paper presents several different approaches to estimate the flood magnitude using hydrological and hydraulic analysis. Akstinas *et al.* (2020a) analyze the causes and effects of two large flood episodes in the Nemunas river basin using data analysis and hydrological modelling. The findings show that snow accumulation was the main factor of the formation of large floods and it also identifies the subcatchments with the largest contributions to the flood volume.

Four of the papers address surface water, groundwater and blue-green solutions in urban areas. Handling surface water in cities is a challenge. With increasing intensity of rainfall for short periods, the challenges will also increase. Events in Copenhagen and elsewhere in Northern Europe in early June 2013 are examples of challenges with headlines like ‘Rain showers in the cities cost far more than floods’ (*Aftenposten*, June 2, 2013). Oslo and other major cities have similar events. Surface water and groundwater must be handled in a comprehensive and not separate manner in order to handle the climate change of the future in a cost-effective and sustainable manner. The drainage and piping of today is not designed for the rainfall of the future, and it is also expensive to clean rainwater that is taken along with graywater. Alternative solutions for surface water management, tailored to future precipitation forecasts and ‘in harmony’ with groundwater, will provide solutions that can withstand heavy rainfall episodes and save society the high costs. For coastal towns and cities, both increasing precipitation and rise of sea level is a potential threat to society. Venvik *et al.* (2020) present a study from the city of Bergen where effects of subsidence and floods are studied to establish risk hazard maps for the city. Through these maps subsidence mitigation and flood water management can be planned. Green roofs are one of the common green strategies for stormwater management. Schärer *et al.* (2020) investigate the suitability of using runoff coefficients for designing green roofs. Based on laboratory and field studies, the applicability and magnitude of runoff coefficients for different types of detention based roofs are assessed. Bian *et al.* (2020) study the flood responses related to changes in precipitation and urbanisation for the Qinhuai basin in China. The catchment has seen a marked development in urbanisation since the start of the 2000s. Based on trend tests on flood data, the impact of urbanization is evaluated by studying periods before and after urbanisation of the catchment. Modelling urban flooding to estimate inundation depths is important in planning urban infrastructure. Li *et al.* (2020) use the SIMWE model to model overland flood for different design rainfall events and spatial resolutions over Oslo to estimate the overland flood. The paper presents the effect the rainfall input and spatial resolution have on flooded areas and the predicted water depths.

ACKNOWLEDGEMENTS

We would like to thank all authors for their interest in *Hydrology Research*, and the reviewers for their help in the review process, which significantly improved the quality of the papers. We are also grateful for the support and patience of Chong-Yu Xu and Bjørn Kløve, *Hydrology Research* Editors, in the preparation of this issue.

Guest Editors
Kolbjørn Engeland
The Norwegian Water Resources and Energy Directorate,
P.O. Box 5091 Majorstua, Oslo N-0301, Norway.
E-mail: *koe@nve.no*

Knut Alfredsen
Department of Civil and Envioronmental Engineering,
The Norwegian University of Science and Technology,
7491 Trondheim, Norway.
E-mail: *knut.alfredsen@ntnu.no*

REFERENCES

Akstinas, V., Meilutytė -Lukauskienė, D., Kriauciunienė, J. & Šarauskienė, D. 2020a Features and causes of catastrophic floods in the Nemunas River basin. *Hydrology Research* **51** (2), 308–321.

Akstinas, V., Jakimavicius, D., Meilutytė -Lukauskienė, D., Kriauciunienė, J. & Šarauskienė, D. 2020b Uncertainty of annual runoff projections in Lithuanian rivers under a future climate. *Hydrology Research* **51** (2), 257–271.

Bhattarai, B. C., Burkhart, J. F., Tallaksen, L. M., Xu, C.-Y. & Matt, F. N. 2020 Evaluation of global forcing datasets for hydropower inflow simulation in Nepal. *Hydrology Research* **51** (2), 202–225.

Bian, G., Du, J., Song, M., Xueliang, Z., Xingqi, Z., Li, R., Wu, S., Duan, Z. & Xu, C.-Y. 2020 Detection and attribution of flood responses to precipitation change and urbanization: a case study in Qinhuai River Basin, Southeast China. *Hydrology Research* **51** (2), 351–365.

Bruland, O. 2020 How extreme can unit discharge become in steep Norwegian catchments? *Hydrology Research* **51** (2), 290–307.

Feiccabrino, J. M. 2020 Precipitation phase uncertainty in cold region conceptual models resulting from meteorological forcing time-step intervals. *Hydrology Research* **51** (2), 180–187.

Grigg, L. D., Feiccabrino, J. & Sherenco, F. 2020 Testing the applicability of physiographic classification methods toward improving precipitation phase determination in conceptual models. *Hydrology Research* **51** (2), 169–179.

Li, H., Gao, H., Zhou, Y., Xu, C.-Y., Ortega, R. Z. & Sælthun, N. R. 2020 Usage of SIMWE model to model urban overland flood: a case study in Oslo. *Hydrology Research* **51** (2), 366–380.

Magnusson, J., Nævdal, G., Matt, F., Burkhart, J. F. & Winstral, A. 2020 Improving hydropower inflow forecasts by assimilating snow data. *Hydrology Research* **51** (2), 226–237.

Massmann, C. 2020 Reproducing different types of changes in hydrological indicators with rainfall-runoff models. *Hydrology Research* **51** (2), 238–256.

Schärer, L. A., Busklein, J. O., Sivertsen, E. & Muthanna, T. M. 2020 The limitations in using runoff coefficients for green and gray roof design. *Hydrology Research* **51** (2), 339–350.

Sivasubramaniam, K., Alfredsen, K., Rinde, T. & Sæther, B. 2020 Can model-based data products replace gauge data as input to the hydrological model? *Hydrology Research* **51** (2), 188–201.

Skaugen, T. & Onof, C. 2014 A rainfall-runoff model parameterized from GIS and runoff data. *Hydrological Processes* **28** (15), 4529–4542. doi:10.1002/hyp.9968.

Tsegaw, A. T., Skaugen, T., Alfredsen, K. & Muthanna, T. M. 2020 A dynamic river network method for the prediction of floods using a parsimonious rainfall-runoff model. *Hydrology Research* **51** (2), 146–168.

Venvik, G., Bang-Kittilsen, A. B. & Boogard, F. C. 2020 Risk assessment for areas prone to flooding and subsidence: a case study from Bergen, Western Norway. *Hydrology Research* **51** (2), 322–338.

Zhang, Y., Hao, Z., Xu, C.-Y. & Lai, X. 2020 Response of melt water and rainfall runoff to climate change and their roles in controlling streamflow changes of the two upstream basins over the Tibetan Plateau. *Hydrology Research* **51** (2), 272–289.

A dynamic river network method for the prediction of floods using a parsimonious rainfall-runoff model

Aynalem Tassachew Tsegaw, Thomas Skaugen, Knut Alfredsen and Tone M. Muthanna

ABSTRACT

Floods are one of the major climate-related hazards and cause casualties and substantial damage. Accurate and timely flood forecasting and design flood estimation are important to protect lives and property. The Distance Distribution Dynamic (DDD) is a parsimonious rainfall-runoff model which is being used for flood forecasting at the Norwegian flood forecasting service. The model, like many other models, underestimates floods in many cases. To improve the flood peak prediction, we propose a dynamic river network method into the model. The method is applied for 15 catchments in Norway and tested on 91 flood peaks. The performance of DDD in terms of KGE and BIAS is identical with and without dynamic river network, but the relative error (RE) and mean absolute relative error (MARE) of the simulated flood peaks are improved significantly with the method. The 0.75 and 0.25 quantiles of the RE are reduced from 41% to 23% and from 22% to 1%, respectively. The MARE is reduced from 32.9% to 15.7%. The study results also show that the critical support area is smaller in steep and bare mountain catchments than flat and forested catchments.

Key words | critical flux (FC), critical supporting area (AC), dynamic river network, floods

Aynalem Tassachew Tsegaw (corresponding author)
Knut Alfredsen
Tone M. Muthanna
Civil and Environmental Engineering,
Norwegian University of Science and Technology,
SP Andersen Vei 5, Trondheim 7491,
Norway
E-mail: *aynalem.t.tasachew@ntnu.no*

Thomas Skaugen
Hydrology Department,
Norwegian Water Resources and Energy Directorate (NVE),
PO Box 5091, Oslo 0301,
Norway

INTRODUCTION

Floods are one of the major climate-related hazards and cause casualties and substantial damage on a global scale every year (Hirabayashi *et al.* 2013; Blaikie *et al.* 2014; Winsemius *et al.* 2015). Floods usually cause damage to agricultural land, infrastructure and buildings (Razi *et al.* 2010). Flood peak is one of the most important variables to be estimated as its magnitude and duration are responsible for the damage (Formetta *et al.* 2017; Gao *et al.* 2017). An accurate estimate of flood peak is a critical requirement for proposing appropriate flood damage mitigation measures in order to reduce social and economic costs (Plate 2009).

The common hydrological tools for flood risk management are flood forecasting models and models used to estimate design floods (Plate 2009). The design flood, where the magnitude of the flood is associated with a return period and hence a level of risk, is important in the planning, design and operation of hydraulic structures and for protection of human life and property (Rahman *et al.* 1998; Reis & Stedinger 2005; Smithers 2012). Methods to estimate design floods are generally classified into three: (a) statistical flood frequency analysis; (b) event-based simulation; and (c) derived flood frequency simulation (Filipova *et al.* 2018).

Derived flood frequency analysis, using continuous rainfall-runoff models, is increasing in use for design flood

doi: 10.2166/nh.2019.003

estimation (Cameron *et al.* 2000; Calver & Lamb 1995; Boughton & Droop 2003; Eschenbach *et al.* 2008). A rainfall-runoff model can be used to simulate several flow values under different conditions for extending and enhancing the observed flow record (Filipova *et al.* 2018). A stochastic weather generator is used to simulate long synthetic series of rainfall and temperature input data for the continuous simulation method. The long series of flow data derived from the simulation is then used to estimate the required return periods, usually using plotting positions (Camici *et al.* 2011; Haberlandt & Radtke 2014). There is a growing interest in continuous simulation method of flood estimation as an alternative to event-based method, and internationally the trend is to adopt the continuous method (Lamb & Kay 2004; Chetty & Smithers 2005; Pathiraja *et al.* 2012). The main advantages of the continuous simulation models are their ability to represent the antecedent moisture condition in the catchment and their capability to model future land use and climate changes impacts on the flood peaks (Brocca *et al.* 2011; Smithers *et al.* 2013). The other reason for using the continuous simulation approach is that precipitation records are more widely available and tend to have longer periods of records than stream flow data (Blazkova & Beven 2002). Continuous simulation can avoid the base flow estimation problem in the event-based method and avoid any need to associate return period of the flood with specific design precipitation because the frequency analysis of floods can be done directly.

Rainfall-runoff models are simplified representations of a complex physical system and therefore carry a certain amount of uncertainty in their applications (Bourdin *et al.* 2012). The performance of rainfall-runoff models depend on several factors which include the quality of precipitation input data and an appropriate model structure capable of simulating floods (Collier 2007). Therefore, the structure and performance of the rainfall-runoff models should be evaluated and improved for their capability in simulating flood peaks before using them in design flood estimation and flood forecasting.

There are several ways to classify rainfall-runoff models (Singh 1995). Rainfall-runoff models can be classified into lumped and distributed models. Lumped models consider the whole catchment as a single unit with state variables that represent the average of the catchment (Beven 2001b). Distributed models make prediction at distributed locations, i.e., by discretizing the catchment into a number of elements with state variables representing local averages (Singh & Frevert 2006). When a rainfall-runoff model is used for design flood estimation, the model could underestimate the design flood. Thomas (1982) evaluated floods estimated by continuous simulation methods on 50 small streams in Oklahoma, and the result showed that the flood peaks were consistently underestimated. Pathiraja *et al.* (2012) used 45 catchments in the Murray–Darling basin in Australia to estimate design floods using the Australian water balance model. They found that the model underestimates the floods from 5% to 30% depending on how reasonably the antecedent moisture condition is simulated. The forecast of floods requires an accurate understanding of catchment characteristics and a precise determination of catchment's initial conditions before flooding (Rusjan *et al.* 2009).

There is a link between catchment morphology and a hydrologic response of a catchment (Rodríguez-Iturbe & Valdés 1979; D'Odorico & Rigon 2003; Rigon *et al.* 2011). Gupta *et al.* (1980) pointed out that the Geomorphic Instantaneous Unit Hydrograph (GIUH) is equivalent to the probability density function of travel times, $f(t)$, from any point in the catchment to the outlet. This permits the formulation of hydrologic response through the geomorphologic width function, $W(x)$. The GIUH and $W(x)$ concepts represent the dependency of peak flows on the geomorphological properties of a catchment and provide a quantitative prediction of peak flows for engineering application (Rinaldo *et al.* 1991; Rinaldo *et al.* 1995; D'Odorico & Rigon 2003; Rigon *et al.* 2011). The form and extent of the stream network reflect the characteristics of the hillslope (Willgoose *et al.* 1991). The stream reflects the ground water dynamics and is often termed as perennial, intermittent and ephemeral streams (Dingman 1978; Bencala *et al.* 2011). Dynamic expansions and contractions of stream networks play an important role for hydrologic processes since they connect different parts of the catchment to the outlet (Nhim 2012). Stream networks in a catchment expand and contract as the catchment wets and dries, both seasonally and in response to individual precipitation events, and this dynamic of stream networks gives an important information to the pattern and process of runoff generation (Godsey & Kirchner 2014; Ward *et al.* 2018). The mean of the distribution of distances from a point in the catchment to the nearest river reach (D_m) and the drainage density (D_d) are among the indexes used to describe a stream network. Horton (1945) defined the traditional D_d as the sum of lengths (L) of all streams in a catchment divided by the catchment area (A). The D_m reflects the spatial characteristics that affect the formation of streams and the response time of a catchment for a particular stream network (Wharton 1994; Tucker *et al.* 2001; Di Lazzaro *et al.* 2014; Skaugen & Onof 2014). The mean distance one has to walk from a random location in a hillslope before encountering a stream, D_m is related to the traditional definition of D_d (Horton 1932; Tucker *et al.* 2001; Di Lazzaro *et al.* 2014).

$$D_m \approx \frac{1}{2D_d} \tag{1}$$

Chorley & Morgan (1962) showed that the maximum flow is related to D_m. Day (1983) studied two catchments of New England (NSW, Australia) and found that the D_m is correlated with discharge. In these two catchments, the D_m was found to decrease for an increase in discharge, indicating that the stream network expands during the flooding events. During the expansion and contraction of streams, the critical supporting area (A_c), which is the area needed to initiate and maintain streams, shows variations within a catchment and is an important variable for assessing geomorphometric characteristics (e.g. D_m) (Papageorgaki & Nalbantis 2017). The relationship between D_d and A_c follows an inverse power law (Moglen *et al.* 1998) as shown in Equation (2):

$$D_d = kA_c^{-n} \tag{2}$$

where k and n are positive numbers. If we insert the value of D_d from Equation (2) into Equation (1), we will get a power relationship between D_m and A_c as shown in Equation (3):

$$D_m = aA_c^b \tag{3}$$

where $a = 1/2k$ and $b = n$.

The Distance Distributions Dynamics (DDD) model is a parsimonious continuous rainfall-runoff model with a small number of calibration parameters recently developed by Skaugen & Onof (2014). Many of the model parameters can be estimated from catchment topography using geographic information system (GIS) and recession characteristics. DDD is a semi-distributed model, i.e., lumped in model parameters and distributed input data (precipitation and temperature). The calibration and validation results for 41 small rural unregulated catchments in Norway (area <50 km^2) with hourly data showed that the DDD model, in most cases, underestimated flood peaks (Tsegaw *et al.* 2019). In the runoff dynamics of the DDD model, there is a single static river network forming the basis for the dynamics of water routing through the hillslopes and in the river network (Skaugen & Onof 2014). However, studies show that the river network has a dynamical nature, being more dense during high flows than at low or medium flows (Godsey & Kirchner 2014). The primary objective of this study is to investigate whether including a dynamic river network model into the DDD model will improve flood prediction in small rural catchments (area <50 km^2). The secondary objective is to improve the understanding of the stream development for different vegetation covers, catchment slopes and climate. The secondary objective helps us to assess whether there is a potential to relate a calibration parameter of the dynamic river network routine with the environmental factors so that there is a possibility for regionalizing the parameter.

METHODOLOGY

Study catchments and data

Fifteen gauged small rural catchments, which show significant underestimation of peak floods during the calibration and validation of 41 small catchments located in Norway (Tsegaw *et al.* 2019), are used in this study for testing the dynamic river network method. The catchments are selected from the Norwegian Water Resources and Energy Directorate (NVE) HYDRA II database. Figure 1 shows locations of the study catchments, and Table 1 shows the catchment characteristics. The definition of small catchment follows that of Fleig & Wilson (2013) with an upper area limit of

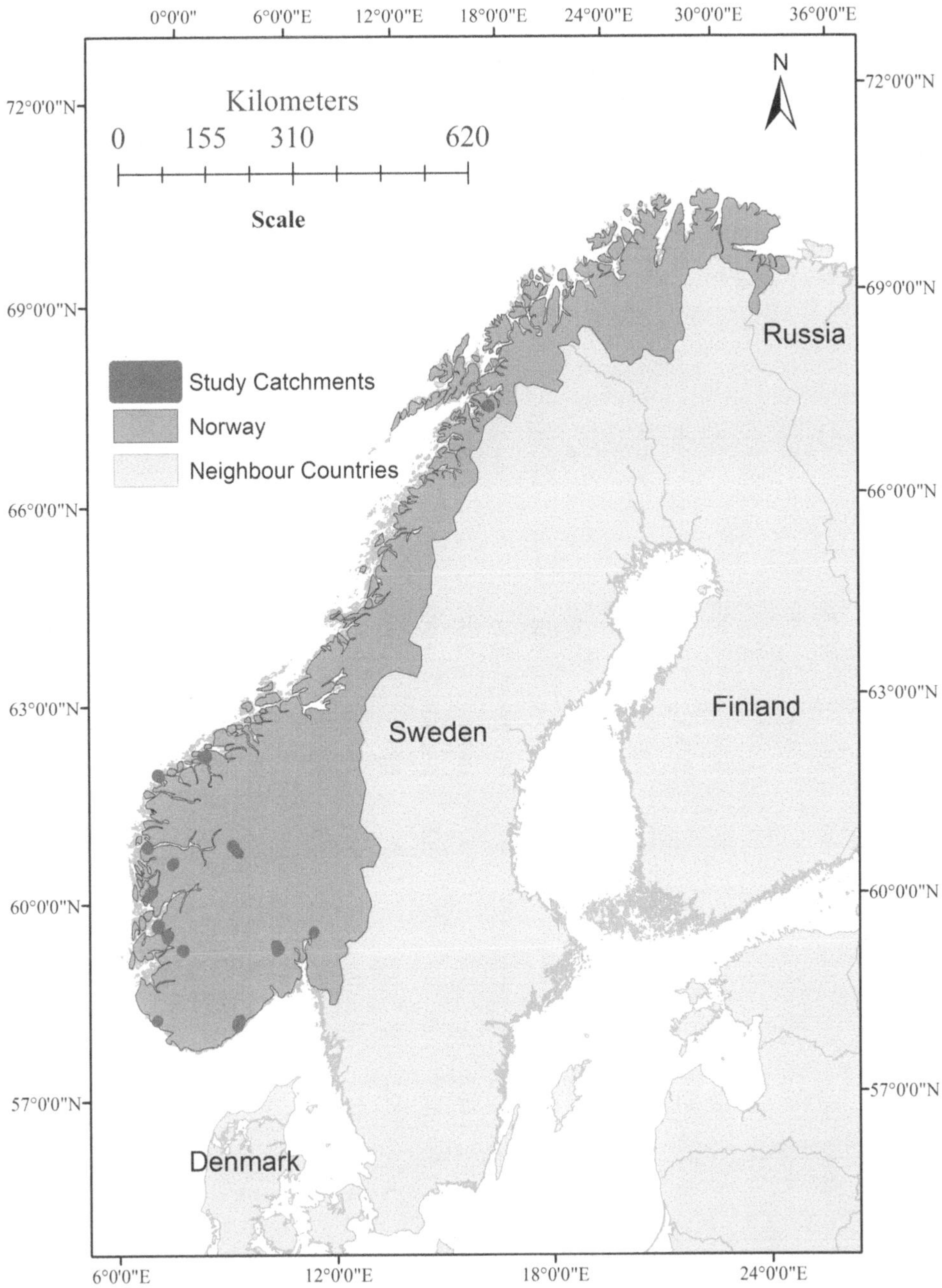

Figure 1 | Locations of study catchments.

50 km^2. We selected where the DDD model had a known history of underestimating floods so that the dynamic river network model could be tested and evaluated.

Precipitation, temperature and discharge are the main input data for running and calibrating the DDD model. We used hourly data of precipitation, temperature and discharge. Precipitation and temperature are based on a 1×1 km gridded product of the Norwegian Meteorological Institute (http://thredds.met.no/thredds/catalog.html) (Lussana *et al.* 2016). We used a total of 5 years of data for calibration and validation. The DDD model uses distributed precipitation and temperature data as input for the model's 10 elevation zones extracted from the hypsographic curve of a catchment. The elevation of the center of each

Table 1 | Catchment characteristics of the study catchments

S.no	Cat_ID	Area (km^2)	Lake (%)	Marsh land (%)	Forest (%)	Bare mountain (%)	Cultivated land (%)	Urban (%)	Mean elvation (m)	Mean annual precipitation (mm)	Mean annual temprature (°C)	Specific runoff ($l/s/km^2$)	Mean hillslope slope (%)
1	6.10	7	2.7	1.6	94.3	0	0	0	302.9	886	4.1	20.6	18.3
2	12.193	50.7	1.3	3	88.4	0	4.3	0.5	306.5	840	3.8	17.5	15.3
3	19.107	41.5	4.8	1.4	86.4	0	4.2	0.8	88.4	1,158	4.2	24.2	14.7
4	26.64	9.7	7.2	2.2	38.8	46.2	1.1	0	203.9	1,688	6.7	45.8	28.3
5	36.32	20.9	2.5	0.9	13.5	81.4	0	0	1,039.4	2,377	0.2	105.2	34.1
6	41.8	27.4	7.7	0.4	8.8	82.2	0	0	836.8	2,955	2.7	126.4	37.5
7	42.2	31.1	2.8	1.4	40.7	52.1	0	0	573.3	2,361	4.8	108.3	40.4
8	55.4	50.6	3.7	1.3	51.8	30.7	2.8	0.2	361.3	2,593	5.4	100.4	41.9
9	63.12	12.8	4.5	0.3	5.8	86.1	0.6	0	886.2	2,579	1.1	94.8	34.4
10	68.2	20.9	4.3	0	20.2	50.3	2.5	0	402.4	2,736	5	125.2	43.6
11	73.21	25.8	7.4	2.2	2.2	88	0	0	1,292.5	946	−0.8	34.7	21.5
12	73.27	30.4	9.1	0.5	0.1	89.4	0	0	1,372	679	−2.5	33.5	14.8
13	91.2	25.8	10.1	3	3.7	66.5	1.3	0	261.9	2,072	6.1	63.5	29.9
14	101.1	40	11.3	1.6	61.3	11.3	6	0.1	232.9	1,704	5.5	54.9	23.9
15	172.8	21.2	10.7	0.5	1.4	82.5	0	0	659	1,465	1	46	17
Minimum		7	1.3	0	0.1	0	0	0	88.4	679	−2.5	17.5	14.7
Maximum		50.7	11.3	3	94.3	89.4	6	0.8	1,372	2,955	6.7	126.4	43.6

temperature and precipitation grid cell has been extracted from the 10×10 m digital elevation model (DEM) of Norway. Discharge data have been obtained from the Norwegian Water Resources and Energy Directorates (NVE) HYDRA II database. The Norwegian Mapping Authority (www.statkart.no) is the source of the topography, observed river network and land use data.

The DDD rainfall-runoff model

The DDD model currently runs operationally with daily and three-hourly time steps at the Norwegian flood forecasting service. It has two main modules: the subsurface and the dynamics of runoff. In DDD, the distribution of distances between points in the catchment and their nearest river reach (distance distributions of a hillslope) is the basis for describing the flow dynamics of the hillslope. The distribution of distances between points in the river network and the outlet forms the basis for describing the flow dynamics of the river network. The hillslope and river flow dynamics of DDD is hence described by unit hydrographs (UHs) derived from distance distributions from a GIS and celerity derived from recession analysis (Skaugen & Onof 2014; Skaugen & Mengistu 2016). When the distance distributions are associated with flow celerity of the hillslope and rivers, we obtain the distributions of travel times which constitute the time area concentration curve (Maidment 1993). The derivative of the time area concentration curve gives the instantaneous UH (Bras 1990), which is basically a set of weights distributing the input (precipitation and snowmelt) in time to the outlet.

Subsurface

The volume capacity of the subsurface water reservoir, M (mm), is shared between a saturated zone with volume S (mm) and an unsaturated zone with volume D (mm). If the volume of the saturated zone is high, the unsaturated volume has to be correspondingly small (Skaugen & Onof 2014; Skaugen & Mengistu 2016). The actual water volume present in the unsaturated zone is described as Z (mm). The subsurface state variables are updated after evaluating whether the current soil moisture, $Z(t)$, together with the input of rain and snowmelt, $G(t)$, represent an excess of water over the field capacity, R, which is fixed at 30% ($R = 0.3$) of $D(t)$ (Skaugen & Onof 2014). If $G(t) + Z(t) > R^*D(t)$, then the excess water $X(t)$ is added to $S(t)$.

Excess water (mm/h)

$$X(t) = \text{Max}\left\{\frac{G(t) + Z(t)}{D(t)} - R,\ 0\right\}D(t) \tag{4}$$

Ground water (mm/h) $$\frac{dS}{dt} = X(t) - Q(t) \tag{5}$$

Soil water content (mm/h) $$\frac{dZ}{dt} = G(t) - X(t) - E_a(t) \tag{6}$$

Soil water zone (mm/h) $$\frac{dD}{dt} = -\frac{dS}{dt} \tag{7}$$

Potential evapotranspiration (mm/h) $$E_p = Cea * T \tag{8}$$

Actual evapotranspiration (mm/h) $$E_a = E_p * \frac{S + Z}{M} \tag{9}$$

$Q(t)$ is runoff, and $E_a(t)$ is the actual evapotranspiration which is estimated as a function of potential evapotranspiration and the level of storage. *Cea* is a degree hour factor which is positive for positive temperature (T) and zero for negative temperature. E_a is drawn from Z. The degree hour approach is a simplification but experiences from Skaugen & Onof (2014) show that the evapotranspiration routine in DDD calculates similar values to the approach used in the well-known rainfall-runoff model HBV (Bergström 1976). A recession analysis of observed runoff from the catchment is used to estimate the catchment-scale fluctuations of storage and the capacity of the subsurface water reservoir (M) (see Skaugen & Mengistu 2016).

Runoff dynamics

The dynamics of runoff in DDD has been derived from the catchment features using a GIS combined with runoff recession analysis. The method for describing the runoff dynamics of a catchment is built on the distance distribution derived from the catchment topography. The distances from the points in the catchment to the nearest river reach are calculated as Euclidean distance for the marsh and soil parts of a hillslope. Previous studies in more than 120 catchments in

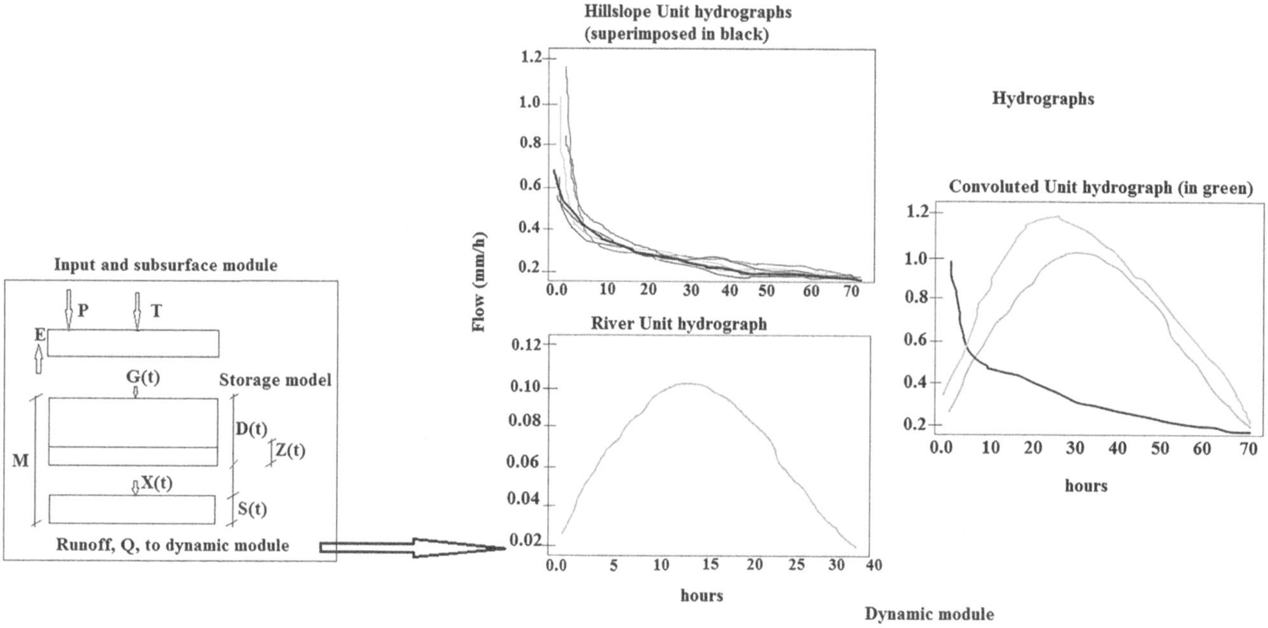

Figure 2 | Structure of distance distribution dynamics (DDD) rainfall-runoff model. Please refer to the online version of this paper to see this figure in colour: http://dx.doi.org/10.2166/nh.2019.003.

Norway showed that the exponential distribution describes the hillslope distance (Euclidean distance from the nearest river reach) distribution well, and the normal distribution describes well the distances between points in the river network and outlet of a catchment (Skaugen & Onof 2014). Figure 2 shows the structure of the DDD model. The model is written in the R programming language. All GIS work is done with ArcGIS 10.3 (ESRI 2014), and the recession analysis is done using an R script (R Core Team 2017).

Water is conveyed through the soils to the river network by waves with celerity determined by the actual storage, $S(t)$, in the catchment (Skaugen & Onof 2014; Skaugen & Mengistu 2016). The celerity associated with the different levels of subsurface storage is estimated by assuming exponential recessions with parameter Λ in the equation $Q(t) = Q_0 \Lambda e^{-\Lambda(t-t_0)}$, where Q_0 is the peak discharge immediately before the recession starts. Λ is the slope per Δt of the recession in the log–log space.

$$\Lambda(t) = \log(Q(t)) - log(Q(t + \Delta t)) \tag{10}$$

The distribution of Λ is modeled using a two-parameter gamma distribution.

The celerity, v, is calculated as a function of Λ using Equation (11):

$$v = \frac{\Lambda \ D_m}{\Delta t} \tag{11}$$

where, D_m is the mean of the distances from points in the catchment (hillslope) to the nearest river. The capacity of the subsurface reservoir M (mm) is divided into five storage levels, i, corresponding to the quantiles of the distribution of Λ under the assumption that the higher the storage, the higher the value of Λ. Each storage level is further assigned a celerity $v_i = \frac{\lambda_i \ D_m}{\Delta t}$ (see Equation (11)), where λ_i is the parameter of the UH for the individual storage level i, and estimated such that the runoff from several storage levels will give a UH equal to the exponential UH with a parameter Λ_i. With the assumption that the recession and its distribution carries information on the distribution of catchment-scale storage, we can consider that the temporal distribution of catchment-scale storage, $S(t)$, is a scaled version to that of Λ. $S(t)$ is calculated using Equation (12), and its distribution is modeled using a two-parameter gamma distribution.

$$S(t) = \frac{Q(t)}{1 - \ e^{-\Lambda(t)}} \tag{12}$$

The DDD model has five storage levels ($i = 1, \ldots, 5$). Four storage levels are subsurface level, whereas the fifth one is an overland flow level with unlimited capacity (Skaugen & Onof 2014; Skaugen & Mengistu 2016). The five levels have five-UHs (four for subsurface flow and one for overland flow) and each of them has different temporal scales as they have been assigned different celerities. The UH is modeled as follows:

$$\mathrm{UH}_i(t) = \lambda_i \mathrm{e}^{-\lambda_i (t - t_o)} \tag{13}$$

where t_o is the time of input, and λ_i is the parameter of the exponential distribution estimated from recession analysis for each level, i.

Model parameters and calibration

The model parameters are divided into three main groups. The first group are those estimated from observed hydro-meteorological data (Table 2), the second group are those estimated by model calibration against observed discharge (Table 3), and the third group are those estimated from digitized maps using a GIS (Table 4). The snow routine in DDD has two parameters estimated from the spatial distribution of observed precipitation data (Skaugen & Weltzien 2016). The shape parameter (a_0) and the decorrelation length (d) of the gamma distribution of snow and snow water equivalent (SWE) are estimated from a previous calibration for 84 catchments in Norway (Skaugen *et al.* 2015). The calibration of the model is performed using the probability particle swarm optimization (PPSO) algorithm (Lu & Han 2011). The Kling–Gupta efficiency criteria (KGE) have been used as objective function for the calibration (Gupta *et al.* 2009), and we used KGE, the BIAS (ratio of the mean of simulated to observed discharge) and visual inspection of hydrographs to evaluate the performance of the model.

Dynamic river network routine

We introduce a dynamic river network concept into the DDD model so that the scale of the overland unit hydrograph (OUH) will be dynamic while keeping the four

Table 2 | List of DDD rainfall-runoff model parameters estimated from observed hydro-meteorological data

Parameters	Description of the parameter	Method of estimation	Unit
d	Parameter for spatial distribution of SWE, decorrelation length	From spatial distribution of observed precipitation	Positive real number
a0	Parameter for spatial distribution of SWE, shape parameter	From spatial distribution of observed precipitation	Positive real number
MAD	Long-term mean annual discharge	From long-term observed mean annual flow data	m^3/s
Gshape	Shape parameter of λ	Recession analysis of observed runoff	Positive real number
Gscale	Scale parameter of λ	Recession analysis of observed runoff	Positive real number
GshInt	Shape parameter of Λ	Recession analysis of observed runoff	Positive real number
GscInt	Scale parameter of Λ	Recession analysis of observed runoff	Positive real number

Table 3 | List of DDD rainfall-runoff model parameters needing calibration

Parameters	Description of the parameter	Method of estimation	Unit	Intervals of calibration
pro	Liquid water in snow	Calibration	fraction	0.03–0.1
cx	Degree hour factor for snow melt	Calibration	mm $^\circ C^{-1}$ h^{-1}	0.05–1.0
CFR	Degree hour factor for refreezing	Calibration	mm $^\circ C^{-1}$ h^{-1}	0.001–0.01
Cea	Degree hour factor for evapotranspiration	Calibration	mm $^\circ C^{-1}$ h^{-1}	0.01–0.1
rv	Celerity for river flow	Calibration	m/s	0.5–1.5

Table 4 | List of DDD rainfall-runoff model parameters estimated from geographical data using GIS

Symbol of parameters	Description of the parameter
Area	Catchment area
maxLbog	Maximum distance of marsh land portion of hillslope
midLbog	Mean distance of marsh land portion of hillslope
bogfrac	Areal fraction of marsh land from the total land uses
zsoil	Areal fraction of DD for soils (what area with distance zero to the river)
zbog	Areal fraction of distance distribution for marsh land (what area with distance zero to the river)
midFl	Mean distance (from distance distribution) for river network
stdFL	Standard deviation of distance (from distance distribution) for river network
maxFL	Maximum distance (from distance distribution) for river network
maxDl	Maximum distance (from distance distribution) of non-marsh land (soils) of hill slope
midDL	Mean distance (from distance distribution) of non-marsh land (soils) of hill slope
midGl	Mean distance (from distance distribution) for glaciers
stdGl	Standard deviation of distance (from distance distribution) for glaciers
maxGl	Maximum distance (from distance distribution) for glaciers
Hypsographic curve	11 values describing the quantiles 0, 10, 20, 30, 40, 50, 60, 70, 80, 90, 100

subsurface UHs constant during the simulation period. The methodology, we used in estimating the dynamic OUHs of a hillslope, is similar to that of GIUH and of the width function, i.e., the travel time probability density function of a unit amount of water draining from a catchment. However, the approaches used in estimating the parameters of the distribution are different, i.e., the approach in calculating the celerity and distances of a flow from the points in the hillslope to the nearest river reach. Further, we assumed that the scale of the travel time distribution in a hillslope is dynamic for generating dynamic OUHs while the shape is held constant. In DDD, the dynamic OUHs are turned on and off according to saturation of the subsurface thus giving a dynamic travel time distribution.

The river network indicates where the subsurface water flow becomes surface water flow. The network system governs the dynamics of runoff for conditions where we have no overland flow from the hillslope in that there is a significant (orders of magnitude) difference in water celerity for flow through the soils and flow in the river network (Robinson *et al.* 1995). In case of overland flow, however, we can imagine a dynamic river network (and hence dynamic distance distributions) as a function of overland flow (OF). We made three assumptions to derive such algorithm.

1. The mean celerity of the overland flow (v_{OF}) is constant, i.e., independent of the subsurface saturation and river network.
2. The overland flow unit hydrograph (OUH) is exponential determined from D_m and v_{OF}.
3. The D_m of a river network is a function of volume of water per area per unit time, i.e., OF. If we assume a critical flux, F_c of 10 m^3/h is necessary to create a stream, then OF of 10 mm for an hour over $A_c = 1000\ m^2$, will provide such a flux, whereas the same flux is obtained for OF of 100 mm over 100 m^2. The two cases have different critical supporting area (A_c), and these cases will provide us with two different river networks where the latter has smaller D_m than the former.

The physical mechanisms underpinning the above three assumptions are:

1. The variable contributions of saturation excess overland flow of a hillslope develops along the existing river network following the concept of Dunne's overland flow (Dunne 1978).
2. The critical supporting area (A_c defines the minimum catchment area from which the generated runoff is sufficient to initiate and maintain river development

(Schaefer *et al.* 1979)). The expansion and contraction of the stream network is governed by the amount of saturation excess overland flow.

3. The hillslope travel time probability density function of overland flow is estimated from the distance distributions at any point from the hillslope to the river network and the celerity of flow in the hillslope (D'Odorico & Rigon 2003; Rigon *et al.* 2011).

In order to compute the OUH, we need the mean (D_m) and the maximum (D_{max}) of the hillslope distance distribution and the mean overland flow celerity, v_{OF}. Using assumption (3), we can derive a dynamic A_c after introducing a critical flux (F_c) as shown in Equation (14), which needs to be determined.

$$F_c\ (m^3/h) = A_c(\mathrm{m}^2) * \mathrm{OF}\ (\mathrm{m/h}) \tag{14}$$

where OF is saturation excess overland flow and is estimated from the DDD model output at each simulation time step. When the subsurface is saturated and there is overland flow (OF > 0), the dynamic river network subroutine is activated and the corresponding A_c will be calculated in the model using Equation (14).

We need to compute the coefficients *a* and *b* of the general power relation between D_m and A_c of each of the study catchments (see Equation (3)) for computing a dynamic D_m during simulation. For computing *a* and *b*, we have used the following procedure:

1. The 10×10 m DEM of a catchment has been reconditioned to the observed river network using the DEM reconditioning tool from Arc Hydro and a raster flow accumulation map has been prepared using GIS.
2. We wrote a python script that can loop through several thresholds of flow accumulation (A_c) to define stream and create several stream networks. From the distance distributions derived from each stream networks, the D_m is calculated.
3. A regression curve is fitted to the synthetically derived A_c and D_m of a catchment to estimate *a* and *b* (Figure 3 shows the fitted curves for six sample catchments). The values of *a* and *b* are unique for each catchment and are listed in Table 5 for all study catchments.

After we have obtained the coefficients of the relationship between A_c and D_m, the A_c estimated from Equation (14) will be used to estimate D_m using Equation (3). We estimated D_{mx} from the exponential distribution with parameter D_m, as a distance where 99% of the catchment area is accounted for. From the recession analysis for estimating celerities, we already have an estimate of v_{OF} in the DDD model (Skaugen & Onof 2014). We estimate a dynamic OUH for every time step when overland flow is estimated.

When the D_m calculated using Equation (3) is greater than the D_m of the observed river network, the dynamic river network degenerates to the observed river network. The observed river network is the basis network for all cases where the subsurface capacity is unsaturated, saturated but no overland flow and when there is overland flow but not sufficient to expand the observed (existing) stream network. When the subsurface capacity is saturated and there is sufficient OF, the observed stream network starts to expand. The extent of expansion is determined by the magnitude of the estimated OF and F_c.

We have tested the performance of the DDD model with and without the dynamic river network routine. We calibrated and validated the DDD model as described in the model parameter and calibration section, and we implemented the dynamic river routine into the model and calibrated F_c. We have calibrated F_c manually after calibrating automatically DDD parameters without dynamic river network. The procedures we have followed in calibrating are as follows:

1. The F_c parameter is adjusted by trial and error to fit the observed flood peaks, which had been underestimated by DDD without dynamic river network.
2. We have visually compared the observed flood hydrographs and flood hydrographs simulated with and without dynamic river network.
3. While calibrating DDD with F_c (with dynamic river network module), the KGE and BIAS values obtained should not be less than the KGE and BIAS values of DDD without dynamic river network method (earlier calibration result).
4. Using the visual inspection of observed hydrographs, KGE and BIAS, the F_c which fits the observed flood peaks well is taken as a calibrated value.

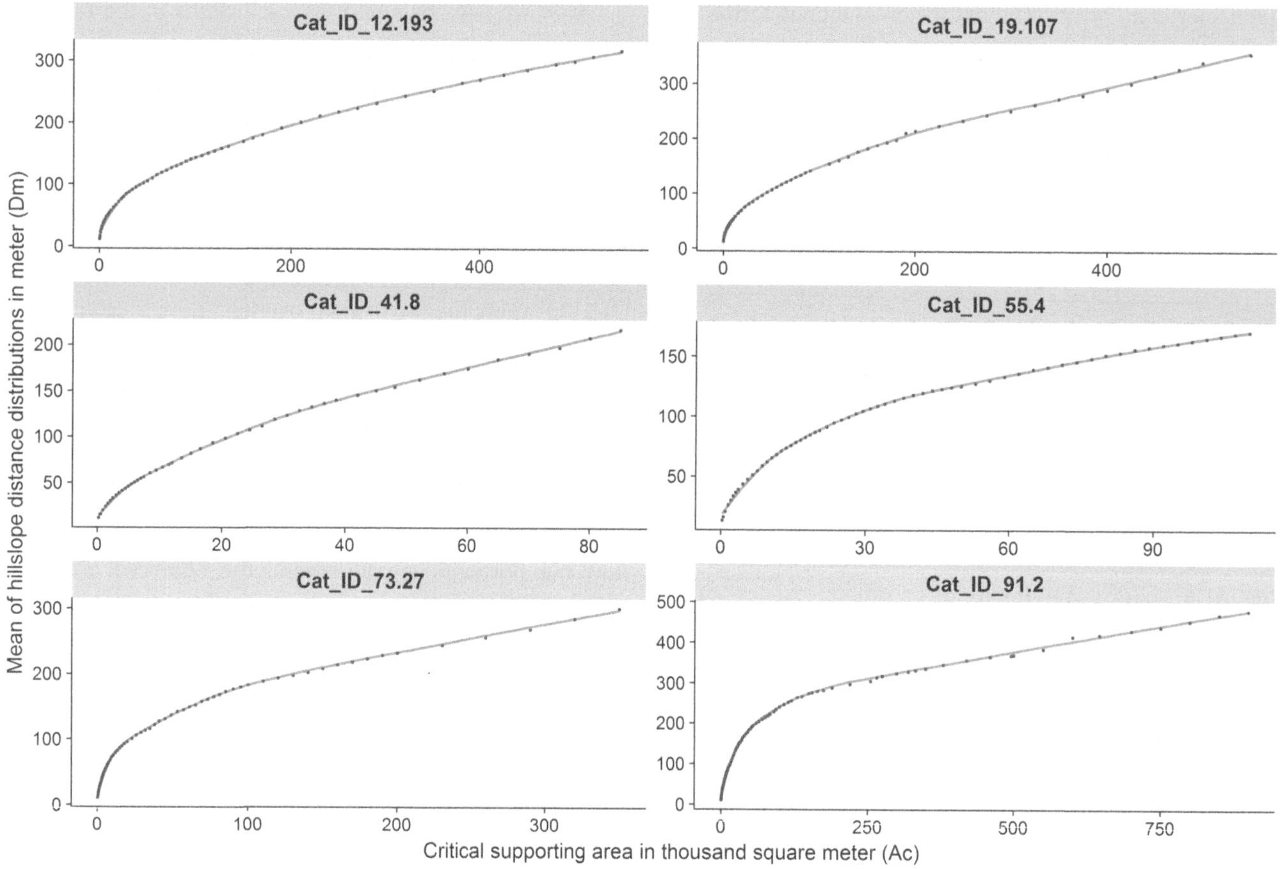

Figure 3 | Curves fitted to the relation between mean distance distribution of hillslope, i.e., D_m and critical supporting area, i.e., A_c for six sample study catchments with a relation, $D_m = aA_c^b$.

As for the previous case, we used KGE, BIAS and hydrographs to evaluate the performance of the model with dynamic river network routine. We have also analyzed the mean absolute relative error (MARE, Equation (15)) of 91 flood peaks with and without river dynamics.

$$\mathrm{MARE}(\%) = \frac{1}{N}\sum_{i=1}^{N}\left|\frac{(O_i - S_i)}{O_i}*100\right| \tag{15}$$

where O_i is the observed flood peak and S_i is the predicted flood peak with and without dynamic river network. N is the number of flood peaks (91 in this study). We have also analyzed the quantiles of the distribution of relative errors (RE, Equation (16)) of the flood peaks prediction.

$$\mathrm{RE}(\%) = \frac{(O_i - S_i)}{O_i}*100 \tag{16}$$

where O_i is the observed flood peak and S_i is the predicted flood peak with and without dynamic river network.

Correlation between A_c and F_c with environmental factors

We have done a correlation analysis between the parameters A_c and F_c and environmental factors to improve the understanding on how the dynamic river network develops and to assess the potential for relating F_c to environmental factors. The environmental factors included in the correlation analysis are mean annual precipitation, mean hillslope slope, bare mountain and forest land covers of the study catchments. We have used the Pearson correlation coefficient for the analysis. The A_c derived from observed river network has a spatial variation within a catchment; therefore, we have estimated the mean A_c for

Table 5 | The coefficient of determination (R-squared) and the coefficients of the power relation between D_m and A_c, i.e., $D_m = aA_c^b$ for all the study catchments

Cat_ID	a	b	R^2
6.1	2.17	0.37	0.99
12.193	1.16	0.42	1
19.107	1.04	0.43	1
26.64	1.23	0.42	1
36.32	1.18	0.42	1
41.8	0.61	0.51	0.99
42.2	0.92	0.45	1
55.4	1.1	0.44	1
63.12	1.05	0.45	1
68.2	0.86	0.49	1
73.21	0.82	0.48	1
73.27	1.19	0.44	0.99
91.2	0.68	0.51	0.99
101.1	1.1	0.45	0.98
172.8	1.08	0.44	1

each of the study catchments before the correlation analysis using the following two steps. First, the D_m of the observed stream network is estimated using GIS. Second, the mean A_c of the observed river network is calculated using Equation (3). To assess the potential for relating F_c with the environmental factors, correlation analysis between the calibrated F_c and the environmental factors has also been performed. We have also done a stepwise method of multiple linear regression between F_c and the four environmental factors mentioned above to see if there is a possibility for regionalizing F_c.

RESULTS

Performance of DDD with and without dynamic river network

The calibration and validation results of the DDD model without a dynamic river network show that the model performs satisfactorily with KGE values between 0.55 and 0.9 and BIAS between 0.75 and 1.25. As stated by Thiemig *et al.* (2013), $0.75 \leq \text{KGE} < 0.9$ is considered good, $0.5 \leq \text{KGE} < 0.75$ is intermediate and $0.0 \leq \text{KGE} < 0.5$ is poor. Seven catchments show good and eight catchments show intermediate performance both during the calibration and validation periods. Even if the KGE performance is satisfactory, the visual inspection of the hydrographs shows that several observed flood peaks are underestimated.

We added the dynamic river network routine into the DDD model and calibrated the critical flux (F_c) parameter of the routine manually for the whole simulation period. The KGE and BIAS performance of the model are similar as before, i.e., without dynamic river network for all study catchments except one, where the KGE is slightly lower. However, the inspection of the hydrographs clearly shows that the predication of several underestimated flood peaks has been improved after the addition of the dynamic river network routine. The dynamic OUHs that resulted from the dynamic river network have higher peaks and narrower width during the flooding events, and these OUHs, added with the subsurface UHs, helped in improving the previously underestimated floods. Figure 4 shows the hillslope distance distributions for variable A_c for catchment 73.27. Figure 5 shows the empirical cumulative distance distributions functions as an example for the dynamic distance distribution presented in Figure 4, and Figure 6 shows the four dynamic OUHs which resulted from the corresponding distance distributions functions. Table 6 shows OF, A_c and D_m for a catchment 12.193 during a flooding event.

Figure 7 shows the hydrographs during the flooding periods with and without the dynamic river network routine for six sample study catchments. Table 7 shows the observed floods, simulated floods, KGE and BIAS performance of DDD model with and without dynamic river network routine for five sample study catchments selected randomly.

The results of the statistical analysis (mean absolute relative errors and quantiles of relative errors), for the 91 observed peak floods from the 15 study catchments, show that the dynamic river network method improved the prediction of peak floods significantly. The 0.75 quantile of the relative errors of the simulated peaks reduced from 41% to 23%, and the 0.25 quantile of the relative errors reduced from 22% to 1%. Figure 8 shows box plots of the relative errors with and without river dynamics. The MARE of the magnitude of the underestimated peak floods is reduced from 32.9% to 15.7%.

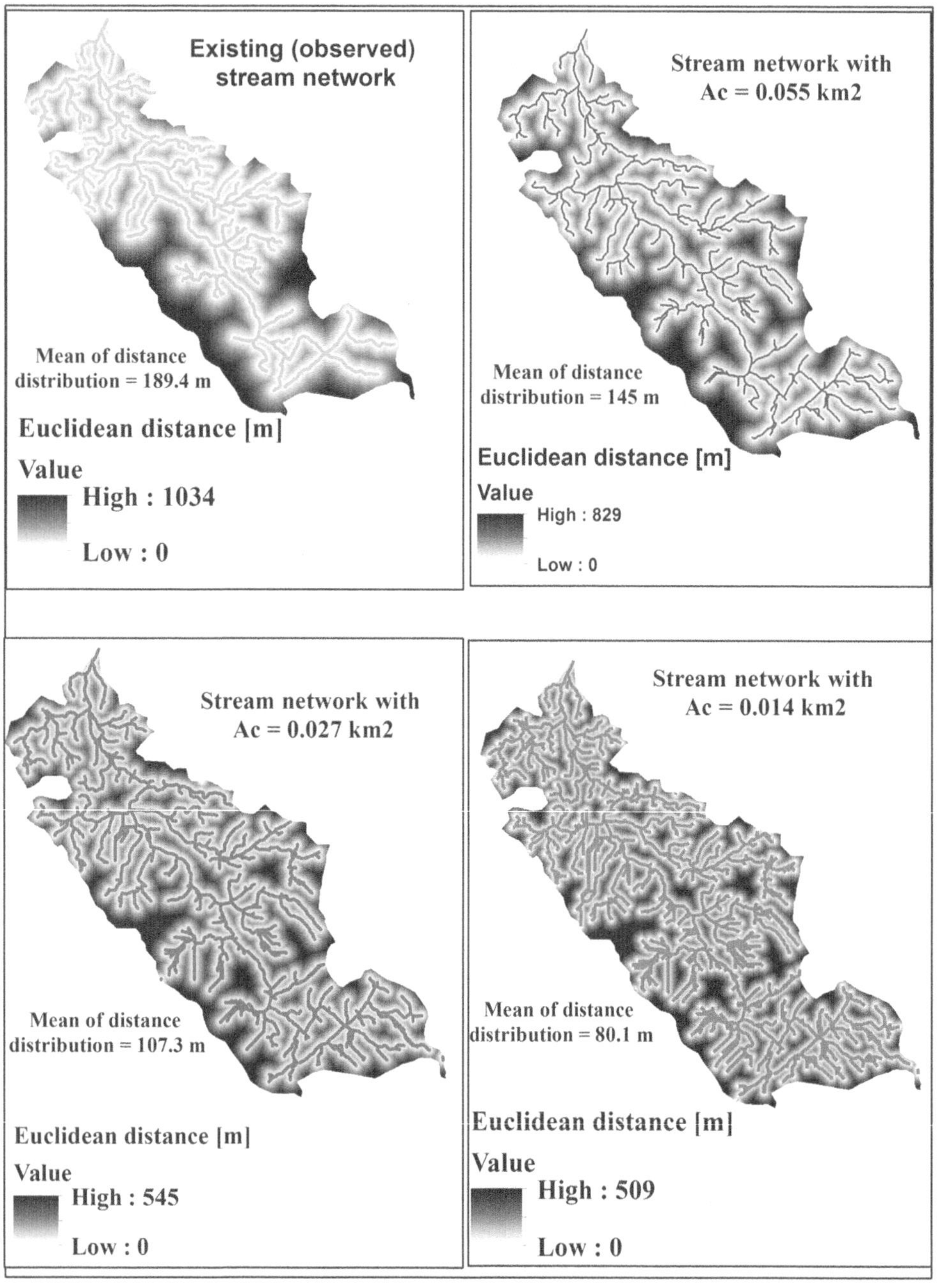

Figure 4 | Map of sample dynamic distance distributions of hillslope generated from dynamic critical supporting area, i.e., A_c during flooding events for catchment 73.27.

Correlation between A_c and F_c with environmental factors

The critical supporting area, A_c, of an observed stream network of a catchment shows spatial variation within the catchment (Figure 9 shows the distributions for five sample catchments); therefore, the mean value of a catchment is used for the correlation analysis. The mean A_c for the observed river networks is correlated with environmental factors, i.e., vegetation cover, topography and climate. The correlation with vegetation cover is stronger than that of topography and climate. The mean

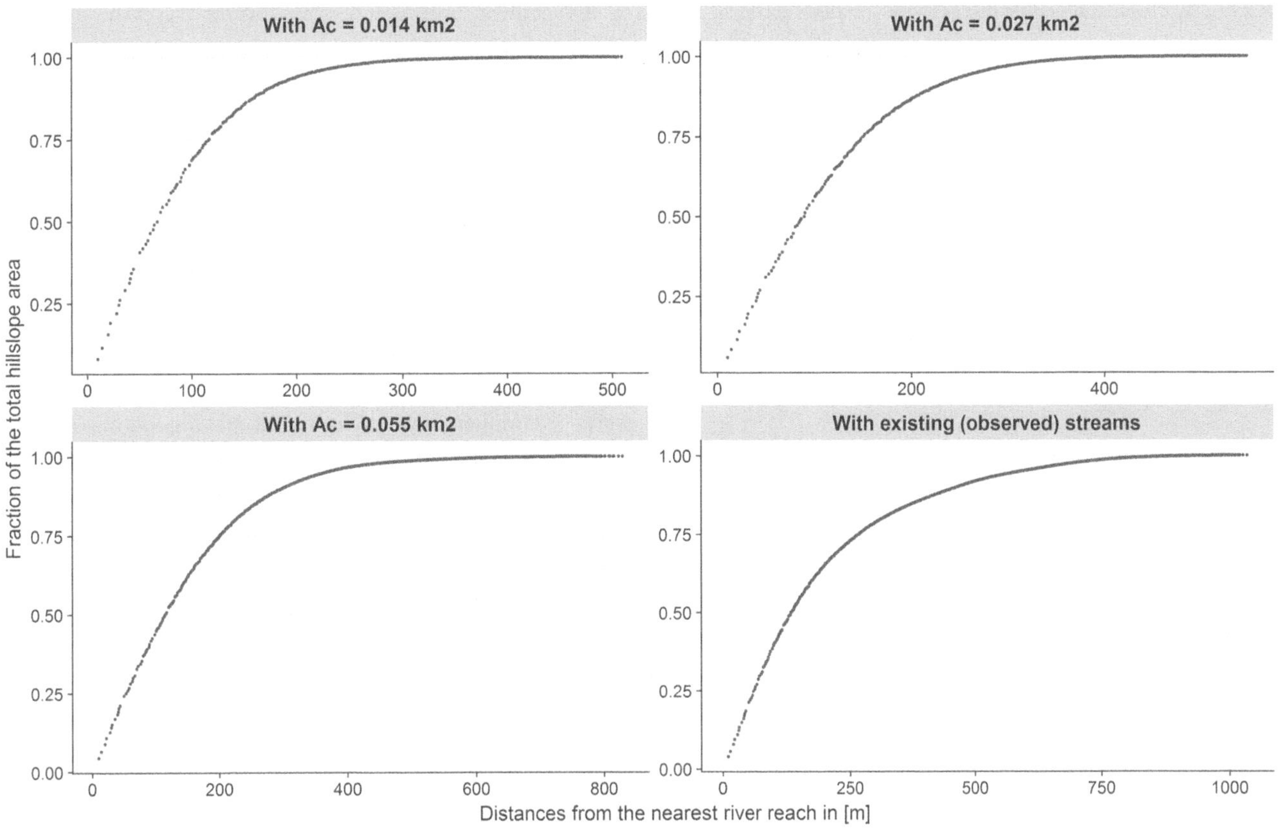

Figure 5 | Cumulative distance distributions functions of the dynamic hillslope distance distributions of Figure 4 for catchment 73.27.

A_c has a positive correlation with the forest cover in a catchment, but it has a negative correlation with mean annual precipitation, bare mountain cover and mean hillslope slope of a catchment. Table 8 shows the correlation between the mean A_c and F_c and the environmental factors.

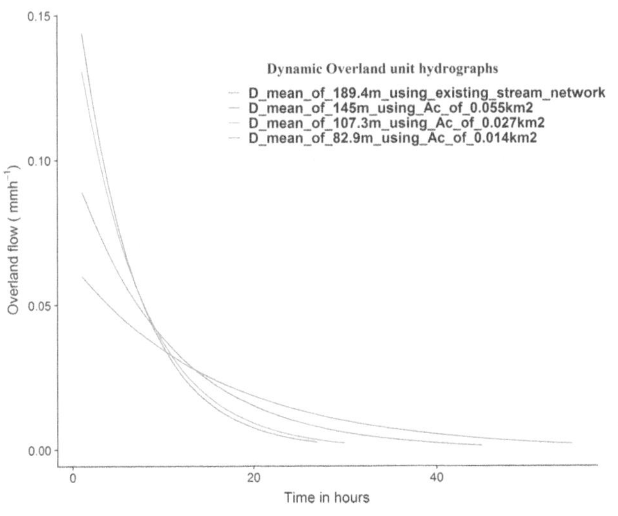

Figure 6 | Dynamic overland unit hydrographs of the cumulative distance distributions functions under Figure 5 during a flooding event for catchment 73.27.

The calibrated critical flux, F_c, of the dynamic river network routine is correlated with the environmental factors. The correlation between F_c and the vegetation cover is stronger than the correlation between F_c and topography and mean annual precipitation. F_c shows positive correlation with forest and negative correlation with bare mountain, mean annual precipitation and mean hillslope slope of a catchment. Table 9 shows the environmental factors used in the correlation analysis, the D_m and mean A_c of observed river network and the calibrated F_c of the dynamic river network routine of the study catchments.

The result of stepwise multiple linear regression shows that there is a potential to estimate F_c from the environmental factors as shown in Equation (17). Bare mountain is the only environmental factor contributing significantly to the regression with a significant level of 0.1. The multiple

Table 6 | Dynamic mean distance of the hillslope distance distributions estimated and used for generating dynamic overland unit hydrograph during flooding event at catchment 12.193 with a calibrated critical flux of 90 m^3/h

OF (mm/h)	A_c (m^2)	D_m (m) estimated using Equation (3)	D_m (m) used in deriving OUH
0.144	1,319,444.4	429.08	*301.1*
2.19	86,758	137.58	137.58
2.22	85,585.6	136.79	136.79
0.94	202,127.7	196.26	196.26
10.9	17,431.2	70.12	70.12
1.4	135,714.3	166.02	166.02
2.7	70,370.4	126	126
1.19	159,663.9	177.75	177.75
0.19	1,027,027	388.44	*301.1*
1.27	149,606.3	172.96	172.96
0.2	950,000	375.93	*301.1*
1.33	142,857.1	169.64	169.64
1.28	148,902.8	172.61	172.61
0.17	1,117,647.1	402.48	*301.1*
0.14	1,319,444.4	431.54	*301.1*
0.64	296,875	230.64	230.64
0.34	558,823.5	302	*301.1*
0.31	612,903.2	312.72	*301.1*
0.83	228,915.7	206.79	206.79

Italic numbers are rounded to two significant figures.

coefficient of determination (R^2) of the multiple regression is 0.3 and the significant level (P) is 0.06.

$$F_c\ (m^3/h) = 160.7 - 1.4 * \text{bare mountain}\ (\%) \tag{17}$$

DISCUSSION

Dynamic river networks

Dynamic river networks and hence dynamic OUHs are introduced and implemented in the DDD rainfall-runoff model to improve the simulation of floods. The dynamic river network method expands the observed river networks during OF events. The expansion means that the A_c required to initiate and maintain a stream decrease. Smaller A_c results in smaller D_m (see Table 5). The smaller D_m value indicates shorter travel times from points in the catchment to the nearest river reach. The shorter travel time distribution generates OUHs with a higher peak and shorter scale for the hillslopes (Figure 6). The dynamic OUHs are superpositioned with the other four subsurface UHs of DDD to give a single dynamic UH of a catchment during flooding events. The results of the method are supported by the previous study of D'Odorico & Rigon (2003) who found that shorter hillslope distances result in shorter travel times and hence higher flood peaks. The smaller D_m results, obtained during the flooding events using the dynamic river network method, are also supported by the study of Humbert (1990) who found that a good correlation exists between the runoff coefficient of flooding events and D_d, and hence the D_m, for 45 French catchments. Lazzaro *et al.* (2015) also found that the variability of runoff due to higher D_d (lower D_m) creates a faster concentration of flow that implies shorter travel times and higher peak floods. The results in this study are also supported by the results of Lee *et al.* (2008) who found that a UH of a catchment is dynamic during different precipitation intensities, i.e., the higher the precipitation intensity, the higher the peak and shorter the temporal scale of the UHs. The results of this study also show that a dynamic river network method could be a solution for rainfall-runoff models which face challenges in predicting flood peaks through continuous simulation. Improving the prediction of peak floods in a continuous simulation is very important because the hydrograph consisting of this peak flow is mainly responsible for the damage caused by floods. Therefore, a dynamic river network is a method to be conceptualized and included as one routine in continuous rainfall-runoff models which underestimate predictions of floods.

We analyzed statistically 91 underestimated flood peaks to evaluate the performance of the dynamic river network. The MARE and quantiles of RE of the prediction with and without dynamic river network show that the overall performance of the method has improved the prediction of the peaks satisfactorily. The dynamic river network overestimated 17 of the 91 flood peaks and still underestimates the remaining 74 floods but with a significant improvement in the prediction of flood peaks compared to the results obtained without a dynamic river network.

A single calibrated critical flux, F_c, improves the prediction of several underestimated floods significantly, but it also

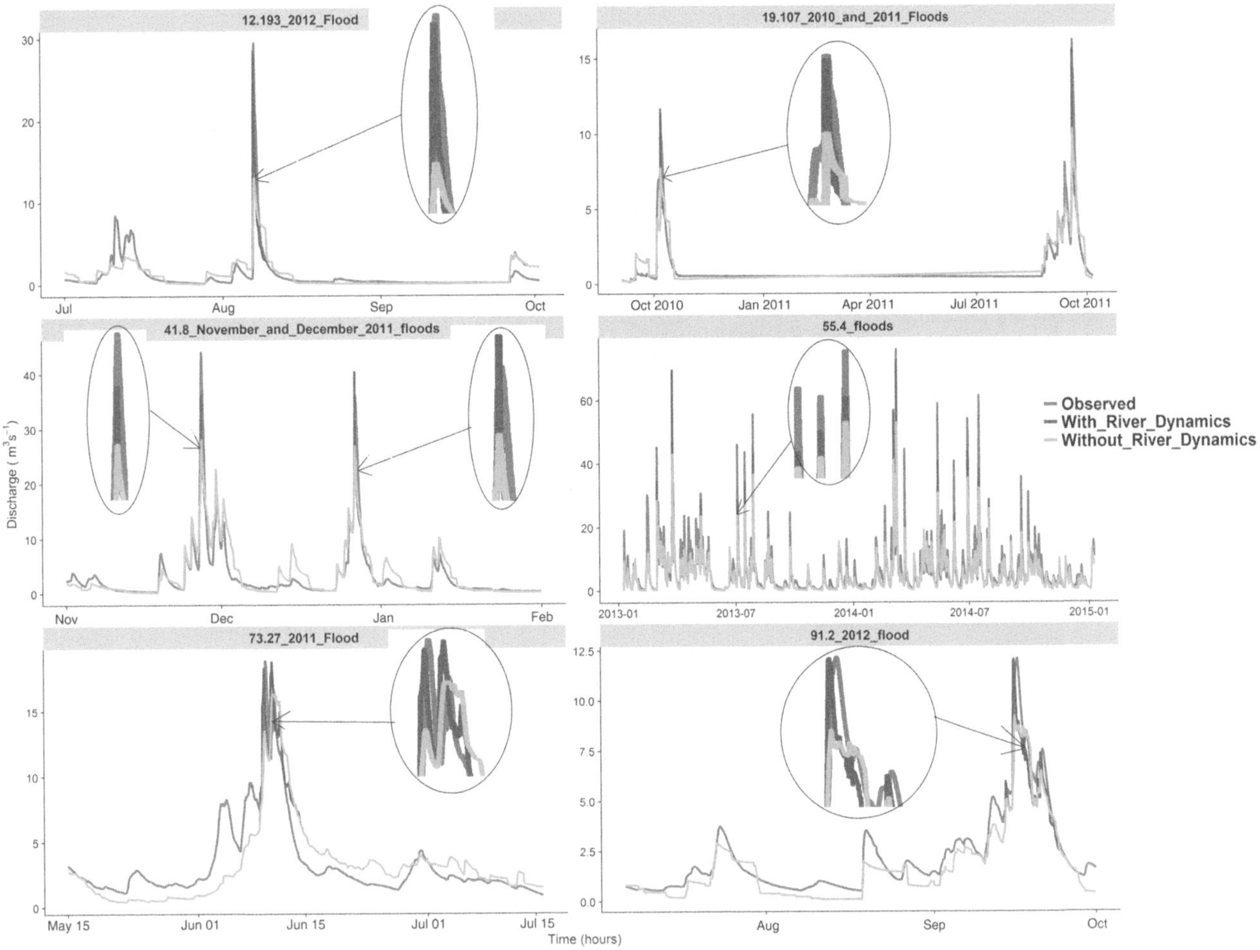

Figure 7 | Hydrographs of continuous simulations results of DDD rainfall-runoff models with flood peaks, i.e. observed, simulated with and without dynamic river network.

Table 7 | Observed and simulated floods using DDD with and without dynamic river network and the corresponding performance of the model for five sample catchments

Cat_ID	Observed flood(s) in m^3/s	Simulation period	Simulated value(s) of flood without river dynamics in m^3/s	Performance of DDD model without river dynamics in calibration: KGE	BIAS	Simulated value(s) of flood with river dynamics in m^3/s	Performance of DDD model with river dynamics: KGE	BIAS
12.193	*29.42*	2 years	*12.97*	0.64	1.2	*28.78*	0.65	1.2
19.107	*11.65* and *16.2*	3 years	*7.64* and *10.3*	0.8	0.93	*9.42* and *16.1*	0.81	0.94
41.8	*43.96* and *36.13*	2 years	*28.02* and *26.74*	0.77	0.84	*36.3* and *40.3*	0.77	0.84
73.27	*18.85*	3 years	*13.3*	0.71	0.76	*18.3*	0.71	0.76
91.2	*12.06*	2 years	*8.34*	0.71	0.8	*12.04*	0.71	0.8

Italic numbers are rounded to two significant figures.

overestimates a few flood peaks (Figure 7). Reasons for overestimation could be that a single calibrated F_c could not represent the different precipitation patterns, overland flow patterns and initial conditions prior to flooding events.

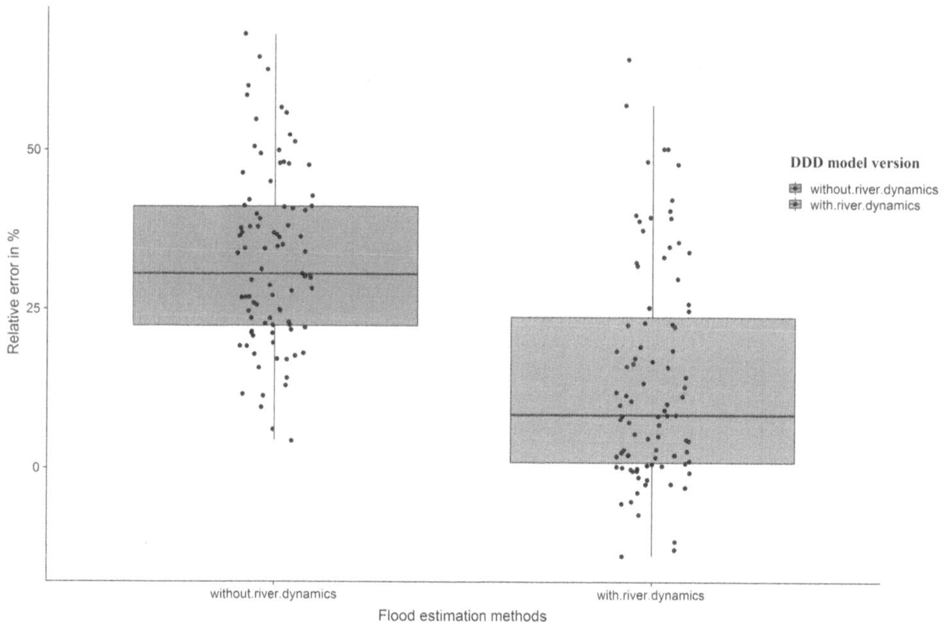

Figure 8 | Distributions of relative errors (%) of prediction of 91 flood peaks with and without dynamic river network.

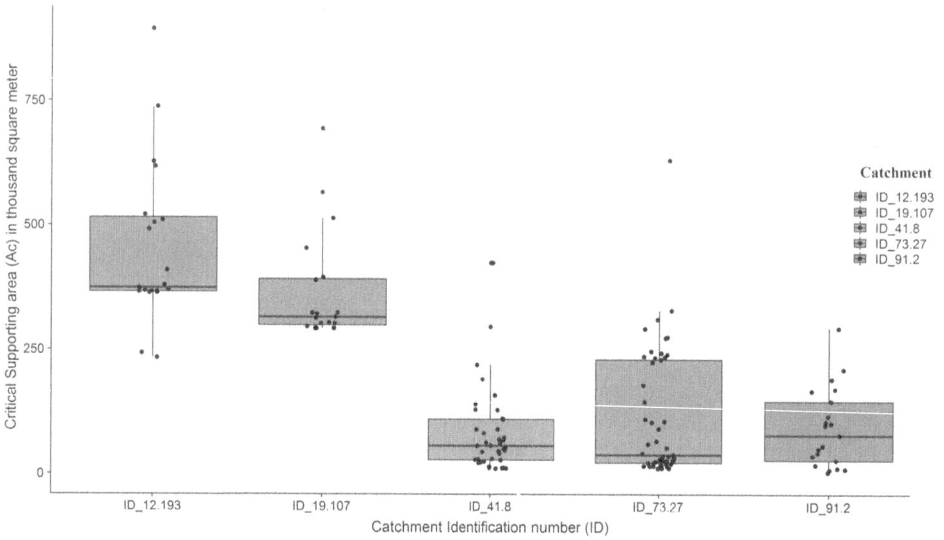

Figure 9 | Distributions of critical supporting area, i.e., A_c of observed stream networks for five of sample study catchments.

Table 8 | Correlation between calibrated critical flux, F_c of the dynamic river network with environmental variables, and correlation between mean critical area, A_c of the observed river network with some environmental variables

	Forest (%)	Bare mountain (%)	Mean hillslope slope (%)	Mean annual precipitation (mm)
Mean A_c (m^2)	0.63	−0.63	−0.56	−0.49
F_c (m^3/h)	0.46	−0.5	−0.29	−0.18

Another reason for such overestimation could be that the dynamic D_m estimated using Equation (3) is not always an accurate representation of the reality during a flood event. The limitations of manual calibration include subjectivity and time consuming. Manual calibration methods are subjective in the evaluation of model fit and the final choice of optimal parameters. Ndiritu (2009) pointed out that

Table 9 | The environmental factors used in the correlation analysis, the mean distance distribution, and mean critical area, A_c of observed river network, and the calibrated critical flux, F_c of the dynamic river network routine

	Environmental factors				Observed river network		Dynamic river network
Cat_ID	Forest (%)	Bare mountain (%)	Mean annual precipitation (mm)	Mean hillslope slope (%)	D_m (m)	Mean A_c (m^2)	F_c (m^3/h)
6.1	94.3	0	886	18.3	149.2	92,406	15
12.193	88.4	0	840	15.3	301.1	560,042	190
19.107	86.4	0	1,158	14.7	336.9	689,123	370
26.64	38.8	46.2	1,688	28.3	181.6	146,146	5
36.32	13.5	81.4	2,377	34.1	168.6	135,213	120
41.8	8.8	82.2	2,955	37.5	157.2	53,418	15
42.2	40.7	52.1	2,361	40.4	175.9	117,477	30
55.4	51.8	30.7	2,593	41.9	155.1	76,715	15
63.12	5.8	86.1	2,579	34.4	181.2	93,596	80
68.2	20.2	50.3	2,736	43.6	211.3	75,573	150
73.21	2.2	88	946	21.5	298.4	216,464	15
73.27	0.1	89.4	679	14.8	189.4	100,964	60
91.2	3.7	66.5	2,072	29.9	283.8	137,481	90
101.1	61.3	11.3	1,704	23.9	334.0	328,354	150
172.8	1.4	82.5	1,465	17	168.1	95,972	10

manual calibration may be more prone to obtaining suboptimal parameter sets than automatic calibration. Studies also show that manual calibration is more subjective than automatic calibration because it largely depends on visual hydrograph inspection and the personal judgment of the hydrologist. Substantial amount of time is also required to adjust F_c so that the observed and simulated flood peaks agreed well. A separate automatic calibration of F_c (which is not included in this study) after defining and writing appropriate objective function could improve the limitations of the manual calibration. However, since we have only one manually calibrated parameter and we have enough experience of using DDD model, the manual calibration result of F_c could be very good. In addition, the results of our study show that the manually calibrated F_c resulted in a significant improvement in predicting flood peaks using dynamic river network method.

The hydrograph in Figure 7 shows that the two floods of the catchment 41.8 could not be estimated well using a single calibrated F_c of magnitude 15 m^3/h even if the overall prediction of the flood peaks is improved. When we look at the flood hydrographs, the 27 November flood of 43.96 m^3/s (at 8 A.M), had been preceded by a 1-day precipitation of 68.9 mm (from 26 November 2011 09:00 to 27 November 2011 08:00). The precipitation was again preceded by a 3-day precipitation of magnitude 86.1 mm, i.e., the 4-day precipitation preceded the flooding event was 154.9 mm. When F_c is fitted to this single flooding event, 5 m^3/h is required. After 1 month, another heavy precipitation event happened (56.9 mm/day) and the event was preceded by 82.2 mm of 3-day precipitation, i.e., the 4-day precipitation preceded the flooding event was 139.7 mm. The magnitude of the flood was 36.13 m^3/s. When F_c is fitted to this single flooding event, 25 m^3/h is required. The variation in the fitted values of F_c for different flooding events in a catchment shows that we could have an overestimation of flooding events when we use a single calibrated F_c for the whole simulation period as a representative for a catchment. Accordingly, a single F_c of 15 m^3/h for the catchment 41.8 has overestimated the December 2011 flood, i.e., 36.13 m^3/s, but it has improved the overall prediction of the flood peaks in the catchment.

The spatial variability of A_c during flooding events, which is not considered in this study, could also be another factor for the overestimation of floods using a single calibration value of F_c for a catchment. We have derived the coefficients of Equation (3) (*a* and *b*) with the assumption of constant A_c using the DEM, which considers only the topography of a catchment. However, the A_c of observed stream networks clearly shows that there is a spatial variability of A_c within a catchment. Geological and land use factors play significant roles in initiating and maintaining a stream, and these factors control the spatial variability of A_c in a catchment (Montgomery & Dietrich 1989; Ogden *et al.* 2011; Sjöberg 2016; Ward *et al.* 2018) in addition to the topography. The correlation results between A_c and vegetation cover, which is done in this study and explained in the next section, also confirm that the land use affects the spatial variability of A_c. Figure 7 shows box plots of the spatial distributions of A_c of observed stream network for five sample catchments. For flooding events preceded by short duration and higher values of OF, Equation (3) gives very low values of D_m. The very low D_m gives OUHs of sharp peak and short scales which overestimated the floods. However, if we had calculated the actual D_m using the spatial variability of A_c, we could have found higher values of D_m than the value calculated using Equation (3) and the overestimation could have been avoided.

For estimating the parameters of travel time distributions of overland flow of a hillslope, we followed the original approach used in DDD (e.g. the distance from any point in the catchment to the nearest river network is calculated using the Euclidean distance and the celerity is determined from recession analysis). The GIUH and of width function estimates the distribution of travel times at the outlet of a catchment combining the hillslope and river network travel times using the steepest descent path from any point in the catchment to the outlet and the shape and scale parameters of the travel time distribution could change with the extent of hillslope saturation. Therefore, further investigation, i.e., comparison assessment is required before concluding as one method is better than the other.

Correlation of A_c and F_c with environmental factors

Environmental factors such as vegetation cover, topography and climate, affect A_c and hence D_m. Land use (e.g. vegetation cover) affects the hydrology and can affect subsurface as well as overland flow which in turn can cause changes in the stream network, i.e., A_c (Jonathan & Dennis 2001). The correlation results show that the denser the vegetation cover in a catchment the higher the A_c is required for initiating and maintaining a stream and vice versa, i.e., positive correlation with forest and negative correlation with bare mountain. The correlation result confirms the findings that a decrease in vegetation causes a decrease in surface resistance and critical shear stress, which result in an increase of drainage density (a reduction in D_m), i.e., streams form easier in less vegetated catchments (Willgoose *et al.* 1991; Prosser & Dietrich 1995; Magnuson *et al.* 1997; Tucker & Slingerland 1997). Field observations also show that higher D_m and hence higher A_c is generally observed in denser vegetation cover (Morisawa 1985).

The steepness of a catchment is one of the topographical factors controlling A_c and hence D_m. In this study, we used the mean hillslope slope of a catchment and found that a catchment hillslope slope has negative correlation with A_c, i.e., the higher the steepness, the lower the A_c required to initiate and maintain stream. This finding is supported by Montgomery & Dietrich (1989) who found that stream initiation on steep slopes shows a negative relationship between valley gradient at the stream head and A_c, i.e., the higher the stream head slope the lower the A_c (lower D_m).

The positive correlation between F_c and vegetation cover in a catchment shows that the denser the vegetation covers, the higher the F_c. F_c shows negative correlation with bare mountain, mean hillslope slope and mean annual precipitation of a catchment. The A_c and hence F_c depend on several factors, which include geology, precipitation, vegetation, morphology, soils and land uses, and one factor may be more important than another (see Table 8). Therefore, a more detailed investigation supported by field work (e.g. mapping of the slope, geology, vegetation cover and soil of a catchment at the head of first-order streams of observed river networks and mapping of the pattern of expansion of first-order streams during flooding events) should be carried out to assess how the combination of these factors control A_c and hence F_c.

We have done a simple multiple linear regression analysis using the four environmental factors as predictors, i.e., forest, bare mountain, slope and mean annual precipitation, to estimate the response variable F_c. The result shows that only bare mountain is contributing significantly in estimating F_c with a significance level of 0.1, and the coefficient of determination of the regression (R^2) is 0.3. The objective of the regression analysis, done in this study, is to assess a preliminary method for regionalization that can predict F_c for ungauged catchments from environmental factors and to lay a foundation for further studies.

A dynamic river network method could be implemented in rainfall-runoff models as shown for DDD for prediction of floods for catchments with a wide range of topography and land uses (Table 1). In this study, the effect of steep hills is reflected in the dynamic river networks as the steepness of a catchment is one of the factors that govern the initiation of streams. As shown in Table 8, the mean hillslope slope has a negative correlation with A_c, i.e., we need a smaller A_c to initiate and maintain streams in steep topography than in a flat topography. Table 9 also shows that the mean A_c of an observed stream network decreases as the mean hillslope slope of a catchment increases. The fundamental theory behind the method is the expansion of river networks during flooding events, i.e., whether the critical flux, F_c, which is required to initiate and maintain a stream, is satisfied or not. The magnitude of F_c depends on the magnitude of saturation excess overland flow, OF, and the critical support area, A_c. The study results also show that the critical support area, required to initiate and maintain a stream, is smaller in steep and bare mountain catchments than flat and forested catchments. Therefore, the method could be applicable at different catchments with different characteristics.

CONCLUSIONS

The dynamic river network method, introduced in Distance Distribution Dynamics (DDD) rainfall-runoff model, can improve the prediction of flood peaks in continuous simulation satisfactorily. The performance of the DDD model is the same with and without dynamic river network in terms of KGE and BIAS. The statistical analysis on 91 flood peaks, underestimated by DDD without dynamic river network method, shows that the MARE of the prediction reduced from 32.9% to 15.7% using the dynamic river network method. With a dynamic river network method, the 0.75 quantile of the relative errors reduced from 41% to 23%, and the 0.25 quantile of the relative errors reduced from 22% to 1%. The visual inspection of the hydrographs also shows an improvement in the prediction of flood peaks for several flooding events. Therefore, we recommend the use of a dynamic river network method in the prediction of floods. The next step in the development of the method is to investigate the applicability of the method from gauged to ungauged catchments and find a way to address the limitations identified in this study.

The critical flux, F_c, the calibration parameter introduced in the method, has been formulated as the product of critical supporting area (A_c) and the saturated excess overland flow (OF). F_c shows stronger correlation with vegetation cover than topographical and climate factors. The parameter shows positive correlation with forest cover of catchments, and negative correlation with bare mountain, mean hillslope slope and mean annual precipitation. The simple multiple linear regression, using the four environmental factors as predictors and F_c as a response variable, shows that there is a potential to estimate F_c from environmental factors and regionalize it for using the method without calibration. The value of the calibrated F_c could be different for the same catchment of different flood magnitudes, and it could be different for the same type of vegetation cover for different catchments. This difference shows that F_c could depend on several environmental factors and further investigations should be carried out.

ACKNOWLEDGEMENTS

The authors would like to acknowledge Cristian Lussana of the Norwegian Meteorological Institute for providing information on how to access and process the 1×1 km spatial and 1-h temporal resolution gridded precipitation and temperature data for Norway. Finally, the authors gratefully acknowledge the financial support by the Research Council of Norway and several partners through the Centre for Research-based Innovation 'Klima 2050' (see www.klima2050.no).

REFERENCES

Bencala, K. E., Gooseff, M. N. & Kimball, B. A. 2011 Rethinking hyporheic flow and transient storage to advance understanding of stream-catchment connections. *Water Resources Research* **47**, 3. doi:10.1029/2010WR010066.

Bergström, S. 1976 *Development and Application of a Conceptual Runoff Model for Scandinavian Catchments*. Department of Water Resources Engineering, Lund Institute of Technology, University of Lund, Lund.

Beven, K. 2001b *Rainfall-Runoff Modelling: The Primer*. John Wiley and Sons, Chichester, UK, pp. 1–360.

Blaikie, P., Cannon, T., Davis, I. & Wisner, B. 2014 *At Risk: Natural Hazards, People's Vulnerability and Disasters*. Routledge, London.

Blazkova, S. & Beven, K. 2002 Flood frequency estimation by continuous simulation for a catchment treated as ungauged (with uncertainty). *Water Resources Research* **38** (8), 14-11–14-14. doi:10.1029/2001WR000500.

Boughton, W. & Droop, O. 2003 Continuous simulation for design flood estimation – a review. *Environmental Modelling & Software* **18**, 309–318. https://doi.org/10.1016/S1364-8152(03)00004-5.

Bourdin, D. R., Fleming, S. W. & Stull, R. B. 2012 Streamflow modelling: a primer on applications, approaches and challenges. *Atmosphere-Ocean* **50** (4), 507–536. doi:10.1080/07055900.2012.734276.

Bras, R. L. 1990 *Hydrology: An Introduction to Hydrologic Science*. Addison-Wesley, Reading, MA.

Brocca, L., Melone, F. & Moramarco, T. 2011 Distributed rainfall-runoff modelling for flood frequency estimation and flood forecasting. *Hydrological Processes* **25** (18), 2801–2813. doi:10.1002/hyp.8042.

Calver, A. & Lamb, R. 1995 Flood frequency estimation using continuous rainfall-runoff modelling. *Physics and Chemistry of the Earth* **20** (5), 479–483. doi:10.1016/S0079-1946(96)00010-9.

Cameron, D., Beven, K., Ja, T. & Naden, P. 2000 Flood frequency estimation by continuous simulation (with likelihood based uncertainty estimation). *Hydology and Earth System Sciences* **4**, 23–34.

Camici, S., Tarpanelli, A., Brocca, L., Melone, F. & Moramarco, T. 2011 Design soil moisture estimation by comparing continuous and storm-based rainfall-runoff modeling. *Water Resources Research* **47**, 5. doi:10.1029/2010WR009298.

Chetty, K. & Smithers, J. 2005 Continuous simulation modelling for design flood estimation in South Africa: preliminary investigations in the Thukela Catchment. *Quarterly Journal of the Royal Meteorological Society* **30**, 3–23.

Chorley, R. J. & Morgan, M. A. 1962 Comparison of morphometric features, Unaka Mountains, Tennessee and North Carolina, and Dartmoor, England. *GSA Bulletin* **73** (1), 17–34. doi:10.1130/0016-7606(1962)73[17:COMFUM]2.0.CO;2.

Collier, C. G. 2007 Flash flood forecasting: what are the limits of predictability? *Quarterly Journal of the Royal Meteorological Society* **133** (622), 3–23. https://doi.org/10.1002/(SICI)1099-1085(19970630)11:8<825::AID-HYP509>3.0.CO;2-G.

Day, D. G. 1983 Drainage density variability and drainage basin outputs. *Journal of Hydrology (New Zealand)* **22** (1), 3–17.

Di Lazzaro, M., Zarlenga, A. & Volpi, E. 2014 A new approach to account for the spatial variability of drainage density in rainfall-runoff modelling. *Boletin Geologico y Minero* **125** (3), 301–313.

Dingman, S. L. 1978 Drainage density and streamflow: a closer look. *Water Resources Research* **14** (6), 1183–1187. doi:10.1029/WR014i006p01183.

D'Odorico, P. & Rigon, R. 2003 Hillslope and channel contributions to the hydrologic response. *Water Resources Research* **39**, 5. doi:10.1029/2002WR001708.

Dunne, T. 1978 Field studies of hillslope flow processes. In: *Hillslope Hydrology*, Vol. 9 (M. J. Kirkby ed.), *Geographical Research Letters* **9**, 227–293.

Eschenbach, D., Haberlandt, U., Buchwald, L. & Belli, A. 2008 Derived flood frequency analysis using rainfall runoff modelling and synthetic precipitation data. *Wasserwirtchaft* **98** (11), 19–23.

ESRI 2014 *ArcGISDesktop Help 10.3 Geostatistical Analyst*. https://www.esri.com/en-us/home.

Filipova, V., Lawrence, D. & Skaugen, T. 2018 A stochastic event-based approach for flood estimation in catchments with mixed rainfall/snowmelt flood regimes. *Natural Hazards and Earth System Sciences Discussion* **2018**, 1–30. doi:10.5194/nhess-2018-174.

Fleig, A. K. & Wilson, D. 2013 *Flood Estimation in Small Catchments: Literature StudyNaturfareprosjektet,Rapport (Norges Vassdrags-og Energidirektorat: Online) (Vol. no. 60-2013)*. Norwegian Water Resources and Energy Directorate, Oslo.

Formetta, G., Prosdocimi, I., Stewart, E. & Bell, V. 2017 Estimating the index flood with continuous hydrological models: an application in Great Britain. *Hydrology Research* **49**, 23–33.

Gao, H., Cai, H. & Duan, Z. 2017 Understanding the impacts of catchment characteristics on the shape of the storage capacity curve and its influence on flood flows. *Hydrology Research* **49** (1), 90–106. doi:10.2166/nh.2017.245.

Godsey, S. E. & Kirchner, J. W. 2014 Dynamic, discontinuous stream networks: hydrologically driven variations in active drainage density, flowing channels and stream order. *Hydrological Processes* **28** (23), 5791–5803. doi:10.1002/hyp.10310.

Gupta, V. K., Waymire, E. & Wang, C. T. 1980 A representation of an instantaneous unit hydrograph from geomorphology. *Water Resources Research* **16** (5), 855–862. doi:10.1029/WR016i005p00855.

Gupta, H. V., Kling, H., Yilmaz, K. K. & Martinez, G. F. 2009 Decomposition of the mean squared error and NSE performance criteria: implications for improving hydrological modelling. *Journal of Hydrology* **377** (1), 80–91. doi:10.1016/j.jhydrol.2009.08.003.

Haberlandt, U. & Radtke, I. 2014 Hydrological model calibration for derived flood frequency analysis using stochastic rainfall and probability distributions of peak flows. *Hydrology and Earth System Sciences* **18** (1), 353–365. doi:10.5194/hess-18-353-2014.

Hirabayashi, Y., Mahendran, R., Koirala, S., Konoshima, L., Yamazaki, D., Watanabe, S. & Kanae, S. 2013 Global flood risk under climate change. *Nature Climate Change* **3**, 816. doi:10.1038/nclimate1911. https://www.nature.com/articles/nclimate1911#supplementary-information

Horton, R. E. 1932 Drainage-basin characteristics. *Eos, Transactions American Geophysical Union* **13** (1), 350–361. doi:10.1029/TR013i001p00350.

Horton, R. E. 1945 Erosional development of streams and their drainage basins: hydrophysical approach to quantitative morphology. Bulletin of the Geological Society of America 56, 275-370. *Progress in Physical Geography: Earth and Environment* **19** (4), 533–554. doi:10.1177/030913339501900406.

Humbert, J. 1990 Interet de la densite de drainage pour regionaliser les donnees hydrologiques en zone montagneuse. In: *Hydrology in Mountainous Regions I, Hydrological Measurements, the Water Cycle*. IAHS Publisher, Strasbourg, pp. 373–380.

Jonathan, L. L. M. & Dennis, P. L. 2001 Effects of forest roads on flood flows in the Deschutes river. *Earth Surface Processes and Landforms* **26**, 115–134, Vol. **26**. Washington.

Lamb, R. & Kay, A. L. 2004 Confidence intervals for a spatially generalized, continuous simulation flood frequency model for Great Britain. *Water Resources Research* **40** (7). doi:10.1029/2003WR002428.

Lazzaro, D. M., Zarlenga, A. & Volpi, E. 2015 Hydrological effects of within-catchment heterogeneity of drainage density. *Advances in Water Resources* **76**, 157–167. doi:10.1016/j.advwatres.2014.12.011.

Lee, K. T., Chen, N.-C. & Chung, Y.-R. 2008 Derivation of variable IUH corresponding to time-varying rainfall intensity during storms/Dérivation d'un HUI variable correspondant à l'évolution temporelle de l'intensité pluviométrique durant les averses. *Hydrological Sciences Journal* **53** (2), 323–337. doi:10.1623/hysj.53.2.323.

Lu, Q. & Han, Q.-L. 2011 A probability particle swarm optimizer with information-sharing mechanism for odor source localization. *IFAC Proceedings Volumes* **44** (1), 9440–9445. doi:10.3182/20110828-6-IT-1002.00507.

Lussana, C., Ole Einar, T. & Francesco, U. 2016 Senorge v2.0: an observational gridded dataset of temperature for Norway. *METreport* **108**, 75–86.

Magnuson, J., Webster, K., Assel, R., Bowser, C., Dillon, P., Eaton, J. & Mortsch, L. 1997 Potential effects of climate changes on aquatic systems: Laurentian Great Lakes and Precambrian Shield region. *Hydrological Processes* **11**. doi:10.1002/(SICI)10991085(19970630)11:8<825::AID-HYP509>3.3.CO;2-7.

Maidment, D. R. 1993 *Developing A Spatially Distributed Unit Hydrograph by Using GIS*. IAHS Publication, Austin, TX, 211, pp. 181–192.

Moglen, G. E., Eltahir, E. A. B. & Bras, R. L. 1998 On the sensitivity of drainage density to climate change. *Water Resources Research* **34** (4), 855–862. doi:10.1029/97WR02709.

Montgomery, D. R. & Dietrich, W. E. 1989 Source areas, drainage density, and channel initiation. *Water Resources Research* **25** (8), 1907–1918. doi:10.1029/WR025i008p01907.

Morisawa, M. 1985 *Rivers. Form and Process*. Longman Inc, New York.

Ndiritu, J. 2009 A comparison of automatic and manual calibration using the Pitman Model. *Engineering* **34**, 62–63.

Nhim, T. 2012 *Variability of Intermittent Headwater Streams in Boreal Landscape: Influence of Different Discharge Conditions. 248 Student Thesis*. Available from: http://urn.kb.se/resolve?urn=urn:nbn:se:uu:diva-183137; DiVA database.

Ogden, F. L., Raj Pradhan, N., Downer, C. W. & Zahner, J. A. 2011 Relative importance of impervious area, drainage density, width function, and subsurface storm drainage on flood runoff from an urbanized catchment. *Water Resources Research* **47** (12). doi:10.1029/2011WR010550.

Papageorgaki, I. & Nalbantis, I. 2017 Definition of critical support area revisited. *European Water* **57**, 273–278.

Pathiraja, S., Westra, S. & Sharma, A. 2012 Why continuous simulation? The role of antecedent moisture in design flood estimation. *Water Resources Research* **48**, 6. doi:10.1029/2011WR010997.

Plate, E. J. 2009 Classification of hydrological models for flood management. *Hydrology and Earth System Sciences* **13** (10), 12. doi:10.5194/hess-13-1939-2009.

Prosser, I. P. & Dietrich, W. E. 1995 Field experiments on erosion by overland flow and their implication for a digital terrain model of channel initiation. *Water Resources Research* **31** (11), 2867–2876. doi:10.1029/95WR02218.

Rahman, A., Hoang, T. M. T., Weinmann, P. E. & Laurenson, E. M. 1998 *Joint Probability Approaches to Design Flood Estimation: A Review Report 98/8*. CRC for Catchment Hydrology, Clayton, VIC.

Razi, M., Ariffin, J., Tahir, W. & Arish, N. A. M. 2010 Flood estimation studies using hydrologic modeling system (HEC-HMS) for Johor river. *Journal of Applied Sciences* **10**, 930–939. doi:10.3923/jas.2010.930.939.

R Core Team 2017 *R: A Language and Environment for Statistical Computing*. Available from: http://www.R-project.org/

Reis, D. S. & Stedinger, J. R. 2005 Bayesian MCMC flood frequency analysis with historical information. *Journal of Hydrology* **313** (1), 97–116. doi:10.1016/j.jhydrol.2005.02.028.

Rigon, R., D'Odorico, P. & Bertoldi, G. 2011 The geomorphic structure of the runoff peak. *Hydrology and Earth System Sciences* **15** (6), 1853–1863. doi:10.5194/hess-15-1853-2011.

Rinaldo, A., Marani, A. & Rigon, R. 1991 Geomorphological dispersion. *Water Resources Research* **27** (4), 513–525.

Rinaldo, A., Vogel, G. K., Rigon, R. & Rodriguez-Iturbe, I. 1995 Can one gauge the shape of a basin? *Water Resources Research* **31** (4), 1119–1127. doi:10.1029/94WR03290.

Robinson, J. S., Sivapalan, M. & Snell, J. D. 1995 On the relative roles of hillslope processes, channel routing, and network geomorphology in the hydrologic response of natural catchments. *Water Resources Research* **31** (12), 3089–3101. doi:10.1029/95WR01948.

Rodríguez-Iturbe, I. & Valdés, J. B. 1979 The geomorphologic structure of hydrologic response. *Water Resources Research* **15** (6), 1409–1420. doi:10.1029/WR015i006p01409.

Rusjan, S., Kobold, M. & Mikoš, M. 2009 Characteristics of the extreme rainfall event and consequent flash floods in W Slovenia in September 2007. *Natural Hazards and Earth System Sciences* **9**, 3. doi:10.5194/nhess-9-947-2009.

Schaefer, M., Elifrits, D. & Barr, D. 1979 *Sculpturing Reclaimed Land to Decrease Erosion, Paper Presented at the Symposium on Surface Mining Hydrology, Sedimentology and Reclamation*. University of Lexington, Kentucky.

Singh, V. P. 1995 *Computer Models of Watershed Hydrology*. Water Resources Publications, Littleton, CO.

Singh, V. P. & Frevert, D. K. 2006 Watershed models. *Experimental Agriculture* **42** (3), 370. doi:10.1017/S0014479706293797.

Sjöberg, O. 2016 *The Origin of Streams: Stream Cartography in Swiss Pre Alpine Headwater. 16003 Student Thesis*. Available from: http://urn.kb.se/resolve?urn=urn:nbn:se:uu:diva-277377 DiVA database.

Skaugen, T. & Mengistu, Z. 2016 Estimating catchment-scale groundwater dynamics from recession analysis – enhanced constraining of hydrological models. *Hydrology and Earth System Sciences* **20** (12), 4963–4981. doi:10.5194/hess-20-4963-2016.

Skaugen, T. & Onof, C. 2014 A rainfall-runoff model parameterized from GIS and runoff data. *Hydrological Processes* **28** (15), 4529–4542. doi:10.1002/hyp.9968.

Skaugen, T. & Weltzien, I. H. 2016 A model for the spatial distribution of snow water equivalent parameterized from the spatial variability of precipitation. *The Cryosphere* **10** (5), 1947. doi:10.5194/tc-10-1947-2016.

Skaugen, T., Peerebom, I. O. & Nilsson, A. 2015 Use of a parsimonious rainfall–run-off model for predicting hydrological response in ungauged basins. *Hydrological Processes* **29** (8), 1999–2013. doi:10.1002/hyp.10315.

Smithers, J. 2012 Methods for design flood estimation in South Africa. *Water SA* **38**, 633–646.

Smithers, J., Chetty, K., Frezghi, M., Knoesen, D. & Tewolde, M. 2013 Development and assessment of a daily time-step continuous simulation modelling approach for design flood estimation at ungauged locations: ACRU model and Thukela Catchment case study. *Water SA* **39**, 00–00.

Thiemig, V., Rojas, R., Zambrano-Bigiarini, M. & De Roo, A. 2013 Hydrological evaluation of satellite-based rainfall estimates over the Volta and Baro-Akobo Basin. *Journal of Hydrology* **499**, 324–338. doi:10.1016/j.jhydrol.2013.07.012.

Thomas, W. O. 1982 An evaluation of flood frequency estimates based on rainfall/runoff modeling1. *JAWRA Journal of the American Water Resources Association* **18** (2), 221–229. doi:10.1111/j.1752-1688.1982.tb03964.x.

Tsegaw, A. T., Alfredsen, K., Skaugen, T. & Muthanna, T. M. 2019 Predicting hourly flows at ungauged small rural catchments using a parsimonious hydrological model. *Journal of Hydrology*. doi:10.1016/j.jhydrol.2019.03.090.

Tucker, G. E. & Slingerland, R. 1997 Drainage basin responses to climate change. *Water Resources Research* **33** (8), 2031–2047. doi:10.1029/97WR00409.

Tucker, G. E., Catani, F., Rinaldo, A. & Bras, R. L. 2001 Statistical analysis of drainage density from digital terrain data. *Geomorphology* **36** (3), 187–202. doi:10.1016/S0169-555X(00)00056-8.

Ward, A. S., Schmadel, N. M. & Wondzell, S. M. 2018 Simulation of dynamic expansion, contraction, and connectivity in a mountain stream network. *Advances in Water Resources* **114**, 64–82. doi:10.1016/j.advwatres.2018.01.018.

Wharton, G. 1994 Progress in the use of drainage network indices for rainfall-runoff modelling and runoff prediction. *Progress in Physical Geography: Earth and Environment* **18** (4), 539–557. doi:10.1177/030913339401800404.

Willgoose, G., Bras, R. L. & Rodriguez-Iturbe, I. 1991 A coupled channel network growth and hillslope evolution model: 1. Theory. *Water Resources Research* **27** (7), 1671–1684. doi:10.1029/91WR00935.

Winsemius, H., Aerts, J., van Beek, L. P. H., Bierkens, M. F. P., Bouwman, A., Jongman, B. & Ward, P. 2015 Global drivers of future river flood risk. *Nature Climate Change* **6**, 381–385.

First received 19 December 2018; accepted in revised form 5 July 2019. Available online 26 August 2019

Testing the applicability of physiographic classification methods toward improving precipitation phase determination in conceptual models

Laurie D. Grigg, James Feiccabrino and Frederick Sherenco

ABSTRACT

Regions with a large percentage of precipitation occurring near freezing experience high percentages (>10%) of misclassified precipitation events (rain versus snow) and necessitate efforts to improve precipitation phase determination schemes through the use of more accurate surface air temperature thresholds (Trs). Meteorological data from 169 sites in Scandinavia were used to test the applicability of using physiographic categories to determine Trs. Three classification methods involving varying degrees of automation were evaluated. The two automated methods tested did not perform as well as when tested on a smaller region, showing only 0.16% and 0.20% reduction in error. A semi-manual method produced the largest average reduction in misclassified precipitation (0.53%) across all sites. Further refinement of classification criteria for mountain and hill stations showed that at mesoscales (>5 km), maximum elevation is a better predictor of Trs (0.89% average reduction in error) than terrain relief (0.22%), but that relief becomes increasingly important at microscales (0.90%). A new method for categorizing mountainous stations based on upslope or downslope air movement increased the average reduction in error up to 0.53%. These results provide a framework for future landscape classification methods and confirm the importance of microscale topography for determining Trs in alpine regions.

Key words | physiographic classification, precipitation phase determination, Scandinavia, snow model, temperature threshold

Laurie D. Grigg
Frederick Sherenco
Department of Earth and Environmental Science,
Norwich University,
158 Harmon Drive, Northfield, VT 05650,
USA

James Feiccabrino (corresponding author)
Department of Water Resources Engineering,
Lund University,
Lund S-221 00,
Sweden
E-mail: *james.feiccabrino@googlemail.com*

INTRODUCTION

The precipitation phase determination scheme (PPDS) is one of the most important parameters in a snow model (Kongoli & Bland 2000), yet remains one of the most difficult tasks for hydrologists and meteorologists (Lackmann *et al.* 2002) in temperatures near freezing. PPDS in conceptual hydrological models often uses a set surface air temperature (Ta) threshold (Trs), assigning all precipitation events with Ta > Trs as rain and all other events as snow. This is a simplistic approach (Daly *et al.* 2000; Harpold *et al.* 2017a) that does not account for the influence of atmosphere or landscape variables such as topography (e.g. Harpold *et al.* 2017b), warm and cold air-mass boundaries (Feiccabrino *et al.* 2012), or ocean temperatures (e.g. Dai 2008). As a result, mid- to high-latitude, topographically complex regions, such as Scandinavia, are associated with high rates (10–40%) of misclassified precipitation events occurring between −3 and 5 °C. The previous work has shown that deriving Trs from groups of physiographically similar sites reduces misclassified precipitation and holds

doi: 10.2166/nh.2020.081

promise as a low-cost method for improving PPDS (Feiccabrino & Grigg 2016). This study builds upon this work by (1) assessing whether this approach can effectively be applied to a wider latitudinal range of meteorological stations and (2) examining the relative importance of elevation and terrain relief at different spatial scales in assigning Trs in mountainous regions.

Correct classification of precipitation into solid and liquid phases is of paramount importance due to vastly different atmospheric, hydrological, and ecological responses to rain and snow (Ye *et al.* 2013). Near freezing precipitation events can be modeled differently depending on the snow fraction (SF) assigned by a PPDS, which impacts snowpack properties such as snow density, albedo, snowpack layering, and water retention capacity (Loth *et al.* 1993). Precipitation mass corrected for gauge undercatch of 2–14% rain and 5–80% snow (Kokkonen *et al.* 2006) is also affected by changes in modeled SF. PPDSs can be based on hydrological, meteorological, or combined approaches. This paper focuses on conceptual hydrologic models, typically using surface temperature and precipitation mass as the meteorological forcing inputs and widely used across many disciplines for their simplicity, data availability, and low computational requirements.

The most commonly used surface temperature measurements for Trs are air temperature (Ta), dew-point temperature (Td), wet-bulb temperature (Tw), or a combination of relative humidity (RH) and Ta (Ye *et al.* 2013). However, Td, Tw, and RH measurements are much less available than Ta, and Trs is typically calibrated to the air temperature resulting in the least misclassified precipitation. This calibration is often conducted over large areas irrespective of terrain, ocean, or seasonal influences (Jennings *et al.* 2018). Trs are known to vary with time and location; however, the practice of validation and calibration of Trs adjusted over different land surfaces is not often applied in models or research (Harpold *et al.* 2017a). The use of broadly established set Trs indirectly forces two notable assumptions: (1) that near surface air is coupled to the atmosphere above, without substantial differences due to physiographic or biophysical processes on the Earth's surface (Aalto *et al.* 2018) and (2) that atmospheric conditions and energy exchanges from precipitation microphysics are invariant. These assumptions are incorrect when the atmospheric lapse rate (the rate of air temperature decrease with height) is greater (more unstable) than normal which has been shown to occur over open ocean water (Dai 2008) due to conductive heat transfer from the water to the lower atmosphere. In mountainous terrain, these assumptions are also often invalidated because of mechanical lifting and cooling of air at the dry adiabatic lapse rate (9.8 °C/km), which is more unstable than the average atmospheric lapse rate (6.5 °C/km) (Feiccabrino *et al.* 2015). These and other landscape-driven changes in environmental lapse rates represent an opportunity to improve high rates of misclassified precipitation (error) rates (>10%) associated with PPDS in near freezing temperatures.

Open-water conductively warms Ta at the ground–atmosphere interface, while the frozen, snow-covered ground has the opposite effect. In the winter, warming of surface air temperatures over ice-free water near Iceland has been shown by Ólafsson & Haraldsdóttir (2003) to increase the environmental lapse rate compared to a model assumed constant lapse rate (e.g. 7.5 °C in CHRM (Fang *et al.* 2013)). The higher atmospheric lapse rate over water gives snow a better chance of reaching the ground at a given Ta than over land, which results in a wide range of Trs in Iceland from 0.5 °C in inland areas to 2.1 °C along the northern coast (Ólafsson & Haraldsdóttir 2003). Dai (2008), using global 3-h data, found average land Trs 1.2 °C and ocean Trs 1.9 °C, a 0.7 °C warm bias over oceans. Similar findings at the regional scale were found in Scandinavia where ocean stations were shown to have a warmer Trs than land stations by ≈ 0.5 °C (Feiccabrino & Grigg 2016).

Precipitation patterns are strongly affected by geographic barriers (hills/mountains) at multiple geographic scales. Air forced to rise over geographic barriers can cool to saturation, allowing water vapor to condense and cause enhanced precipitation on the windward side (Roth *et al.* 2018). This orographically enhanced precipitation causes a thicker melting layer, a lower 0 °C isotherm, and a decrease in the snow elevation compared to upwind areas (Minder *et al.* 2011). Descending air on the lee side of mountains dries resulting in lower atmospheric RH, more sublimation, and decreased precipitation totals compared to the windward side (Jennings *et al.* 2018; Roth *et al.* 2018). Geographic barriers also cause large variance in precipitation mass at local scales (Henn *et al.* 2018) and increased

PPDS errors when fixed lapse rates or Trs are used (e.g. Harpold *et al.* 2017b). The impacts of topography on Trs are evident in data from Northern Hemisphere land stations which show the coolest Trs in lowland and maritime climates and the warmest Trs in continental mountain climates with a maximum of 4.5 °C on the Tibetan Plateau (Jennings *et al.* 2018). At local to regional scales, Trs from Scandinavia show a similar trend with lowland stations in Scandinavia having cooler Trs (1.0–1.1 °C) than hill and mountain stations Trs (1.2–1.6 °C) (Feiccabrino & Grigg 2016).

This study aims to verify and improve upon the methods developed by Feiccabrino & Grigg (2016) with the end goal of decreasing precipitation phase uncertainty in conceptual hydrological models. The previous study developed a geographic information system (GIS)-based landscape classification method for a 7° latitude wide cross-section in Scandinavia from the North Sea over the Scandinavian mountains to the Bay of Bothnia and resulted in a reduction in error of 0.59% and 1.26%, using Ta and Tw-based Trs, respectively. The limited latitudinal range of the Feiccabrino & Grigg's (2016) study enabled the simplification of windward versus leeward stations based on the prevailing westerlies but restricted the future applicability of this method as a broader country-wide approach to assigning Trs. In this study, an expanded data set of 169 meteorological stations from across Norway and Sweden was used, and the simplified windward versus leeward designation was replaced with the GIS-derived classification of the upslope or downslope movement of air that can be modified based on the regional prevailing wind direction. This study also introduces a semi-manual classification scheme that considers the more varied wind sources and substrates of the larger geographic area.

Another aspect of the previous study's results which is addressed is the difficulty in assigning Trs for mountain and hill stations. The mountain and hill categories had both the largest Trs variability between stations and the highest percent of misclassified precipitation. This study tests which elevation variables, in addition to or in replace of relief, are most predictive of Trs at multiple spatial scales. The original study used a 15 km buffer surrounding each station in order to assign landscape categories, which may have been too coarse to capture microscale (<10 km) differences in topography. Harpold *et al.* (2017a) suggest that in mountain and coastal environments, extreme changes in atmospheric dynamics over short distances may drive microscale differences in optimal Trs.

STUDY AREA

A meteorological observation data set for the years 1995–2011 from 84 Norwegian stations and 85 Swedish stations was selected for the analysis (Figure 1(a)). This data set was obtained freely from the websites of the Swedish Meteorological and Hydrological Institute (SMHI 2018) and the Norwegian Meteorological Institute (NMI 2018). Data collected after 2011 were not used because of multiple changes to data collection methods made between 2011 and 2014. The Scandinavian peninsula is an ideal setting to develop new physiographic-based PPDS because of its latitude, proximity to the ocean on three sides, prevailing wind direction, and topographic diversity over a relatively small region.

Much of the weather impacting Scandinavia travels over the North Atlantic Ocean and the Norwegian Sea and is subject to an increase in the environmental lapse rate causing higher Trs over water. The prevailing westerly winds then experience rising/cooling followed by sinking/warming, as they travel over the Scandinavian mountains which run SSE–NNW along the Norway/Sweden border. These orographic effects change the moisture content and precipitation intensity both of which impact Trs. The stations range in latitude from ≈ 55°N in southern Sweden to 77°N in Svalbard and in elevation from the sea level to 723 m. 20% to 60% of total precipitation in Scandinavia occurs between −3 and 5 °C, and the average rate of misclassified precipitation at these temperatures is 11.00% (Figures 1(b) and 1(c)).

METHODS

Original-automated classification – all sites

The original-automated (GIS)-classification method (Table 1) by Feiccabrino & Grigg (2016) was tested, but this time on an expanded and more geographically complex

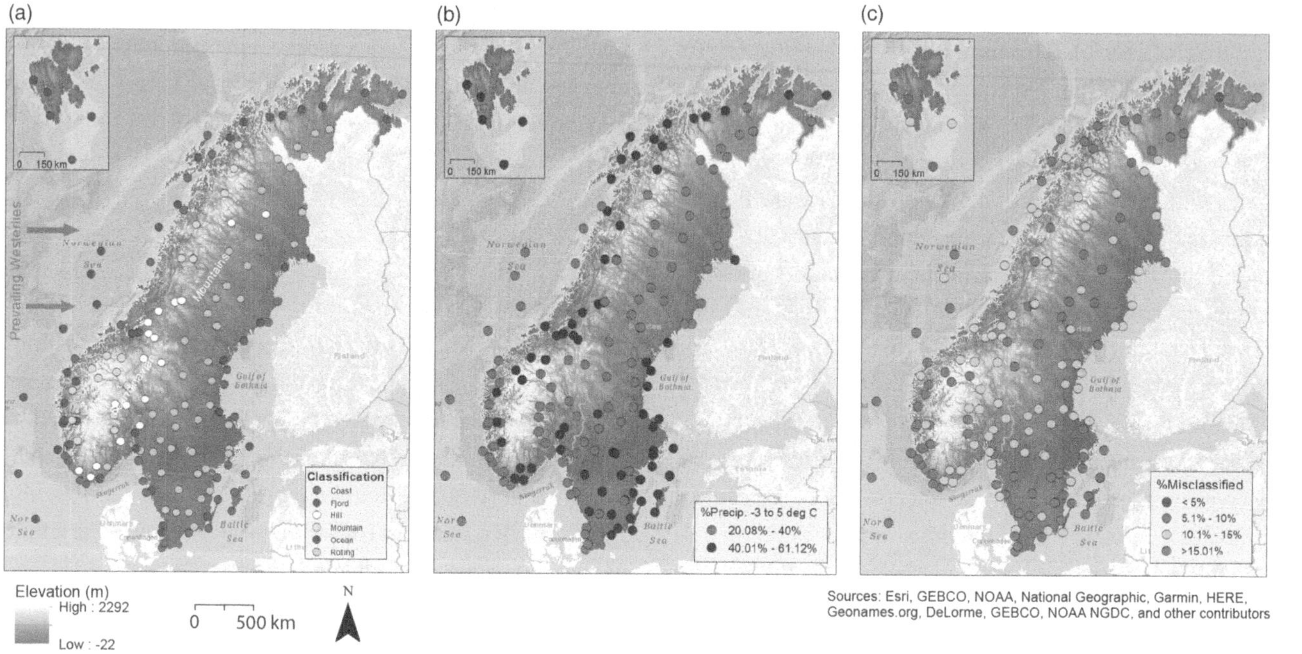

Figure 1 | (a) Map showing the locations of Scandinavian meteorological stations classified using the original-automated method. Topography and the general direction of the prevailing westerly winds are also shown. (b) Stations symbolized based on the percent of precipitation occurring between −3 and 5 °C. (c) Stations symbolized by the percent of misclassified precipitation when using the single country-wide Trs.

set of meteorological stations. Basic physiographic categories were determined by user-defined criteria to classify sites as the ocean, coast, fjord, rolling, hill, or mountain (Table 1; Figure 1(a)). Key aspects of this method are the differentiation of ocean-influenced versus land sites within a 15 km radius of the station, the differentiation of land sites based on maximum relief within a 15 km station radius, and the assumption that prevailing westerlies caused all Norwegian stations to be windward and all Swedish stations to be leeward regardless of local terrain. Further

Table 1 | Physiographic categories used in Feiccabrino & Grigg (2016), the starting point for all three classification methods in this study (original-automated, new-automated, and semi-manual)

Physiographic category	% Water within 15 km	Elevation change within 15 km
Ocean	90–100	N/A
Coast	60–90	
Fjord	40–10	
Rolling	<10%	0–499 m
Hill		500–999 m
Mountain		Above 1,000 m

details on the method can be found in Feiccabrino & Grigg (2016).

New-automated method – all sites

As in the previous study, ArcGIS was used to develop a 'new-automated' method that attempted to improve on the simplistic country-based classification of windward and leeward sites. This approach uses a measured upslope or downslope classification based on the location of the maximum elevation within each 15 km station buffer relative to the station (Supplemental Table S-1 in Supplementary Materials). The orientation of the maximum elevation point relative to the station was determined using the Euclidean Direction tool within the Spatial Analyst extension. Sites with a maximum elevation located to the west of the station were classified as downslope sites, while sites with a maximum elevation located to the east were classified as upslope. Elevation data were obtained from a 30-arc second European digital elevation model (GTOP30; USGS 1996). This upslope or downslope distinction was then combined with the 15 km maximum relief-based categorization of land sites from the original-automated method (Table 1; Figure 1(a)).

Semi-manual method – all sites

With the expansion of the study area (Figure 1), the number of impacting climatic variables increased. Sites from southern Scandinavia could have weather coming from Europe over the Baltic Sea which would not experience the same orographic effect as sites further north, while Svalbard stations are much further north, have frozen ground and little relief. To address this variability, a semi-manual classification (Supplemental Table S-1) was done using the original-automated method to first determine basic physiographic categories (Table 1). Further categorizations of the land sites were then determined manually based on the relative location of the station to mountains or high-ground and the potential for different air mass source region effects on advecting air. Ocean, coastal, and fjord stations were similarly classified based on the direction of the expected mean wind flow and substrate (i.e. frozen ground, deep or shallow ocean; see Supplemental Table S-2).

Relief versus elevation – mountain and hill sites only

In Scandinavia, relief of the terrain is believed to have a stronger effect than the elevation on changing atmospheric conditions due to the windward ascent of air and the leeward decent of air during precipitation events. However, since the elevation is readily available and often used in hydrological models (Lehning *et al.* 2011; Henn *et al.* 2018; Roth *et al.* 2018), this study tests the use of station, maximum and average elevation from within a 15 km station radius as classification criteria (Supplemental Table S-3). These trials were conducted only on the 37 stations originally classified as mountain or hill sites, where relief was large enough to potentially impact Trs. The methods followed the new-automated classification scheme except maximum and average elevation within the 15 km radius or station elevation replaced maximum relief.

Station radius size – mountain and hill sites only

A final variation in the classification methodology used 5 and 1 km station radii to determine the different GIS-derived elevation parameters for mountain and hill sites (Table 2) in order to better characterize site-specific microscale topography. This approach used the new-automated methodology at 5 and 1 km radii to determine relief and upslope versus downslope. Further tests using 5 and 1 km radii were performed for the different elevation parameters described in the previous section.

Table 2 | Percent reduction in error in mountain and hill sites relative to single country-wide Trs when different elevation parameters and station radii were used to calculate landscape-based Trs

Station radius size	Relief or elevation parameter	% Reduction in misclassified precipitation
15	Maximum relief	0.22
5		0.64
1		**0.90**
15	Maximum elevation	**0.89**
5		0.72
1		**0.89**
15	Average elevation	0.36
5		0.34
1		**1.09**

Bold values highlight the radius with the largest percent reduction in error for each elevation parameter.

Calculations

Using the same method as Feiccabrino & Grigg (2016), non-precipitation observations for each station were removed along with precipitation events occurring in air temperatures warmer than 5 °C and cooler than −3 °C. Mixed phase observations were removed, as were freezing rain/drizzle observations, which can be characterized as rain or snow and consisted of less than 1% of the remaining precipitation observations. The total sum of misclassified precipitation events for each 0.1 °C interval of possible Trs between −3 and 5 °C was calculated as the total of PPDS assigned snow events observed as rain and PPDS assigned rain events observed as snow divided by the total number of precipitation events in the observation data set. See Feiccabrino & Grigg's (2016) study for further details on calculating misclassified precipitation.

Optimum Trs values assigned for physiographic groups in a PPDS were set at the air temperature with the lowest group average of station percent misclassified precipitation. This is not necessarily the Trs that would result in the lowest number of misclassified events, as the sample size was different for each station. However, this allowed stations with small and large sample sizes to be given equal weighting toward the Trs for the physiographic groups within a PPDS.

RESULTS

All sites – comparison of three methods

Optimal Trs assigned by all three landscape-based PPDS classification methods resulted in slight reductions in the average percent of misclassified precipitation relative to the country-wide thresholds (Figure 2; Supplemental Table S-1). The semi-manual classification method showed the lowest average percent of misclassified precipitation at 10.49% which amounted to a 0.53% average reduction in error compared to the single country-wide threshold. The results from the two automated methods were very similar with the new-automated method performing slightly better than the original-automated method, 0.20% versus 0.16% average reduction in misclassified precipitation, respectively. In all three methods, the largest average reduction in error was seen in Norway, with the semi-manual method showing the largest difference between countries, 0.88% for Norway, and 0.19% for Sweden.

An examination of the spatial distribution of the results from all three methods grouped by the major physiographic category shows that the average percent reduction in error was greatest for sites categorized as the ocean (0.68%) and mountain (0.67%). However, these results also show that mountain and hill sites have the highest average percent of misclassified precipitation, at 12.02% and 11.81%, respectively. The rolling and coastal categories showed the least improvement at 0.17% and 0.11% but along with the fjord sites have the lowest percentages of misclassified precipitation (9.60–10.73%).

Mountain and hill sites – relief versus elevation

The higher rates of misclassified precipitation events from all three methods for mountain and hill categories were the motivation to further refine the classification of topographically complex sites. The first set of calculations compared the use of maximum relief versus various elevation parameters from within a 15 km station radius (Supplemental Table S-3). When no additional classification of sites based on the prevailing wind direction was used, station elevation produced the greatest (0.75%) reduction in error of misclassified precipitation, while average elevation produced the smallest reduction in error (0.19%). However, when the categories were expanded to include upslope and downslope wind designations, relief and all elevation parameters showed some increase in the reduction of error (Figure 3). The use of maximum elevation within the 15 km station combined with upslope versus downslope yielded the largest reduction in error at 0.89%. All the elevation parameters outperformed relief when using the 15 km station radius, with the exception of average elevation alone (no prevailing wind direction) (Figure 3, Supplemental Table S-3).

Mountain and hill sites – station radius size

The comparison of using 15, 5, or 1 km station radii to determine maximum relief showed a steady increase in the reduction in error of misclassified precipitation with decreasing station radii (Figure 3, Table 2). The use of maximum relief within a 1 km radius resulted in the largest reduction in error at 0.90%. When the smaller station radii were used to determine maximum and average elevation, the results showed no improvement between 15 and 5 km radii. However, the use of a 1 km radius improved these results, with the largest reduction in misclassified precipitation (1.09%) occurring with average elevation calculated for a 1 km radius.

DISCUSSION

Automated versus semi-manual methods

The three basic landscape classification methods (original-automated, new-automated, and semi-manual) tested in this study all resulted in an average percent reduction in misclassified precipitation events relative to the use of a single country-wide threshold (Supplemental Table S-1, Figure 2).

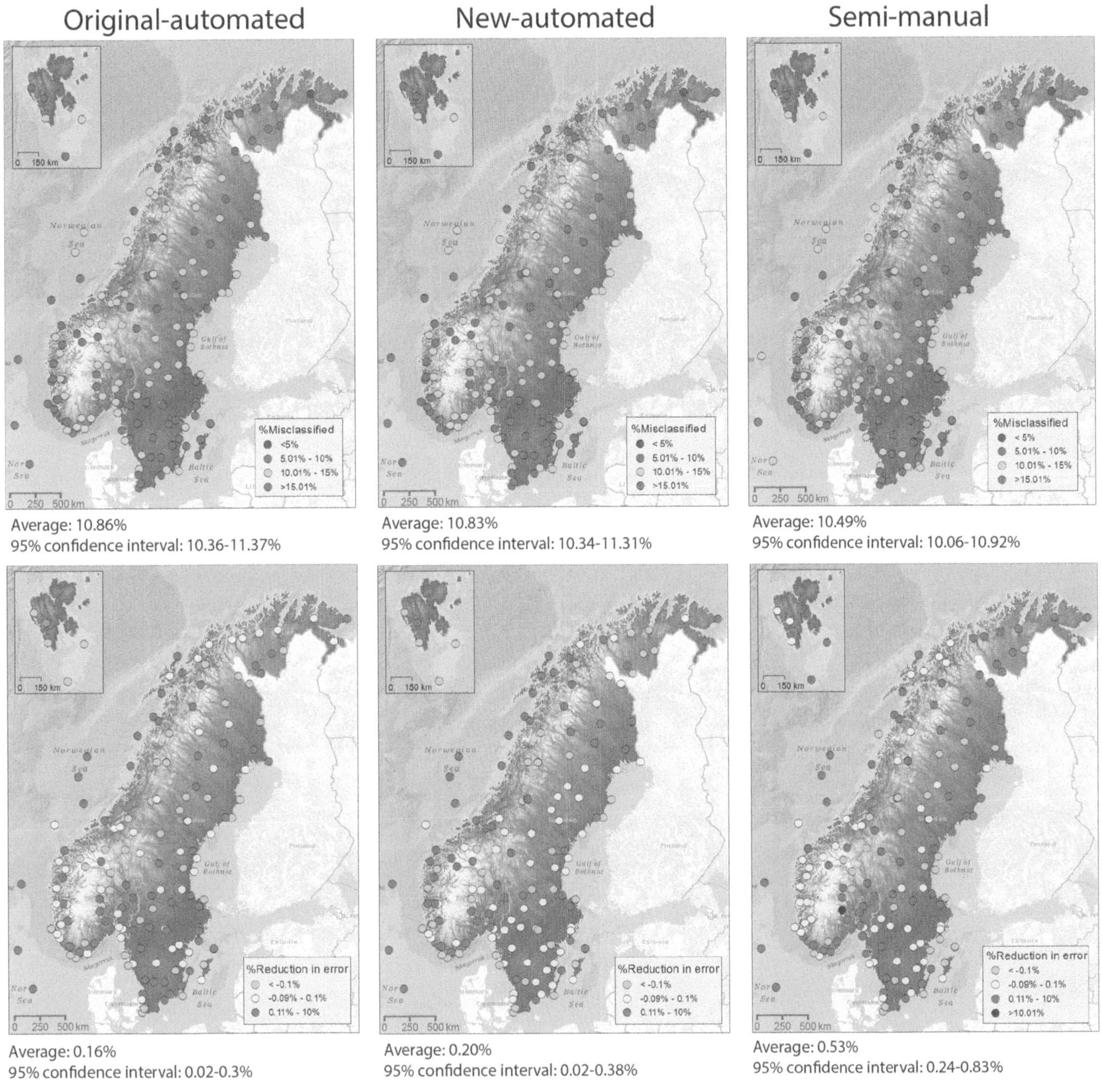

Figure 2 | Results of the three methods tested. The upper set of maps shows the resulting percent misclassified precipitation of each site. The lower set of maps shows the percent reduction in error of each classification method by site relative to a single country-wide Trs.

However, both automated methods resulted in less improvement (0.16% and 0.20% versus 0.59% reduction) than the Feiccabrino & Grigg (2016) study, which used the original-automated method on a smaller, geographically constrained set of stations. The difference in the performance of the same method between this and the previous study can be explained by the larger and more climatically heterogeneous set of stations used in the current study. The semi-manual method addresses the expected shortcomings of the larger data set and results in a similar reduction in error (0.53%) as the original study (0.59%). One of the key variables accounted for by the semi-manual method and not the automated methods is the changing precipitation source area from north to south along the Scandinavian peninsula

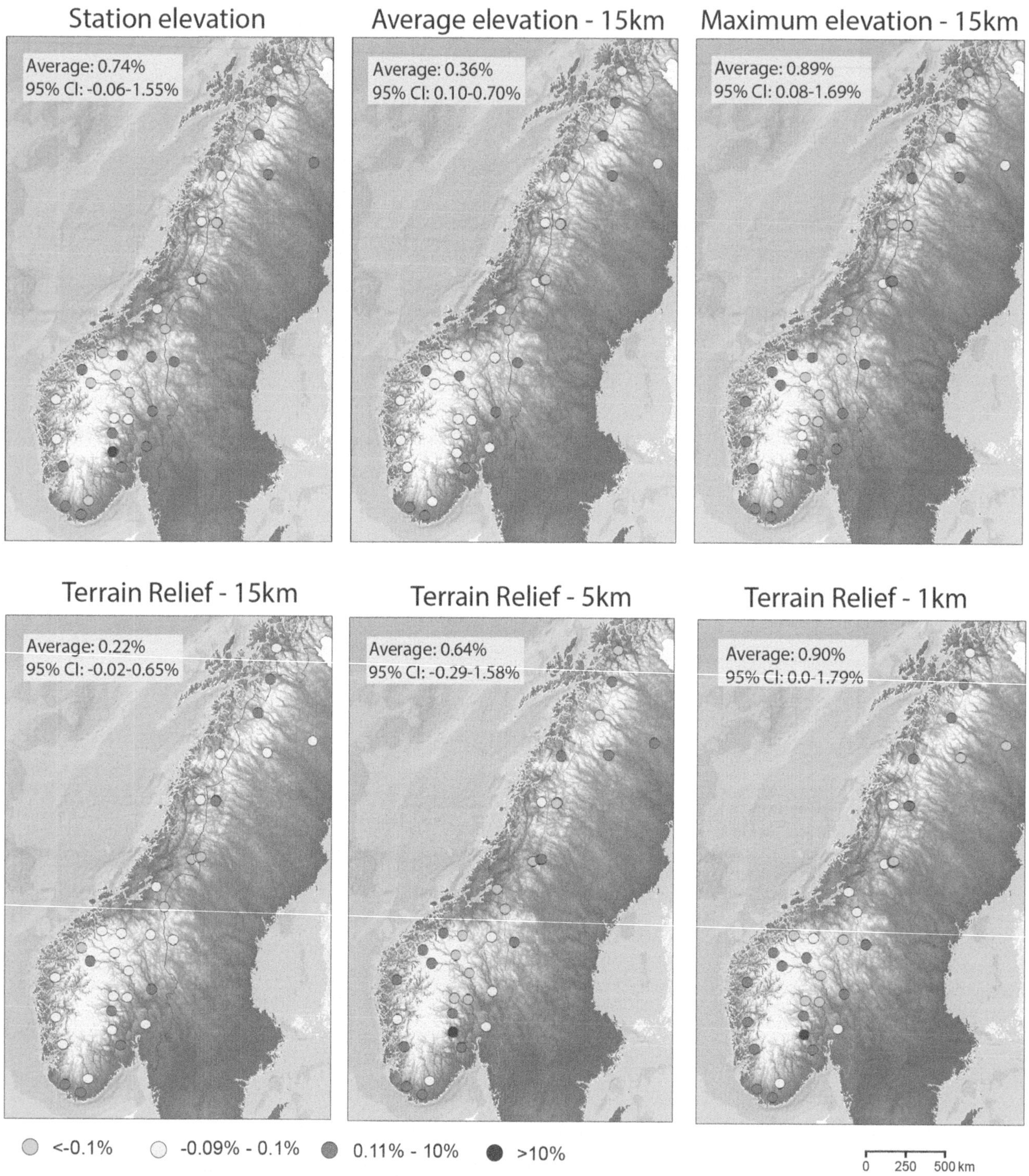

Figure 3 | Maps comparing percent reduction in error of misclassified precipitation for hill and mountain sites relative to a country-wide Trs using different elevation parameters (top panel) and different station radii (bottom panel). For both panels, the average percent reduction in error for all sites increases from left to right.

(Supplemental Table S-2). For example, coastal stations in northern Norway often receive snow advected from the colder Barents Sea which is less likely to melt while falling to the ground, regardless of surface temperatures, and results in a higher Trs (1.5–1.4 °C) than those for central Norway coastal sites (1.2–1.3 °C), which receive weather from the Gulf Stream-warmed Norwegian Sea. Likewise, southern coastal sites have the lowest Trs (0.6 °C) because of the impact of a modified continental air mass from the European mainland with less maritime influence. A future approach to objectively account for diverse precipitation source areas is to use wind and storm track data to first delineate regions with similar precipitation source areas that can then be classified separately by other landscape variables.

Although the new-automated method with GIS-derived upslope or downslope designations showed only a slight improvement relative to the original-automated method (0.16% versus 0.20%), these results would likely be improved when used on a smaller subset of stations or modified to accommodate known geographic changes in the prevailing winds. This method was most effective relative to the other methods at reducing average % error for the hill stations (0.5%) and was least effective (0.1%) at Norway mountain stations (Figure 2). These results suggest that small and previously unrecognized changes in upslope and downslope conditions in regions of low-to-moderate relief may impact Trs, but that in areas with multiple mountain ridges and passes such as the southern Norwegian mountains, other physiographic factors are important. In intermountain areas, the air loses moisture after each successive mountain pass, and the atmospheric changes caused by upslope and downslope winds become less effective than the first coastal mountains. Additionally, dense cold air can get trapped in the valleys between mountains causing warm air to pass over the mountains without experiencing orographic lifting.

PPDSs have broad societal applications across multiple disciplines including weather forecasting, transportation safety, and snowpack modeling. Conceptual PPDSs rely on a best estimate of Trs that can be applied across both spatially and temporally changing atmospheric conditions. The methods examined in this study provide the basis for an automated approach to deriving Trs that accounts for spatial, landscape-driven atmospheric changes averaged through time using widely available meteorological data that are a better starting point for the development of conceptual PPDS than broadly standardized Trs.

Refinement of classification for topographically complex regions

Higher rates (>11%) of misclassified precipitation at mountain stations across all three methods were the motivation for the further refinement of landscape classification for mountain and hill stations. Results show that the use of elevation parameters (station elevation, maximum elevation, and average elevation) instead of relief as a classification criteria yielded better results when using the 15 km station radius (Figure 3). The largest reduction (average: 0.89%) occurred using maximum elevation and upslope/downslope wind direction (Supplemental Table S-3). These results indicate that atmospheric conditions within 15 km are impacted by the up- or downslope movement of air caused by topographic highs. Additionally, Jennings *et al.* (2018) suggest that at higher elevations, less dense air may allow snow to persist at higher temperatures, and thus local high points could warm Trs and may also explain the better performance of maximum elevation.

A reduction in the station radii from 15 km to 5 km and to 1 km improved the performance of relief and average elevation as criteria for classification (Figure 3; Table 2). Maximum elevation did not improve between the 15 and 1 km radii, which is consistent with the explanation of lower air pressure at higher elevation impacting Trs (e.g. Jennings *et al.* 2018) because altitudinal changes in air pressure are laterally consistent. The increased importance of relief and average elevation at microscales suggests that orographically enhanced precipitation and lowering of both the zero-degree isotherm and snow lines (Minder *et al.* 2011) on the windward side of a mountain begin to affect Trs within 5 km of the station.

The findings from the mountain and hill data set contribute to other efforts to better predict snowfall in topographically complex regions (e.g. Marks *et al.* 2013). This study confirms the importance of microscale landscape parameters and could be combined with the microscale snow hydrology work by Cristea *et al.* (2017) which uses the relative elevation position and aspect to better characterize the snow cover in alpine regions. Future efforts of landscape classification should include the use of maximum elevation at macroscales (>10 km) to develop a first order of classification. Mountain and hill sites (maximum elevations

>499 m) can then be further classified using a microscale (<5 km) analysis of terrain relief and/or average elevation. In both stages of classification, the interaction between topography and the prevailing wind direction should be accounted for by an analysis of upslope versus downslope conditions.

CONCLUSIONS

This study contributes to an expanding body of work on the use of physiographic variables to derive more accurate Trs at global, hemispheric, and regional scales (Ólafsson & Haraldsdóttir 2003; Dai 2008; Harpold *et al.* 2017a; Jennings *et al.* 2018). These efforts are crucial to improving conceptual PPDS, which are widely used across meteorology, hydrology, and ecology. The methods tested reveal both the potential for and limitation of automated, GIS-based landscape classification at regional scales. The original classification method when applied to a climatically diverse region did not perform as well as it did when applied to the smaller latitudinal range used by Feiccabrino & Grigg (2016), although it still resulted in a small improvement (0.16% reduction in error) in misclassified precipitation relative to the set country-wide Trs. The modification of the original method to include the GIS-derived upslope versus downslope air movement showed a 0.20% reduction in error and also fell short of improvements made in the original study. Another modification tested was the manual sub-classification of stations following the initial automated landscape classification which resulted in a 0.53% average reduction in error and highlights the need to better integrate precipitation source area and intermountain physiographic effects into future automated methods.

A closer examination of mountain and hill stations indicates that with the 15 km station buffer, the greatest average reduction in misclassified precipitation (0.89%) occurred when using maximum elevation and the upslope and downslope movement of air. When smaller station buffers were used, maximum elevation results remained unchanged, but the use of terrain relief and average elevation improved average reduction in error rates up to 0.73%.

Based on the results of this study, several recommendations can be made for future work on landscape classification. The first is that the use of GIS-based, automated methods in climatically diverse regions could be improved by an initial classification of the precipitation source area based on the analysis of wind direction data during precipitation events. Once similar precipitation source regions are established, stations can be further classified using a 15 km station buffer to determine: (1) proximity to the ocean, (2) maximum elevation, (3) and whether the station is up- or downslope relative to the precipitation source. For stations with maximum elevations >499 m, a third tier of classification that uses terrain relief and/or average elevation within a 1 km station buffer is recommended. This and future work aimed at improving the landscape classification of Trs provide needed and practical approaches to decreasing the error currently associated with conceptual PPDS in mid- and high-latitude regions.

SUPPLEMENTARY MATERIAL

The Supplementary Material for this paper is available online at https://dx.doi.org/10.2166/nh.2020.081.

REFERENCES

Aalto, J., Scherrer, D., Lenoir, J., Guisan, A. & Luoto, M. 2018 Biogeophysical controls on soil-atmosphere thermal differences: implications on warming Arctic ecosystems. *Environmental Research Letters* **13** (7), 1–11. https://doi.org/10.1088/1748-9326/aac83e.

Cristea, N. C., Breckhelmer, I., Raleigh, M. S., HilleRisLanbers, J. & Lundquist, J. D. 2017 An evaluation of terrain-based downscaling of fractional snow covered area data sets based on LiDAR-derived snow data and orthoimagery. *Water Resources Research* **53** (8), 6802–6820. https://doi.org/10.1002/2017WR020799.

Dai, A. 2008 Temperature and pressure dependence of the rain-snow phase transition over land and ocean. *Geophysical Research Letters* **35** (12). https://doi.org/10.1029/2008GL033295

Daly, S. F., Davis, R., Ochs, E. & Pangburn, T. 2000 An approach to spatially distributed snow modelling of the Sacramento and San Joaquin basins, California. *Hydrological Processes* **14** (18), 3257–3271. http://dx.doi.org/10.1002/1099-1085(20001230)14:18 < 3257:AID-HYP199 > 3.0.CO;2-Z.

Fang, X., Pomeroy, J. W., Ellis, C. R., MacDonald, M. K., DeBeer, C. M. & Brown, T. 2013 Multi-variable evaluation of hydrological model predictions for a headwater basin in the Canadian Rocky Mountains. *Hydrology and Earth System Sciences* **17** (4), 1635–1659. https://doi.org/10.5194/hess-17-1635-2013.

Feiccabrino, J. M. & Grigg, L. D. 2016 A new GIS landscape classification method for rain/snow temperature thresholds in surface based models. *Hydrology Research* **48** (4), 902–914. https://doi.org/10.2166/nh.2016.055.

Feiccabrino, J., Lundberg, A. & Gustafsson, D. 2012 Improving surface-based precipitation phase determination through air mass boundary identification. *Hydrology Research* **43** (3), 179–191. http://dx.doi.org/10.2166/nh.2012.060.

Feiccabrino, J., Graff, W., Lundberg, A. & Sandström, N. 2015 Meteorological knowledge useful for the improvement of snow rain separation in surface based models. *Hydrology* **2** (4), 266–288. https://doi.org/10.3390/hydrology2040266Harpold.

Harpold, A. A., Kaplan, M. L., Klos, P. Z., Link, T., McNamara, J. P., Rajagopal, S., Schumer, R. & Steele, C. M. 2017a Rain or snow: hydrologic processes, observations, prediction, and research needs. *Hydrology and Earth System Sciences* **21** (1), 1–22. https://doi.org/10.5194/hess-21-1-2017.

Harpold, A. A., Rajagopal, S., Crews, J. B., Winchell, T. & Schumer, R. 2017b Relative humidity has uneven effects on shifts from snow to rain over the Western U.S. *Geophysical Research Letters* **44** (19), 9742–9750. https://doi.org/10.1002/2017GL075046.

Henn, B., Newman, A. J., Livneh, B., Daly, C. & Lundquist, J. D. 2018 An assessment of differences in gridded precipitation datasets in complex terrain. *Journal of Hydrology* **556**, 1205–1219. https://doi.org/10.1016/j.jhydrol.2017.03.008.

Jennings, K. S., Winchell, T. S., Livneh, B. & Molotch, N. P. 2018 Spatial variation of the rain-snow temperature threshold across the Northern Hemisphere. *Nature Communications* **9** (1148). doi:10.1038/s41467-018-03629-7.

Kokkonen, T., Koivusalo, H., Jakeman, A. & Norton, J. 2006 Construction of a degree-day snow model in the light of the 'ten iterative steps in model development'. In: *Proceedings of the iEMSs Third Biennial Meeting: Summit on Environmental Modelling and Software*, July, Burlington, VT, USA. Available from: http://www.iemss.org/iemss2006/papers/w4/Kokkonen.pdf (accessed 24 July 2018).

Kongoli, C. E. & Bland, W. L. 2000 Long-term snow depth simulations using a modified atmosphere-land exchange model. *Agricultural and Forest Meteorology* **104** (4), 273–287. https://doi.org/10.1016/S0168-1923(00)00169-6.

Lackmann, G. M., Keeter, K., Lee, L. G. & Ek, M. B. 2002 Model representation of freezing and melting precipitation: implications for winter weather forecasting. *Weather Forecasting* **17** (5), 1016–1033. http://dx.doi.org/10.1175/1520-0434(2003)017<1016:MROFAM > 2.0.CO;2.

Lehning, M., Grunewald, T. & Schimer, M. 2011 Mountain snow distribution governed by an altitudinal gradient and terrain roughness. *Geophysical Research Letters* **38** (19). https://doi.org/10.1029/2011GL048927.

Loth, B., Graf, H. F. & Oberhuber, J. M. 1993 Snow cover model for global climate simulations. *Journal of Geophysical Research* **98** (D6), 10451–10464. https://doi.org/10.1029/93JD00324.

Marks, D., Winstral, A., Reba, M., Pomeroy, J. & Kumar, M. 2013 An evolution of methods for determining during-storm precipitation phase and the rain/snow transition elevation at the surface in a mountain basin. *Advances in Water Resources* **55**, 98–110. https://dx.doi.org/10.1016/jadvwatres.2012.11.012.

Minder, J. R., Durran, D. R. & Roe, G. H. 2011 Mesoscale controls on the mountainside snow line. *Journal of the Atmospheric Sciences* **68** (9), 2107–2127. http://dx.doi.org/10.1175/JAS-D-10-05006.1.

Norwegian Meteorological Institute (NMI) 2018 *Meteorological Data*. Available from: http://eklima.met.no (accessed 20 December 2018).

Ólafsson, H. & Haraldsdóttir, S. H. 2003 Diurnal, seasonal, and geographical variability of air temperature limits of snow and rain. In: *Proceedings of the International Conference on Alpine Meteorology (ICAM 2003)*, May 19–23, Brig, Switzerland, pp. 473–476. Available from: www.map.meteoswiss.ch/map-doc/icam2003/741.pdf (accessed 23 September 2016).

Roth, A., Hock, R., Schuler, T. V., Bieniek, P. A., Pelto, M. & Aschwanden, A. 2018 Modeling winter precipitation over the Juneau Icefield, Alaska, using a linear model of orographic precipitation. *Frontiers in Earth Science* **6**. https://doi.org/10.3389/feart.2018.00020.

Swedish Meteorological and Hydrological Institute (SMHI) 2018 *Meteorological Data*. Available from: http://opendatadownload-metobs.smhi.se/explore/?parameter=8 (accessed 20 December 2018).

United States Geological Survey (USGS) 1996 *Global 30 Arc-Second Elevation (GTOP30)*. Available from: https://www.usgs.gov/centers/eros/science/usgs-eros-archive-digital-elevationglobal-30-arc-second-elevation-gtopo30?qtscience_center_objects=0#qt-science_center_objects.

Ye, H., Cohen, J. & Rawlins, M. 2013 Discrimination of solid from liquid precipitation over northern Eurasia using surface atmospheric conditions. *Journal of Hydrometeorology* **14** (4), 1345–1355. http://dx.doi.org/10.1175/JHM-D-12-0164.1.

First received 28 May 2019; accepted in revised form 5 December 2019. Available online 14 February 2020

Precipitation phase uncertainty in cold region conceptual models resulting from meteorological forcing time-step intervals

James M. Feiccabrino

ABSTRACT

Precipitation phase determination is a known source of uncertainty in surface-based hydrological, ecological, safety, and climate models. This is primarily due to the surface precipitation phase being a result of cloud and atmospheric properties not measured at surface meteorological or hydrological stations. Adding to the uncertainty, many conceptual hydrological models use a 24-h average air temperature to determine the precipitation phase. However, meteorological changes to atmospheric properties that control the precipitation phase often substantially change at sub-daily timescales. Model uncertainty (precipitation phase error) using air temperature (AT), dew-point temperature (DP), and wet-bulb temperature (WB) thresholds were compared using averaged and time of observation readings at 1-, 3-, 6-, 12-, and 24-h periods. Precipitation phase uncertainty grew 35–65% from the use of 1–24 h data. Within a sub-dataset of observations occurring between AT −6 and 6 °C representing 57% of annual precipitation, misclassified precipitation was 7.9% 1 h and 11.8% 24 h. Of note, there was also little difference between 1 and 3 h uncertainty, typical time steps for surface meteorological observations.

Key words | conceptual models, precipitation phase, snow, snow model, temperature threshold

James M. Feiccabrino
Department of Water Resources Engineering,
Lund University,
Lund S-221 00,
Sweden
E-mail: *james.feiccabrino@googlemail.com*

INTRODUCTION

An essential question for all cold region hydrological models is: was the precipitation mass in the form of rain, snow, or a mix of the solid and liquid phase (Harpold *et al.* 2017b)? To answer this question, hydrologists have used precipitation phase determination schemes (PPDS) in their surface-based conceptual models. These PPDS often apply a single rain/snow temperature threshold (T_{RS}) where all precipitation occurring in the same and cooler temperatures are assigned to the snow phase, and rain if warmer (e.g., Bartlett *et al.* 2006). Many conceptual hydrological models use either a 1-h or 24-h time steps, but the precipitation phase uncertainty related to the choice of a time step is a relatively unexplored gap in the research (Harpold *et al.* 2017b). This study intends to quantify precipitation phase uncertainty related to the time steps of 1, 3, 6, 12, and 24 hours, and explains some meteorological reasoning to support the findings.

Snowfall and rainfall have different effects on water and energy fluxes, and when misclassified can result in streamflow, surface albedo, or winter snow water storage calculation errors (Jennings *et al.* 2018). The significance of PPDS uncertainties depends on many factors, such as the intended model application, with a magnitude that can vary between precipitation events and or locations (Harpold

doi: 10.2166/nh.2020.080

et al. 2017b). In the Scandinavian Peninsula, 41.77% of annual precipitation occurred with air temperatures (AT) −3 to 5 °C in Norway, and 38.49% in Sweden with station maximum and minimum of 61.12% and 20.08%, respectively (Grigg *et al.* 2020). This abundance of precipitation occurring in near-freezing temperatures leads to a fair amount of precipitation phase uncertainty within conceptual models. Therefore, it is logical to seek ways to decrease this precipitation phase uncertainty.

Incorrect precipitation phase determination can have a cascading negative effect on both rapid response (e.g., flooding and road maintenance) and longer-term (e.g., water supply and ecosystem response) hydrological models (Harpold *et al.* 2017a). For instance, a massive snow event misclassified as rain in a model could result in: (1) a rapid response flood model indicating higher water levels due to a significant melt event which would not be observed or (2) energy loss and wetting of a modeled snowpack which would be unrepresentative of actual conditions. Rapid response models for flooding can be on a 15-min, 1-, 3-, or 6-h timescale to allow quick responses to heavy precipitation events. Other models do not require such high-temporal resolution and may have 24 h or even monthly timescales.

Conceptual hydrological models often use a set T_{RS} calibrated over a large area regardless of changes in physiography, vegetation, or other characteristics that may affect local/regional average atmospheric conditions (Grigg *et al.* 2020). Precipitation phase at the ground surface is a result of microphysical processes (melting, freezing, condensation, evaporation, ice condensation, and sublimation) between hydrometers and the atmosphere they fall through (see Stewart 1992; Thériault & Stewart 2010). The use of a set T_{RS} assumes that atmospheric conditions acting on hydrometeors falling through the lower atmosphere are invariant (Feiccabrino *et al.* 2015) and is, therefore, a source of precipitation phase uncertainty (Feiccabrino 2020). However, the use of 24-h average temperatures also assumes that atmospheric conditions over an area are static for a full day.

A 24 h time step does not account for many regular atmospheric changes two of which are: (1) diurnal changes in temperature which are affected by clear skies, overcast skies, partly cloudy skies, or changes in the cloud cover through the day and (2) frontal boundaries and troughs which separate air-masses with often vastly different atmospheric properties.

Typically, on a cloud-free day, the near-surface air and the boundary layer are warmed by incoming short-wave solar radiation and cooled overnight as long-wave radiation is emitted from Earth. On an overcast day, the incoming short-wave radiation is reduced by cloud cover, which in turn reduces daytime high temperatures, and nighttime long-wave radiation is reradiated by the clouds moderating nightly low temperatures. If clouds move over an area in the morning after a full night of cooling, the typical daily pattern will be disrupted by a cold night and a cool day. If clouds move over an area in the evening after a full day of warming, the daily pattern will be disrupted by a warm day and a cool night. Both examples above run a chance of not being adequately represented by a 24-h average temperature.

Many meteorological changes take place on sub-daily time steps, therefore making 24-h averaged meteorological inputs into a model unrepresentative (e.g., Figures 1–4). Cyclones, described in the Norwegian cyclone model (Bjerknes 1919) and many updates to this model, e.g., the conveyor belt model (Browning 1986), describe the atmospheric interactions at air-mass boundaries causing a majority of winter precipitation (Stewart *et al.* 1994). Fraedrich *et al.* (1986) showed that 82.7% of winter precipitation mass in Germany was generated by cyclones. The most common of these air-mass boundaries are cold fronts and warm fronts, but there are also troughs, occlusions, and arctic fronts (reviewed in Feiccabrino *et al.* 2012).

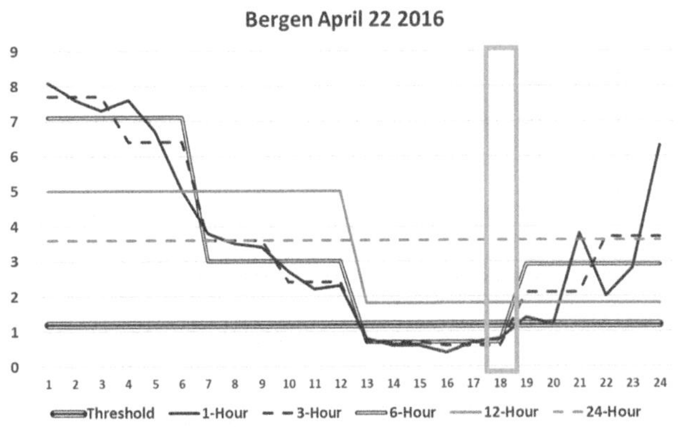

Figure 1 | 2.4 mm water equivalent/4 cm snow (blue box) misclassified as rain in 12- and 24-h time steps. Please refer to the online version of this paper to see this figure in colour: http://dx.doi.org/10.2166/nh.2020.080.

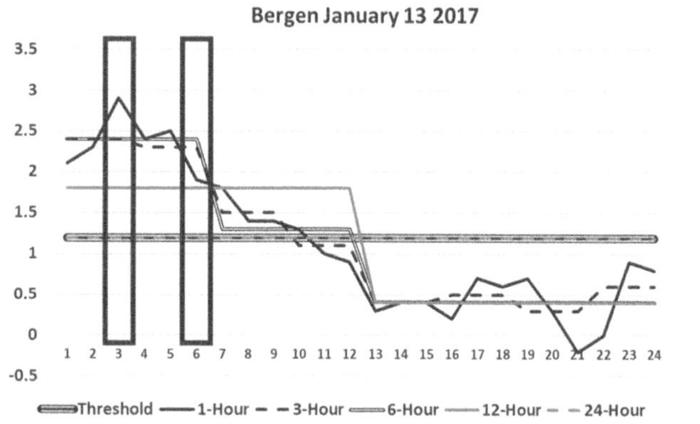

Figure 2 | 0.8 and 0.4 mm rain events (green boxes) misclassified as snow by a 24-h time step. Please refer to the online version of this paper to see this figure in colour: http://dx.doi.org/10.2166/nh.2020.080.

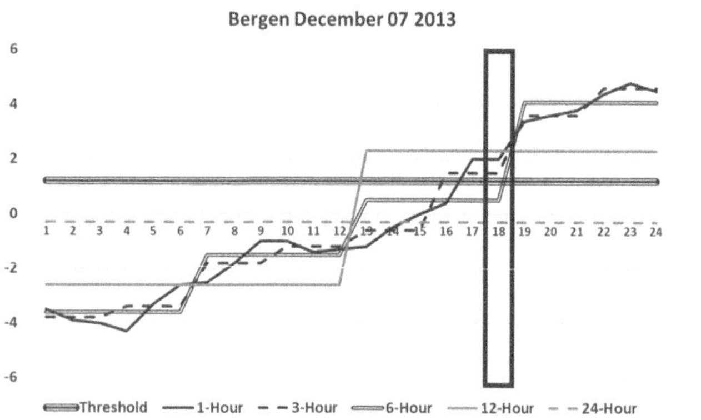

Figure 3 | 7.3 mm warm frontal rain event (green box) misclassified as snow by 6- and 24-h time steps. Please refer to the online version of this paper to see this figure in colour: http://dx.doi.org/10.2166/nh.2020.080.

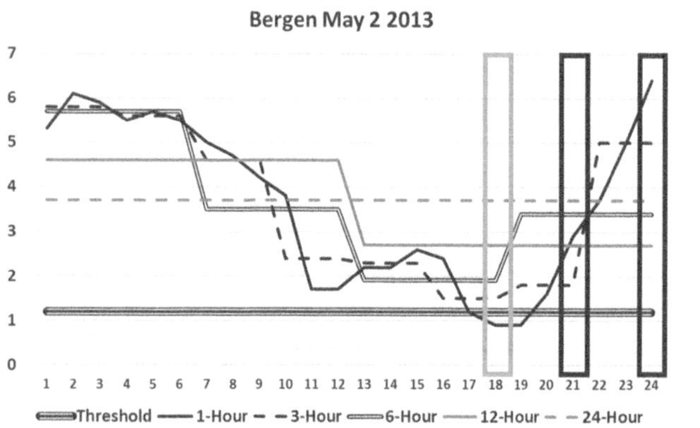

Figure 4 | 5.8 mm water equivalent snow event (blue box) misclassified by all but 1-h time step on a day with multiple precipitation phases (green rain boxes) observed. Please refer to the online version of this paper to see this figure in colour: http://dx.doi.org/10.2166/nh.2020.080.

Occlusions and warm fronts characteristically have shallow frontal slopes with warm air rising over denser cold air near the surface. This causes a broader precipitation field (see Hanesiak *et al.* 1997; Venne *et al.* 1997) allowing hours of steady precipitation before warm frontal passage. Cold fronts have a much steeper frontal slope than other fronts causing stereotypically heavier precipitation and showers due to increased lift (Browning 1986; Bjerknes 1919; Venne *et al.* 1997). Cold fronts, troughs, and arctic fronts with steeper frontal slopes typically take less time than warm fronts to pass over a station. All including showers typically occur well under the daily timescale unless the storm path drags an air-mass boundary parallel to a station.

With many winter precipitation events occurring on a sub-daily time step, this paper will compare misclassified precipitation rates for 1-, 3-, 6-, 12-, and 24-h time steps for all stations analyzed together. This addresses misclassified precipitation events resulting from the assumption of constant atmospheric conditions over a daily time step. The misclassified precipitation findings from this climatological study should translate to model uncertainty when applied in a cold region hydrological model.

METHOD

Hourly meteorological observations from 48 publicly available Norwegian weather stations (NMI 2019; Figure 5) representing diverse physiographic categories (Grigg *et al.* 2020) of ocean platform (6), island (5), Svalbard (3), coast (9), fjord (7), rolling (4), hill (5), and mountain (9) stations having 98,849 liquid or solid precipitation events/observations used in this study.

Observations with AT, dew-point temperature (DP), and wet-bulb temperature (WB) between −6 and 6 °C accounted for 57% AT and 63% DP and WB precipitation observations in each of the 1-, 3-, 6-, 12-, and 24-h datasets. This is the temperature range over which PPDS is most uncertain. Mixed-phase precipitation (4,782 observations) was excluded from the datasets due to a lack of information on rain/snow ratios and many prior studies disregarding this phase, e.g., Bartlett *et al.* (2006). Frozen precipitation (247 observations) was included as the liquid in this analysis. However, frozen precipitation can be considered either

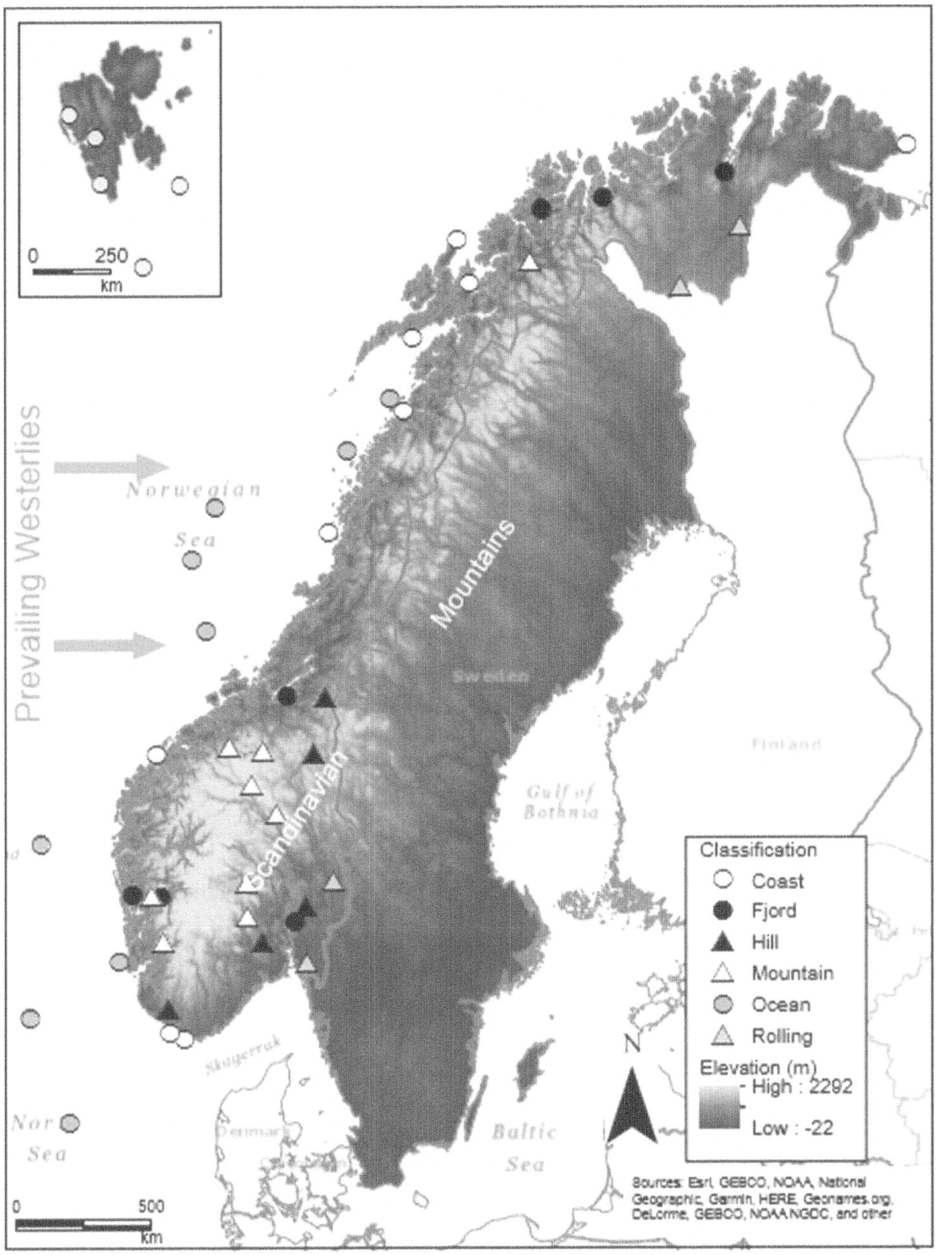

Figure 5 | Map of meteorological stations and their physiographic classification created by Laurie Grigg.

rain or snow, depending on the intended use of a model (Feiccabrino 2020).

The average of all hourly temperature values within a 24-, 12-, 6-, 3-, and 1-h temporal resolution was calculated and assigned to each observation within the period. Observation datasets for each time resolution were then created. The phase of precipitation from the manual observation was compared to the assigned precipitation phase for AT, DP, and WB PPDS for each possible T_{RS} at 0.1 °C intervals between −3 and 5 °C. A precipitation event was considered erroneous in a PPDS if the liquid or solid phase assigned by the T_{RS} did not match the manually observed phase. The sum of errors divided by the total number of observations gave a station's percent misclassified precipitation for each

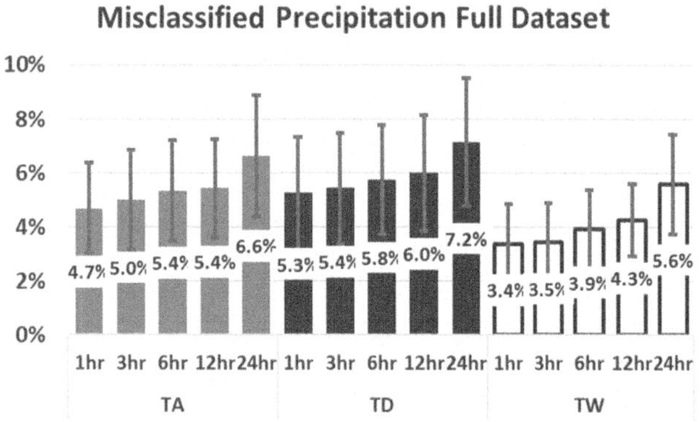

Figure 6 | Average misclassified precipitation per station for each full dataset with standard deviation error bars.

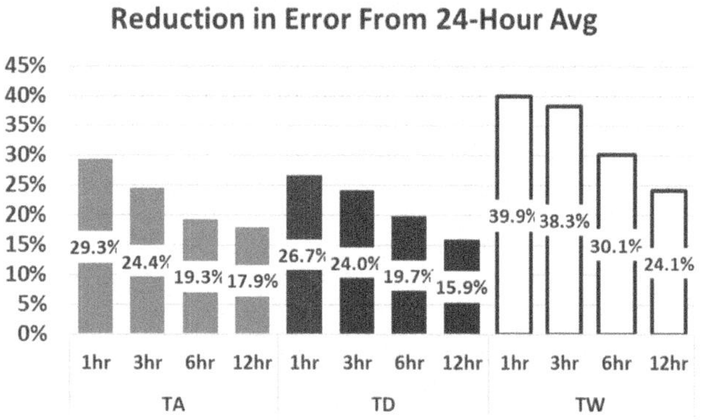

Figure 7 | Average reduction in misclassified precipitation when comparing the labeled dataset to the matching 24-h dataset for air temperature (AT), dew-point temperature (DP), and wet-bulb temperature (WB).

possible T_{RS}. The temperature corresponding to the lowest average station misclassified precipitation was the T_{RS} value for all stations analyzed together. Results for decreases in error between time steps refer to the absolute difference in precipitation points. Reduction in error results represents the relative change in error between two time-step options.

RESULTS AND DISCUSSION

Harpold *et al.* (2017b) stated in their review that PPDS accuracy is generally increased at finer timescales and or with the inclusion of RH data. There has not been a hydrological study focused on model PPDS uncertainties related to the temporal resolution of meteorological inputs. However, there have been some discussion comments, e.g., Harder & Pomeroy (2013) mentioning that there was little change in uncertainty between 15-min and 1-h time steps, but a vast difference between those datasets and a 24-h dataset. Here (Figure 6), misclassified precipitation for AT, DP, and WB all decreased for each shortening of the time resolution from 24 to 12, 6, 3, and 1 h. The misclassified precipitation results for AT and DP are similar for all temporal resolutions. However, misclassified precipitation resulting from

Table 1 | Temperature range (Range) between 90% snow fraction (90% SF) and 10% snow fraction (10% SF) and the percent misclassified precipitation occurring with temperatures cooler than −6, warmer than 6, and between −6 and 6 for air temperature (AT), dew-point temperature (DP), and wet-bulb temperature (WB) thresholds in 1, 3, 6, 12, and 24 h time-step datasets

	Air temperature			**Dew-point temperature**			**Wet-bulb temperature**		
	90% SF	**10% SF**	**Range**	**90% SF**	**10% SF**	**Range**	**90% SF**	**10% SF**	**Range**
1 h	0.1 °C	2.7 °C	2.6 °C	−2.4 °C	0.3 °C	2.7 °C	−0.8 °C	1.2 °C	2.0 °C
3 h	0.1 °C	2.8 °C	2.7 °C	−2.6 °C	0.2 °C	2.8 °C	−0.9 °C	1.1 °C	2.0 °C
6 h	0.0 °C	3.0 °C	3.0 °C	−2.8 °C	0.1 °C	2.9 °C	−1.0 °C	1.2 °C	2.2 °C
12 h	0.0 °C	3.0 °C	3.0 °C	−3.0 °C	0.1 °C	3.1 °C	−1.1 °C	1.2 °C	2.3 °C
24 h	−0.2 °C	3.1 °C	3.3 °C	−3.6 °C	0.1 °C	3.7 °C	−1.6 °C	1.2 °C	2.8 °C
Percent misclassified precipitation in each category									
	AT < −6	**AT > 6**	**−6 ≤ AT ≤ 6**	**DP < −6**	**DP > 6**	**−6 ≤ DP ≤ 6**	**WB < −6**	**WB > 6**	**−6 ≤ WB ≤ 6**
1 h	0.3%	1.2%	98.6%	1.2%	0.2%	98.6%	0.6%	0.5%	99.0%
3 h	0.3%	1.2%	98.6%	1.4%	0.2%	98.4%	0.6%	0.5%	98.9%
6 h	0.2%	1.1%	98.7%	1.2%	0.2%	98.7%	0.5%	0.4%	99.1%
12 h	0.3%	0.9%	98.9%	1.4%	0.2%	98.5%	0.5%	0.3%	99.2%
24 h	0.5%	1.1%	98.5%	2.1%	0.1%	97.8%	0.8%	0.3%	98.9%

the use of WB is much lower than AT and DP in all temporal resolutions. All three temperature measurements have similar reductions in misclassified precipitation as time resolutions are decreased (Figure 7). For example, AT, DP, and WB all had 60% of the error reduction from 24 to 1 h occur with a time step decrease to 12 h.

These results indicate that a majority of the daily variability in average temperature measurements affecting misclassified precipitation using T_{RS} was eliminated by cutting a 24-h time period in half. Interestingly, for AT, DP and WB, every reduction in time step produced a reduction in misclassified precipitation (Figures 6 and 7).

These results (Figure 6) along with many previous studies, e.g., Matsuo *et al.* (1981), have found WB to be a better indicator of the surface precipitation phase than AT. Other studies, e.g., Marks *et al.* (2013), have found DP to be a better precipitation phase indicator than AT alone. However, AT is still used in many models and is available at almost every station reporting environmental measurements. WB and DP require RH, and other observation elements not always measured by stations for their calculation. Due to availability issues for RH and the continued use of AT in many models, improvement of AT methods have elevated importance. However, RH methods to include WB, consistently identify precipitation phase better than AT in the model PPDS.

98% misclassified precipitation occurred in AT, WB, and DP temperatures between −6 and 6 °C, for 1-, 3-, 6-, 12-, and 24-h datasets (Table 1). The daily time resolution has the greatest T_{RS} difference (Figure 8(a)), largest misclassified precipitation percentages (Figures 6 and 9), and greatest mixed precipitation range (90% SF–10% SF) (Table 1) for AT DP and WB. As time resolution increased toward 1 h (Figure 8(a)–8(e)), the T_{RS} for DP warms from −1.9 to −0.7 °C and T_{RS} for AT cool slightly from 1.4 to 1.2 °C, bringing AT and DP T_{RS} closer to WB ≅ 0.0 °C. As time resolutions increase from daily to hourly, misclassified precipitation decreases in each time step (Figure 9), while the mixed-phase temperature range (Table 1) stays steady or decreases. This leads to noticeable decreases (Figure 10) and reductions in error (Figure 11) for AT, DP, and WB.

Interestingly, the proportion of observations in each −6 and 6 °C dataset for AT (57%), DP (63%), and WB (63%) remained constant while the percent misclassified

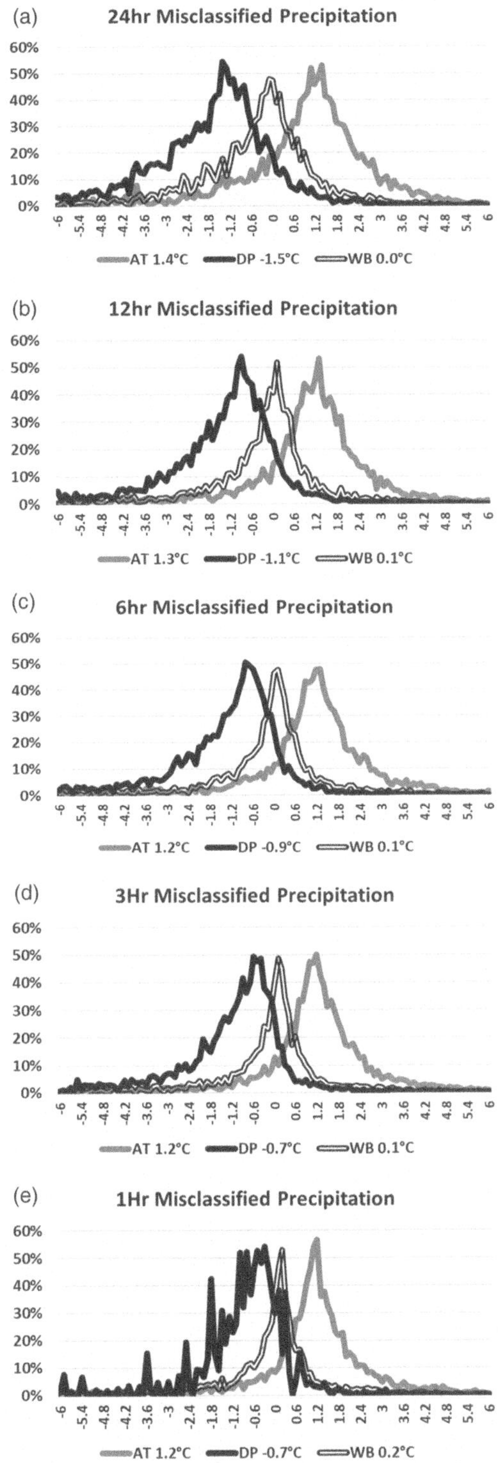

Figure 8 | (a–e) Percent misclassified precipitation occurring at each 0.1 °C between −6 and 6 °C with labeled threshold temperatures for air temperature (AT), dew-point temperature (DP), and wet-bulb temperature (WB) for datasets of 24, 12, 6, 3, and 1 h.

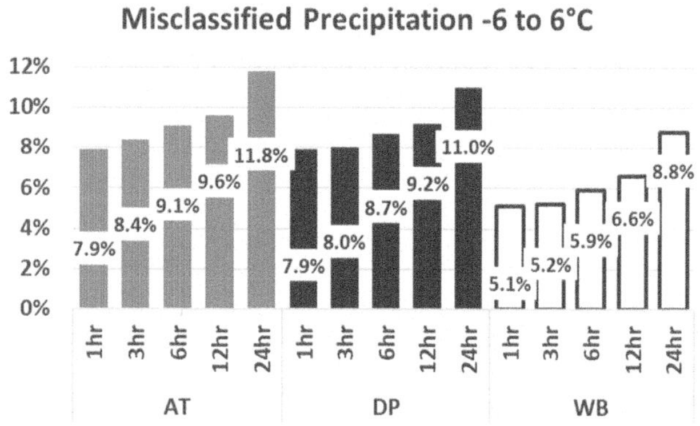

Figure 9 | Average misclassified precipitation for precipitation observations between −6 and 6 °C.

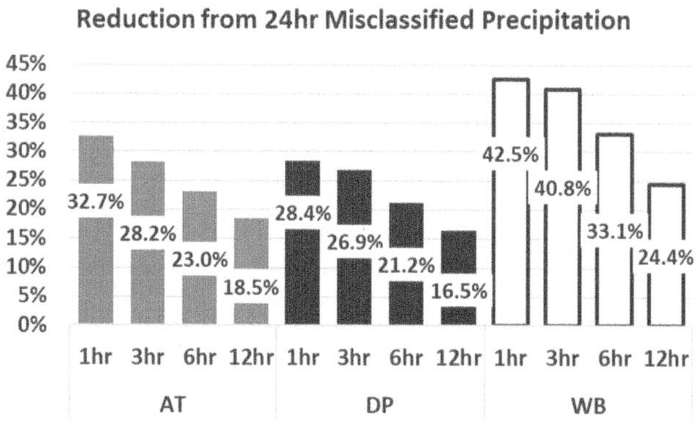

Figure 11 | Average reduction in misclassified precipitation when comparing the labeled dataset to the matching 24-h dataset for air temperature (AT), dew-point temperature (DP), and wet-bulb temperature (WB).

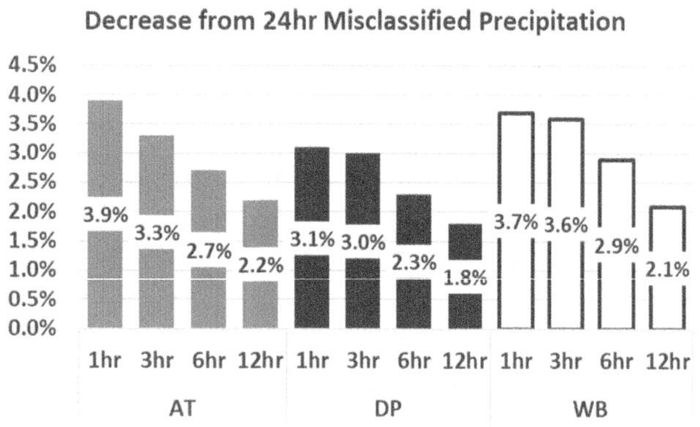

Figure 10 | Average decrease in misclassified precipitation when comparing the labeled dataset to the matching 24-h dataset for air temperature (AT), dew-point temperature (DP), and wet-bulb temperature (WB).

precipitation (Figure 9), and the mixed-phase temperature range (90% SF–10% SF) (Table 1) increased with decreasing time resolutions. The fact that the sample size was relatively constant, while misclassified precipitation and the mixed-phase temperature range increased with time steps suggests that precipitation occurring within the −6 to 6 °C range become more poorly characterized as the time resolution becomes more course.

One result of concern for further studies is the relatively large standard deviations ranging from one-third to one-fourth of the averaged misclassified precipitation for each temperature measurement in all time steps (1.7–2.3%) (Figure 6). These large standard deviations could be a result of averaging all station results together without regard for expected changes in the lower atmosphere caused by, for example, ocean heating of near-surface air, diurnal maritime cloud changes affecting daily heating and cooling patterns, or other physiographic effects on locations (Grigg *et al.* 2020).

CONCLUSIONS

- Reducing the temporal resolution of meteorological forcing data in hydrological models from 24 to 1 h greatly reduced misclassified precipitation for air temperature (AT) (29.3%), dew-point temperature (DP) (26.7%), and wet-bulb temperature (WB) (39.9%) thresholds.
- In almost all cases, reducing the temporal resolution between 24, 12, 6, 3 and hourly meteorological forcing reduced misclassified precipitation. However, the most significant decreases were between 24 and 12 h. Surprisingly, 60% of the decrease between 24 and 1-h time resolutions could be attained for AT, DP, and WB by only cutting the daily temporal resolution in half.
- It is here suggested that if attempting to reduce precipitation phase uncertainty in a cold region hydrological model with a daily air temperature time step, the best two options would be to either switch to using wet-bulb temperature or reduce the time step to 3 or 1 h for more representative meteorological forcing.

REFERENCES

Bartlett, P. A., MacKay, M. D. & Verseghy, D. L. 2006 Modified snow algorithms in the Canadian land surface scheme: model runs and sensitivity analysis at three Boreal forest stands. *Atmosphere-Ocean* **43**, 207–222. https://doi.org/10.3137/ao.440301.

Bjerknes, J. 1919 On the structure of moving cyclones. *Geophysical Publications* **1** (2), 1–8. https://doi.org/10.1175/1520-0493(1919)47<95:OTSOMC > 2.0.CO;2.

Browning, K. 1986 Conceptual models of precipitation systems. *Weather and Forecasting* **1**, 23–41. https://doi.org/10.1175/1520-0434(1986)001<0023:CMOPS > 2.0.CO;2.

Feiccabrino, J. 2020 Reducing misclassified precipitation phase in conceptual models using cloud base heights and relative humidity to adjust air temperature thresholds. *Hydrology Research* (in press).

Feiccabrino, J., Lundberg, A. & Gustafsson, D. 2012 Improving surface-based precipitation phase determination through air mass boundary identification. *Hydrology Research* **43** (3), 179–191. http://dx.doi.org/10.2166/nh.2012.060.

Feiccabrino, J., Graff, W., Lundberg, A. & Sandström, N. 2015 Meteorological knowledge useful for the improvement of snow rain separation in surface based models. *Hydrology* **2** (4), 266–288. https://doi.org/10.3390/hydrology2040266.

Fraedrich, K., Bach, R. & Naujokat, G. 1986 Single station climatology of central European fronts: number, time and precipitation statistics. *Physical Atmosphere* **59** (1), 54–65.

Grigg, L. D., Feiccabrino, J. M. & Sherenco, F. 2020 Testing the applicability of physiographic classification methods towards improving precipitation phase determination in conceptual models. *Hydrology Research* (in press).

Harder, P. & Pomeroy, J. 2013 Estimating precipitation phase using a psychrometric energy balance method. *Hydrological Processes* **27**, 1901–1914. https://doi.org/10.1002/hyp.9799.

Harpold, A. A., Kaplan, M. L., Klos, P. Z., Link, T., McNamara, J. P., Rajagopal, S., Schumer, R. & Steele, C. M. 2017a Rain or snow: hydrologic processes, observations, prediction, and research needs. *Hydrology and Earth System Sciences* **21** (1), 1–22. https://doi.org/10.5194/hess-21-1-2017.

Harpold, A. A., Rajagopal, S., Crews, J. B., Winchell, T. & Schumer, R. 2017b Relative humidity has uneven effects on shifts from snow to rain over the Western U.S. *Geophysical Research Letters* **44** (19), 9742–9750. https://doi.org/10.1002/2017GL075046.

Hanesiak, J., Stewart, R., Szeto, K., Hudak, D. & Leighton, H. 1997 The structure, water budget, and radiational features of a high-latitude warm front. *Journal of Atmospheric Sciences* **54** (12), 1553–1573. https://doi.org/10.1175/1520-0469(1997)054<1553:TSWBAR>2.0.CO;2

Jennings, K. S., Winchell, T. S., Livneh, B. & Molotch, N. P. 2018 Spatial variation of the rain-snow temperature threshold across the Northern Hemisphere. *Nature Communications* **9** (1148). doi:10.1038/s41467-018-03629-7.

Marks, D., Winstral, A., Reba, M., Pomeroy, J. & Kumar, M. 2013 An evolution of methods for determining during-storm precipitation phase and the rain/snow transition elevation at the surface in a mountain basin. *Advances in Water Resources* **55**, 98–110. https://dx.doi.org/10.1016/jadvwatres.2012.11.012.

Matsuo, T., Sato, Y. & Sasyo, Y. 1981 Relationship between types of precipitation on the ground and surface meteorological elements. *Journal of the Meteorological Society of Japan* **59**, 462–476. https://doi.org/10.2151/jmsj1965.59.4_462.

Norwegian Meteorological Institute (NMI) 2019 *Meteorological Data*. Available from: http://eklima.met.no (accessed 22 May 2019).

Stewart, R. E. 1992 Precipitation types in the transition region of winter storms. *Bulletin of the American Meteorological Society* **73**, 287–296. https://doi.org/10.1175/1520-0477(1992)073<0287:PTITTR > 2.0.CO;2.

Stewart, R. E., Marwitz, J. D., Pace, J. C. & Carbone, R. E. 1994 Characteristics through the melting layer of stratiform clouds. *Journal of Atmospheric Science* **41**, 3227–3237. https://doi.org/10.1175/1520-0469(1984)041<3227:CTTMLO > 2.0.CO;2.

Thériault, J. M. & Stewart, R. E. 2010 A parameterization of the microphysical processes forming many types of winter precipitation. *Journal of Atmospheric Sciences* **67**, 1492–1508. https://doi.org/10.1175/2009JAS3224.1.

Venne, E., Jasperson, W. & Venne, D. 1997 *Difficult Weather: A Review of Thunderstorms, Fog, and Stratus, and Winter Precipitation Forecasting*. Technical Report #A246633. AIRFM Command http://DTIC.mil.

First received 31 May 2019; accepted in revised form 10 January 2020. Available online 17 March 2020

Can model-based data products replace gauge data as input to the hydrological model?

K. Sivasubramaniam, K. Alfredsen, T. Rinde and B. Sæther

ABSTRACT

Hydrological models require accurate and representative meteorological inputs for better prediction of discharge and hence, the efficient management of water resources. Numerical weather prediction model-based reanalysis data products on the catchment scale are becoming available, and they could be an alternative input data for hydrological models. This study focuses on the applicability of a set of model-based data as input to hydrological models used in inflow predictions for operational hydropower production planning of three hydropower systems in middle Norway. First, the study compared the data products with gauge measurements. Then, Hydrologiska Byråns Vattenbalansavdelning (HBV) models of the three catchments were calibrated with three different meteorological datasets (model-based, gauge and observational gridded) separately using a Monte Carlo approach. It was found that the correlation between the model-based and gauged precipitation was highly variable among stations, and daily values showed a better correlation than hourly. The performance of model-based input data with daily timestep was nearly as good as the gauge or gridded data for the model calibration. Further, the annual simulated flow volume using the model-based data was satisfactory as similar to the gauge or gridded input data, which indicate that model-based data can be a potential data source for long-term operational hydropower production planning.

Key words | HBV model, hydropower production planning, inflow prediction, meteorological reanalysis, Monte Carlo calibration, numerical weather prediction (NWP) model

K. Sivasubramaniam (corresponding author)
K. Alfredsen
Department of Civil and Environmental Engineering,
Norwegian University of Science and Technology (NTNU),
7491 Trondheim,
Norway
E-mail: *kuganesan.sivasubramaniam@ntnu.no*

K. Sivasubramaniam
T. Rinde
Norconsult AS,
Postboks 626, 1303 Sandvika,
Norway

B. Sæther
NTE Energi AS,
Sjøfartsgt. 3, 7736 Steinkjer,
Norway

INTRODUCTION

Today, precipitation-runoff models are employed as standard tools and routinely used for various hydrological applications (e.g. flood estimation, real-time flood forecasting, prediction of design flood and investigation of climate change and land use variability) (Wagener *et al.* 2004). Hydrological models combined with meteorological forecasts can provide a quantitative forecast of inflow to reservoirs and power plants, and it helps increase power production by reducing water spill and improving water management. Such models have been in operational use by hydropower companies in Norway since the 1970s, and they have proved to be cost-effective tools for hydropower operation and optimization (Killingtveit & Sælthun 1995). Calibration and updating of the states in a model are required before the model is used in an operational inflow forecast. The primary input data for precipitation-runoff models are typically time series of precipitation and air temperature with daily or hourly temporal resolution. Traditionally, *in situ* gauge observations are used as inputs for the models.

Hydrological models require accurate and representative meteorological inputs for better prediction and hence, the efficient management of water resources (Kirchner 2009; Beven 2012).

doi: 10.2166/nh.2020.076

Precipitation is an intermittent variable and various difficulties exist in obtaining quantitative precipitation precisely (Hwang *et al.* 2012). The measurements of precipitation using *in situ* gauges are subject to several error sources, such as wind-induced undercatch, wetting and evaporation losses (Førland *et al.* 1996; Taskinen & Söderholm 2016). The gauge measurement of solid precipitation (snow) in high latitudes and mountainous areas exhibits significant undercatch due to high wind conditions (Wolff *et al.* 2015). Further, the traditional *in situ* gauge observations represent point measurements and require a dense network of gauges to measure representative input on the catchment scale. However, in many areas, dense gauge networks are not common. In cases where existing sparse gauges do not capture the local precipitation distribution, the measured precipitation is not representative of the concerned catchment. Errors in the air temperature measurements are normally smaller, and the spatial variation of air temperature is also less; hence, air temperature observations from a station are generally more representative than precipitation (Ledesma & Futter 2017).

In some places (e.g. natural reserves, sanctuaries and remote mountainous areas), there are also restrictions and difficulties in operating *in situ* gauges. The Børgefjell national park in Norway which is located within the present study area is a typical example. Water draining from a 700 km^2 natural catchment is exploited for hydropower production. The power company is not able to install gauges within the nature reserve, and the hydrological model for inflow forecasting for Børgefjell is based on a single gauge located outside the area. Moreover, the operation and maintenance of precipitation gauges in remote mountainous areas incur considerable expenses.

Due to various challenges associated with the traditional approach of obtaining meteorological input data for hydrological models used in inflow predictions for operational hydropower production planning, hydropower companies in Norway seek alternative data sources for these purposes. Observational gridded datasets, remote sensing (weather radar and satellite) and numerical weather prediction (NWP)-based meteorological reanalysis data on the catchment scale can be potential alternative data sources to overcome challenges associated with traditional station data (Te Linde *et al.* 2008; Oke *et al.* 2009; Vu *et al.* 2012; Lauri *et al.* 2014; Ledesma & Futter 2017).

Observational gridded datasets are increasingly obtainable from the national and regional institutes (Haylock *et al.* 2008; Lussana *et al.* 2016; Lussana *et al.* 2018). Several studies have evaluated the observational gridded precipitation and air temperature datasets as model input compared to station data for medium- and large-scale river basins (Photiadou *et al.* 2011; Vaze *et al.* 2011; Essou *et al.* 2016a) and for small catchments (Ledesma & Futter 2017). Even though the gridded datasets have a continuous spatial coverage over the catchment and relatively fewer missing data compared to gauges, these datasets are generally derived from the available gauge measurements by spatial interpolation, and they have little additional information other than elevation (Essou *et al.* 2016a). Further, limitations in different interpolation techniques can also be a source of uncertainty (Vu *et al.* 2012; Lauri *et al.* 2014).

Precipitation measurements using remote sensing techniques (weather radar and satellite) are existing with high spatio-temporal resolution; however, these measurements of precipitation are indirect and subject to many sources of errors and uncertainties (Oke *et al.* 2009; Villarini & Krajewski 2010). Because of errors and uncertainties, the data from remote sensing techniques have not been widely used in operational hydrology so far (Berne & Krajewski 2013). Errors in the remote sensing are often corrected using ground-based gauge observations (Hasan *et al.* 2016; Sivasubramaniam *et al.* 2018); however, such corrections can only be possible in densely gauged regions.

In recent years, NWP model-based data products on the catchment scale with the increasing spatio-temporal resolution are becoming increasingly available as free and site-specific commercial products. The first guess forecasts from the NWP model are assimilated with the available past observations to make initial conditions for the next forecast. The same analysis for a fixed period produces meteorological reanalysis datasets with high spatio-temporal resolution (Talagrand 1997; Takahashi *et al.* 2010). Reanalysis datasets have been used in weather and climate studies (Takahashi *et al.* 2010) and used as atmospheric forcing data in hydrological models (Essou *et al.* 2016b). Compared with gridded data, the advantage of reanalysis data is that the dataset is updated regularly and available almost near real time (Essou *et al.* 2016b). However, errors and uncertainties related to reanalyses have not been understood well enough compared to those associated with gauge measurements (Parker 2016).

Previous studies have assessed the global and regional reanalysis datasets from different institutes and evaluated the use of them with hydrological models for runoff simulation (Te Linde *et al.* 2008; Lorenz & Kunstmann 2012; Vu *et al.* 2012; Lauri *et al.* 2014; Yang *et al.* 2014; Essou *et al.* 2016b; Roth & Lemann 2016). The focus of these studies was to use the reanalysis dataset as an alternative atmospheric forcing where the lack of gauge measurements exists. This study investigates the use of model-based data as input to hydrological models used in inflow predictions for operational hydropower production planning.

A typical inflow forecasting chain consists of the following components: (1) historical data to calibrate the models, (2) real-time data to update the current model states and (3) meteorological forecasts to generate inflow forecasts. Hydropower companies are involved in short-term and long-term operational planning. For short-term inflow forecasting, a calibrated and updated model is forced by 1–10 days of meteorological forecasts. Long-term predictions are normally run on average precipitation values taken from historical years to simulate a range of likely outcomes for the coming season or hydrological year. While short-term inflow forecasting is important for hydropower systems with low regulation capacity (runoff river schemes), long-term operational planning is required for well-regulated hydropower systems that consist of reservoirs with large volume relative to annual inflow, and snow-fed catchments where inflow prediction depends on snow storage in the catchments.

Hydropower companies usually buy the meteorological forecast data, used for daily inflow forecasting, from commercial weather service providers. The use of a gauge calibrated hydrological model with spatially defined prognosis data can also provide uncertainties in the predicted flow. The same providers now also produce meteorological reanalysis data as commercial products, and it is, therefore, of particular interest for hydropower companies to use them as a substitute for traditional gauge measurements since the model will be calibrated on data with the same spatial representation as the prognosis data.

The present study aims to answer two main research questions. First, can NWP model-based meteorological reanalysis datasets (precipitation and air temperature) replace traditional gauged precipitation and air temperature in the context of inflow predictions? Second, how do the model parameter and simulation uncertainty due to input data vary for the model-based data compared to the gauge and observational gridded data? To answer these questions, the study compares the time series of model-based data products with gauge observations at available gauge locations. Then, the study evaluates the performance of data products as an input to the hydrological model compared to the gauge and observational gridded datasets as an input. Further, the study analyses the uncertainty in the model parameters and the model response with the three forcing datasets.

STUDY AREA AND DATA

Study area

The model-based data were assessed over the Trøndelag region of central Norway. A Norwegian power company, Nord-Trøndelag Elektrisitetsverk (NTE) owns and operates more than 20 hydropower stations in this region, and its annual production is nearly 4,500 GWh. Three test catchments (Namsvatn, Follavatn and Tevla) with areas of 700, 200 and 350 km^2, respectively, are used in the setup of the Hydrologiska Byråns Vattenbalansavdelning (HBV) model in order to evaluate the performance of model-based data as an input. These are the major catchments in the NTE production system. The three catchments are shown in Figure 1, and their basic characteristics are presented in Table 1.

Based on the climatology for the period from 1961 to 2017, the mean annual precipitation in the study region is 2,000–4,000 mm along the coast and 750–2,000 mm inland. The annual mean temperature is in the range of 2–8 °C along the coast, and it is −4 to 2 °C in the inland mountainous areas (http://www.senorge.no/).

Data

Gauge, observational gridded and model-based precipitation and air temperature data and river flow records from January 2010 to December 2016 were used in the present study.

NTE operates its own meteorological stations and uses the data (precipitation and air temperature) from them for

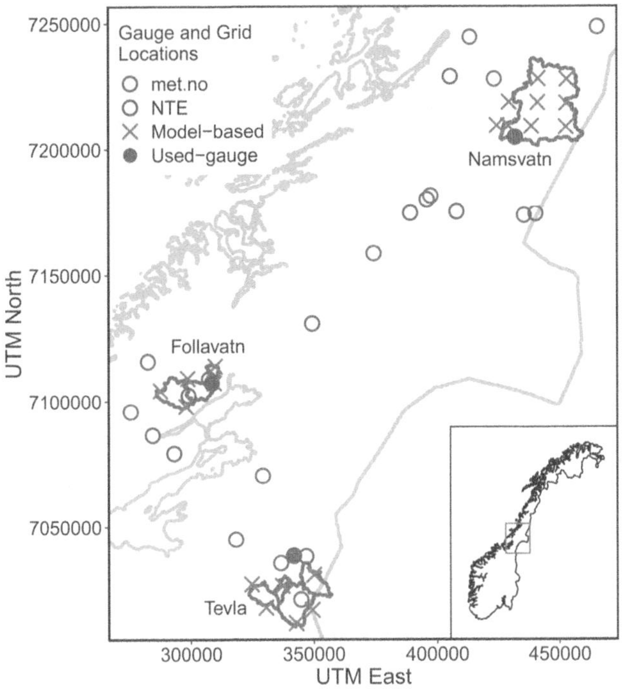

Figure 1 | The weather stations operated by met.no (green circles), NTE (blue circles) and grid points of model-based data and three catchments (purple polygons) used in the study. The weather station (NTE), used for hydrological modelling of each catchment, is marked with a filled blue circle. Please refer to the online version of this paper to see this figure in colour: http://dx.doi.org/10.2166/nh.2020.076.

Table 1 | Characteristics of the study catchments (source – http://nevina.nve.no/)

Description	Namsvatn	Follavatn	Tevla
Area (km^2)	701.5	202.7	345.9
Elevation range (m.a.s.l)	439–1,675	180–660	110–343
River slope (m km^{-1})	9.5	12.5	24.2
Forest (%)	19.0	37.2	42.3
Wetland (%)	6.0	14.1	24.8
Agriculture (%)	0.1	0.9	0.7
Bare mountain (%)	62.2	34.3	27.3
Lake (%)	12.5	13.5	4.7
Glacier (%)	0.2	0	0
Urban (%)	0	0	0.2

inflow simulations. Within the study area, there are 12 NTE stations with the available hourly observations for the study (Figure 1). Besides, observations from the 14 weather stations operated by the Norwegian Meteorological Institute (met.no) were also used for the comparison with model-based data. Out of the 14 met.no stations, two of them are available with hourly precipitation and four with hourly temperature, and the rest are with daily observations.

The hourly time series of inflow data for the three catchments were obtained from NTE. The discharge values are back calculated. NTE calculated the net outflow draining from the catchments using the measurements of water level in the reservoirs and intake and transfer of water from and to the catchments. It can be noted that NTE has used these flow data in its operational HBV model.

StormGeo (http://www.stormgeo.com) commercially distributes meteorological forecasts to hydropower companies including NTE in Norway. Currently, StormGeo generates and distributes NWP model-based meteorological reanalysis using the MESAN (Mesoscale analysis model) (Häggmark *et al.* 2000) from the Swedish Meteorological and Hydrological Institute (SMHI) to its customers as site-specific end-user data products. MESAN assimilates NWP with ground observations (gauge and weather radar) to generate a meteorological reanalysis dataset. Here, NWP from the High-Resolution Limited Area Model (HIRLAM) were used. HIRLAM uses NWP from the European Centre for Medium-Range Weather Forecasts (ECMWF) as boundary conditions. The MESAN analysis model provides a dataset with 11 km × 11 km spatial resolution.

From MESAN analysis, StormGeo provided model-based hourly precipitation and air temperature data on the representative grid locations that spatially covers each of the catchments (Figure 1). It can be noted that StormGeo has distributed daily meteorological forecasts to NTE at these grid locations for operational model runs for the study catchments. In addition, StormGeo derived hourly precipitation and air temperature at the nearest model grids to the 26 meteorological stations (Figure 1) in order to compare the time series of model-based data with gauge observations in this study.

The Norwegian Meteorological Institute spatially interpolated the past observed precipitation and air temperature records from meteorological stations to develop the daily gridded (1 km × 1 km) precipitation (Lussana *et al.* 2018) and hourly and daily gridded air temperature (Lussana *et al.* 2016) datasets covering Norway. These datasets are freely available to the public through met.no's thredds server (http://thredds.met.no/thredds/catalog.html). The gridded precipitation and air temperature were downloaded for each study catchment. Hereafter, the NWP model-based reanalysis dataset from StormGeo is referred to as 'model-based' and observational gridded data as 'gridded' throughout the study.

METHODS

Data comparison

At each gauge location, the time series of model-based hourly precipitation and air temperature data were compared with the available hourly gauged observations. In addition, model-based hourly datasets were aggregated to daily and then compared with daily gauged data.

HBV model

The HBV precipitation-runoff model is a semi-distributed conceptual model. A detailed description of the HBV model structure can be found in the literature (Bergström 1976; Bergström 1992; Killingtveit & Sælthun 1995; Sælthun 1996). The HBV model has been widely used in the Nordic region and other parts of the world for various hydrological studies (Steele-Dunne *et al.* 2008; Te Linde *et al.* 2008; Lawrence & Haddeland 2011). Most of the hydropower companies in Norway use a version of the HBV model for inflow forecasting.

In this study, PINEHBV (Rinde 1999), a variant of HBV, was used. The PINEHBV is in a structure similar to the model used by NTE. The model consists of four main storage components such as snow and soil moisture routines and two linear response tanks, upper and lower. The upper and lower zones generate the surface runoff and base flow, respectively. An illustration of the structure of the HBV model is added to Supplementary Figure S1 in Supplementary Materials. In the snow routine, the catchment is divided into ten elevation zones in order to account for the elevation-dependent variability in the type and amount of precipitation and snow storage. Further, among the ten zones, the lowest zones below the forest line based on the topography are defined as forested, and the remaining zones are non-forested. Determining the type of precipitation (snow or rain) and calculation of snowmelt and snow accumulation in each of the ten zones are the main processes in this component. The processes in the rest of the storage components are lumped at the catchment scale. Input to the PINEHBV model is the time series (daily or hourly) of precipitation and air temperature and monthly average potential evaporation.

Performance evaluation of datasets

HBV uses a single input series of areal precipitation and temperature. We spatially averaged the model-based data from StormGeo grid points (Figure 1), and areal precipitation was estimated for each catchment. The operational HBV model at NTE uses observations from a single gauge for each catchment, and the same gauges (Figure 1) were used in this study and considered as the reference model. In addition, a spatial average of daily observational gridded precipitation of a regular grid (1 km × 1 km) was computed and used as a third input alternative.

For all three catchments, the same monthly average potential evaporation values were used, as shown in Table 2.

For each of three catchments, the HBV model was calibrated separately using gauge, observational gridded and model-based precipitation, and air temperature datasets. Since the study catchments are snow-fed, the start of the simulation was set to September to ensure no initial snow storage. Four years of data from September 2010 to August 2014 was used for the model calibration, and the model performance was evaluated for the three forcing datasets using a two-year verification period (September 2014–August 2016).

The Nash–Sutcliffe efficiency (NSE) (Nash & Sutcliffe 1970) was used as an objective performance criterion to evaluate model performance. The NSE is the most commonly used performance measure in hydrology (Essou

Table 2 | Monthly values of daily potential evaporation

Month	January	February	March	April	May	June	July	August	September	October	November	December
Daily potential evaporation (mm/day)	0.1	0.2	0.7	1.0	2.3	3.5	3.5	2.3	1.0	0.7	0.2	0.1

Table 3 | Parameter ranges used in MC calibration

Parameter	Description	Unit	Minimum	Maximum
RCORR	Precipitation correction factor, rainfall	–	0.3	1.6
SCORR	Precipitation correction factor, snowfall	–	0.3	2.5
TX	Threshold temperature for rain/snow	°C	–4.0	8.0
CX	Melt index (degree day factor)	mm/°C day	0.3	25.0
CXN	Melt index – forest zones	mm/°C day	0.3	25.0
TS	Threshold temperature for melt/freeze	°C	–4.0	8.0
TSN	Threshold temperature for melt/freeze – forest zones	°C	–4.0	8.0
FC	Field capacity	mm	5.0	1,500.0
BETA	Relative contribution to upper zone from soil storage	–	0.1	12.0
FCDEL	Threshold value for potential evapotranspiration in soil moisture	–	0.1	1.0
KUZ2	Upper recession coefficient, upper zone	mm/day	0.1	5.0
KUZ1	Middle recession coefficient, upper zone	mm/day	0.1	1.0
KUZ	Lower recession coefficient, upper zone	mm/day	0.01	0.6
KLZ	Recession coefficient, lower zone	mm/day	0.001	0.15
UZ2	Upper threshold, upper zone	mm	5.0	500.0
UZ1	Lower threshold, upper zone	mm	5.0	100.0
PERC	Percolation constant upper to lower zone	mm	0.0	5.0

et al. 2016a). In addition, accumulated flow difference (AccDiff) was used as an additional measure.

It is often shown that many different parameter sets can give similar good NSE (Beven & Binley 1992), and it is not given that the parameter set with the best NSE during the calibration provides good performance outside the calibration period (Seibert 1997). Therefore, a Monte Carlo (MC) approach of the model calibration was used to investigate how the uncertainty of the HBV model parameters varies for the three forcing datasets. The advantage of the MC is that the resulting parameter sets are not only a basis for investigating the model parameter uncertainty but also the simulated flow, and other model responses can be provided as a range instead of a single value (Steele-Dunne *et al.* 2008). Using the MC approach, Ledesma & Futter (2017) assessed the observational gridded data product compared to gauge measurements as the hydrological model input. Steele-Dunne *et al.* (2008) applied the MC method to generate an ensemble of simulated flows to assess the impacts of climate change on hydrology.

Each of the free parameters (17 parameters) in the PINEHBV model was given a range of reasonable values, as suggested in earlier studies (Killingtveit & Sælthun 1995; Sælthun 1996; Rinde 1999) and shown in Table 3. An MC model calibration with uniform sampling (Seibert 1997; Seibert 2003; Steele-Dunne *et al.* 2008; Ledesma & Futter 2017) was undertaken to generate an ensemble of 100,000 parameter sets for each of the three catchments using the three forcing datasets separately. From the 100,000 parameter sets, the best 100 parameter sets with the highest NSE were chosen, and then, from those 100 parameter sets, the best 50 parameter sets which also give the highest NSE during the verification period were finally selected. An ensemble of the 50 simulated model responses with the 50 best parameter sets were used for the analysis.

RESULTS

Data comparison

For comparing model-based precipitation and temperature with gauge observations, a linear regression analysis was carried out. Pearson correlation coefficient between model-based and gauge data was calculated at each gauge location. Figure 2 shows the box plot of the estimated

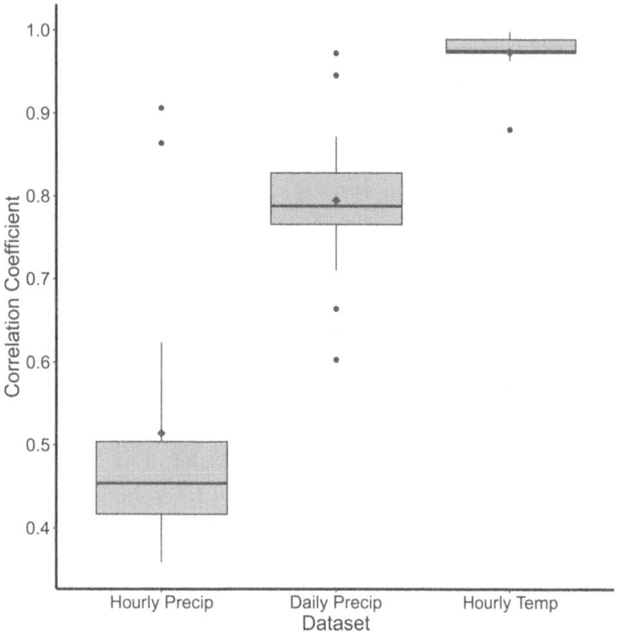

Figure 2 | Box plot of the correlation coefficient between model-based and gauge precipitation and air temperature data with an hourly and daily resolution, estimated at gauge locations. The values outside $1.5 \times$ IQR are represented by the whiskers.

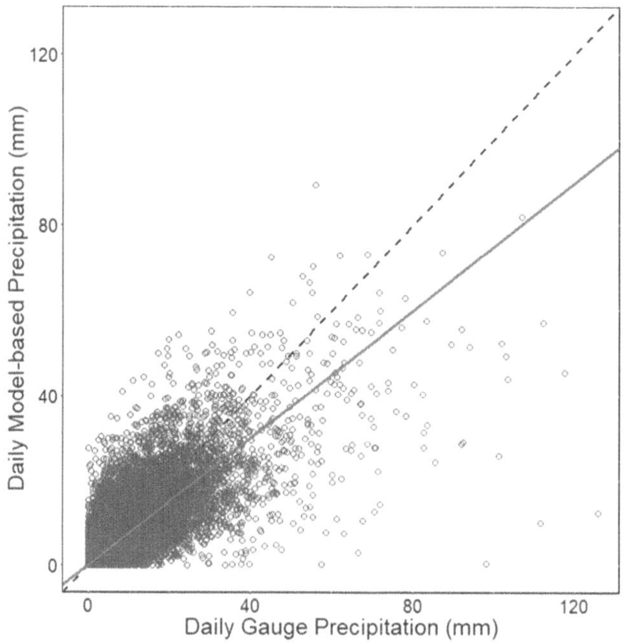

Figure 3 | Scatter plot of model-based and gauge precipitation data, pooled from all gauge locations. The dashed line denotes the perfect fit 45-degree line, and the red solid line shows the regression.

correlation coefficient between the model-based and gauge datasets.

Looking at Figure 2, the model-based hourly precipitation shows a poor correlation with the gauge observations. For the two met.no stations available with hourly measurements, the correlation is relatively high (shown as outliers in Figure 2). For all NTE stations, the hourly precipitation data show a poor correlation. However, daily precipitation data show a reasonably good correlation with a few exceptions. For all gauge locations, the hourly model-based temperature correlated well with the gauge measurements.

We prepared scatterplots and compared the model-based and gauged datasets at each gauge location. Figure 3 shows a single scatterplot of all data pooled together. Looking at Figure 3, it is particularly seen that the model-based data in most cases underestimate high-intensity daily precipitation events observed by gauges.

To investigate how the precipitation volume of the three forcing datasets vary and how this variation influences the model performance, we compared the accumulated annual precipitation of the three forcing datasets over the three catchments. Figure 4 shows that the model-based and

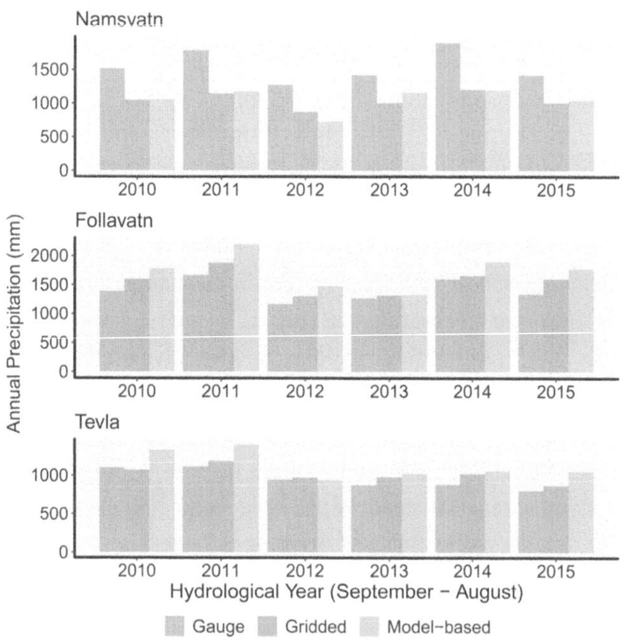

Figure 4 | Annual precipitation from the gauge, gridded and model-based precipitation input data for the three catchments during the calibration (September 2010–August 2014) and verification period (September 2014–August 2016).

gridded precipitation are similar in volume for Namsvatn. It is also seen that for the Namsvatn catchment, the model-based and gridded precipitation are lower than the gauge precipitation, while for Follavatn and Tevla catchment, they are higher than the gauge value, and the model-based yielded the highest annual volume for almost all years for these two catchments.

Performance in simulating the hydrological response

The performance of flow simulation (NSEs) of the three forcing datasets (gauge, gridded and model-based) is shown in Figure 5. The NSE value above 0.6 is generally considered as an acceptable model by hydrologists (Essou *et al.* 2016b). As presented in Figure 5, the NSEs for the 50 optimum parameter sets during the calibration and verification period for all three forcing datasets were mostly above 0.6 in this study except for model-based data on the Follavatn catchment (NSEs during the verification period in the range of 0.52–0.66). Looking at Figure 5, the performance of gauge and gridded during the calibration period is higher than model-based for Namsvatn and Follavatn, while the performance of model-based was superior to that of the gauge and gridded for the Tevla catchment. The best NSEs for the gridded dataset during the calibration period for Namsvatn resulted in a relatively lower NSE during the verification period. It was found that a parameter set with the best NSE during the calibration did not give the best NSE during the verification; a parameter set slightly lower NSE than the best one gave a better NSE during the verification.

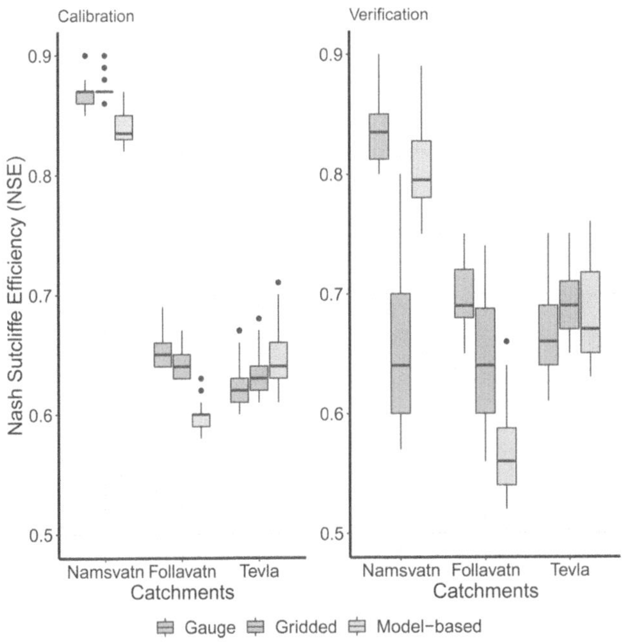

Figure 5 | Box plot summary of the NSEs of 50 optimum parameter sets for the three forcing datasets (gauge, gridded and model-based) for the three catchments during the calibration and verification period.

While Figure 5 presents a summary of NSEs for daily flow, Table 4 presents the median values of NSEs, estimated from the simulated flows with daily and weekly timescale using the 50 optimum parameter sets. Looking at Table 4, the performance (NSE) on a weekly temporal scale is higher than the daily, and it is over 0.67 for all three datasets. Like daily timescale, the performance of the three forcing datasets showed a nearly similar variation on the weekly timescale.

An ensemble of 50 simulated flows using the 50 optimum parameter sets for each of the forcing datasets for the three catchments is shown in Figure 6. Here, the verification period is presented to illustrate the responses for the three forcing datasets; a plot for the calibration period (Supplementary Figure S2) is added to the supplementary material. In Figure 6, three colour bands represent the ensemble of simulated flows for the three forcing datasets. If blue or yellow are not visible, they are within the model-based simulation (green). A subplot of Figure 6, where a shorter period is zoomed, is added to Supplementary Figure S3.

Table 4 | Median values of calculated NSEs of the simulated flow (daily and weekly) using the 50 optimum parameter sets for the three forcing datasets (gauge, gridded and model-based) for the three catchments during the calibration (a) and verification (b) period

	Daily			Weekly		
	Gauge	Gridded	Model-based	Gauge	Gridded	Model-based
a) Calibration period						
Namsvatn	0.87	**0.87**	0.84	0.90	**0.91**	0.87
Follavatn	**0.65**	0.64	0.60	**0.76**	0.73	0.71
Tevla	0.62	0.63	**0.64**	0.70	0.73	**0.73**
b) Verification period						
Namsvatn	**0.84**	0.64	0.80	**0.89**	0.73	0.86
Follavatn	**0.69**	0.64	0.56	**0.79**	0.71	0.67
Tevla	0.66	**0.69**	0.67	0.74	**0.81**	0.72

Maximum NSE among the three forcing datasets is marked in bold.

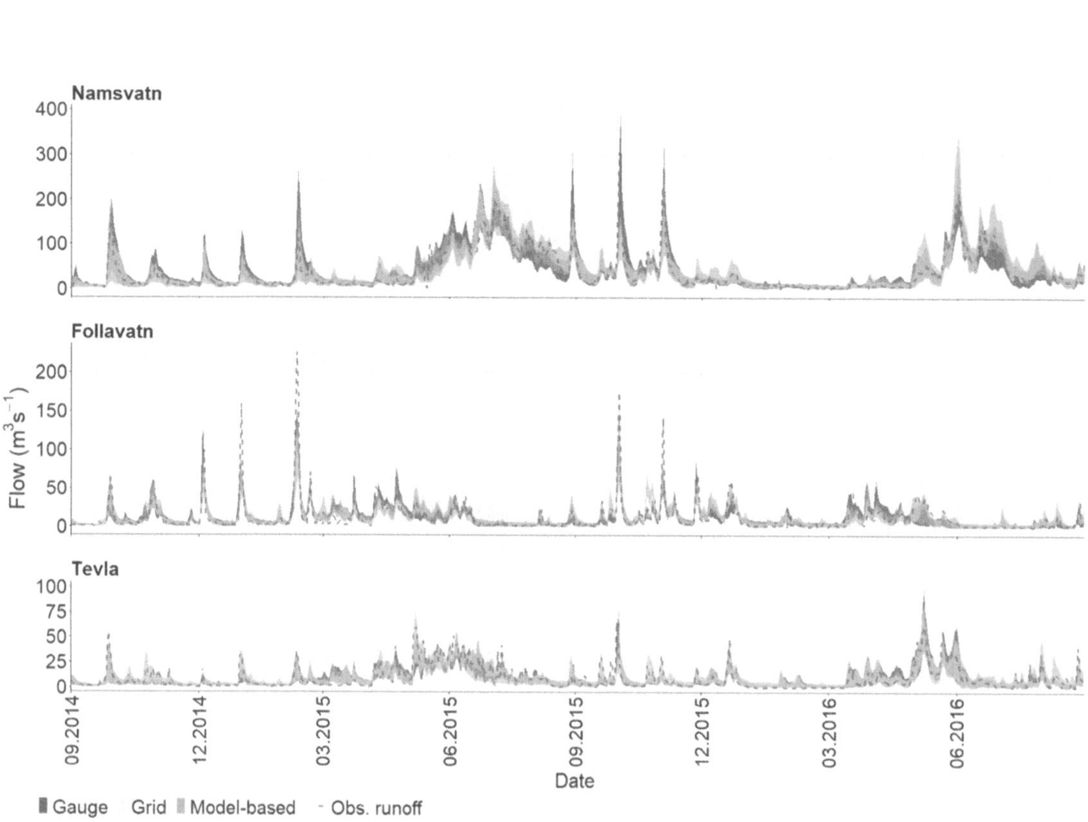

Figure 6 | Ribbon plot showing three colour bands of maximum and minimum of the 50 simulated flow ensembles for three different forcing datasets for the three catchments during the verification period. Observed runoff is denoted by a red dashed line. Please refer to the online version of this paper to see this figure in colour: http://dx.doi.org/10.2166/nh.2020.076.

Looking at Figure 6, it is visible that the observed runoff mostly falls within the band of maximum and minimum for all three datasets except a few extremes. The bandwidth denotes the uncertainty in the simulated daily river flow due to model parameter uncertainty. A visual inspection of Figure 6 shows that the uncertainty in the simulated flow for the three datasets is generally in the same extent.

Looking at Figure 7, the average simulated annual runoff volume for all three forcing datasets are nearly the same as the observed annual runoff volume, and the difference is less than 15% of the observed runoff volume except for Namsvatn in the year 2012 for model-based data (24%) and Tevla in the year 2014 for gauge data (17%). Further, except Namsvatn in the year 2012 for model-based data, the observed runoff volume falls within the lower and upper value of the simulated flow volume from the 50 parameter sets for all three datasets for all three catchments.

The gridded dataset yielded the best-simulated flow volume compared to observed runoff. The percentage mean absolute error for the annual simulated flow volume with reference to the observed runoff volume is 4–6% for gridded and 5–10% for model-based, while it is 5–9% for gauge dataset.

In Figure 7, the range of 50 simulated flow volumes is shown using the error bar, which represents the uncertainty in the simulated flow volume due to the model parameter uncertainty.

The length of the error bars is between 20 and 30% of observed runoff for Namsvatn and Follavatn, while it is 35–55% for Tevla. Looking at Figure 7, the size of the error bar does not differ largely for the three forcing datasets. The error bars for the gridded dataset are relatively smaller for Namsvatn and Tevla catchments, and they are smaller for model-based data for Follavatn catchment.

Parameter uncertainty

We investigated the HBV model parameter values of the 50 optimum parameter sets to see how they vary with the forcing dataset. Apart from precipitation correction factor – rainfall (PCORR) and precipitation correction

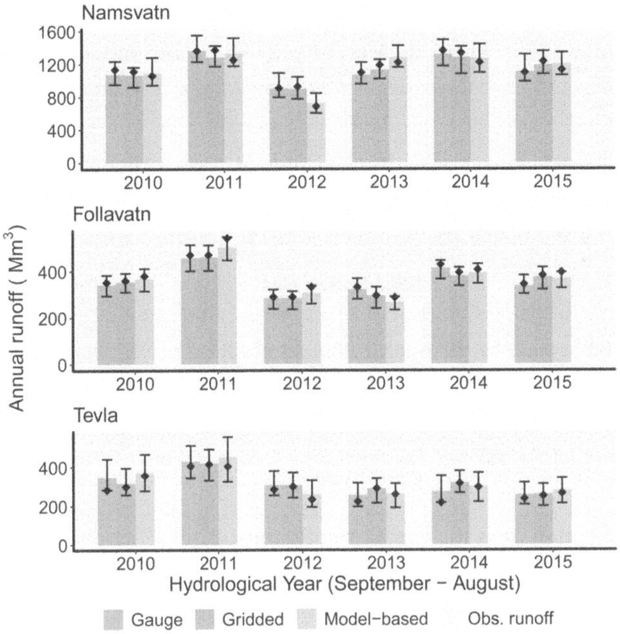

Figure 7 | Bar plot of the average simulated annual runoff volume of the 50 best MC parameter sets for the three forcing datasets (gauge, gridded and model-based) and observed runoff for the three catchments during the calibration (September 2010–August 2014) and verification period (September 2014–August 2016). The error bar denotes the lower and upper value of the simulated volume from the 50 parameter sets. The simulated volume using the parameter set with the best NSE during the calibration period is denoted by a black diamond point.

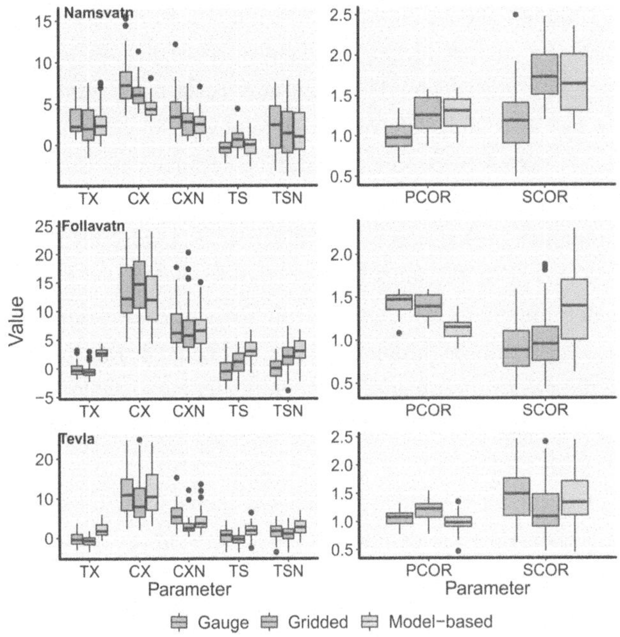

Figure 8 | Summary of the resulting range of values for the HBV model parameters of the 50 optimum parameter sets for the three forcing datasets for the three catchments. Here, the parameters which showed variation with forcing datasets are only displayed.

factor – snowfall (SCORR), parameters in the snow routine, threshold temperature for rain/snow (TX), degree day factor (CX), degree day factor – forest zones (CXN), threshold temperature for melt/freeze (TS) and threshold temperature for melt/freeze – forest zones (TSN) showed differences in the range of values depending on the forcing dataset for a given catchment, and these parameters are shown in Figure 8. The rest of the calibration parameters assumed a similar range of values for three different forcing datasets.

The calibration parameters, such as PCORR and SCORR, correct the rainfall and snowfall input to the HBV model. This correction for the precipitation input covers several factors, including catch errors and lack of representativeness of gauges (Sælthun 1996). For observational gridded and model-based datasets, a need for the correction can also be due to under/overestimation by the interpolation techniques and data assimilation in NWP models.

In Figure 8, PCORR and SCORR for model-based and gridded data for Namsvatn catchment are similar, and they are higher than the values associated with gauge data. This agrees with the underestimation of model-based and gridded precipitation input compared to gauged precipitation for Namsvatn catchment, as shown in Figure 4. For the Follavatn catchment, PCORR and SCORR for model-based assumed a considerably different range of values to the gauge and gridded dataset, while they were similar to the gauge data for the Tevla catchment, but here different compared to the gridded dataset.

To check whether the estimates of actual evapotranspiration in the model influence the variation of PCORR and SCORR, we plotted the ratio of actual evapotranspiration to precipitation input for the three forcing datasets for the three catchments during the calibration for the best 50 parameter sets (Supplementary material Figure S4). It appears that the ratio of actual evapotranspiration to precipitation is almost the same for all three datasets. It shows that PCORR and SCORR were not influenced by the estimation of actual evapotranspiration in the model, and they vary depending on the forcing dataset.

PCORR and SCORR are of primary concern when a different source of precipitation input is used. The results show that snow routine parameters also highly depend on the forcing dataset. Looking at Figure 8, TX, TS and TSN values for the model-based data were higher than the gauge and gridded for Follavatn and Tevla catchments. It can be seen that CX assumed relatively higher values (10–20) for all three forcing datasets, compared to the traditional range of values for CX (1–5) in the literature (Killingtveit & Sælthun 1995; Lawrence & Haddeland 2011).

Kuczera & Williams (1992) demonstrated that the parameter uncertainty increases with the uncertainty in the areal precipitation input. Looking at Figure 8, the uncertainty (size of the range) in the parameters associated with model-based data is nearly similar in extent to the gauge and gridded datasets.

The calibrated models for each catchment using the three input datasets (50 optimal parameter sets of each atmospheric forcing) were forced with model-based data during the verification period, and the computed NSE is shown in Figure 9. The HBV model calibrated with model-based data performed noticeably better than the model calibrated with gauge or gridded dataset for all three catchments.

With reference to Figure 8, for the Namsvatn catchment, parameters 'PCOR' and 'SCOR' are nearly the same for the gridded and model-based datasets. When the model calibrated with gridded was forced with the model-based dataset, the performance is nearly as good as the model calibrated using the model-based dataset. In contrast, for Follavatn and Tevla, 'PCOR' and 'SCOR' associated with model-based dataset are different from the values obtained using the gauge and gridded datasets and the performance was poorer as shown in Figure 9. This underlines the importance of using the same dataset for the model calibration as is later used in the operational forecasting of inflow.

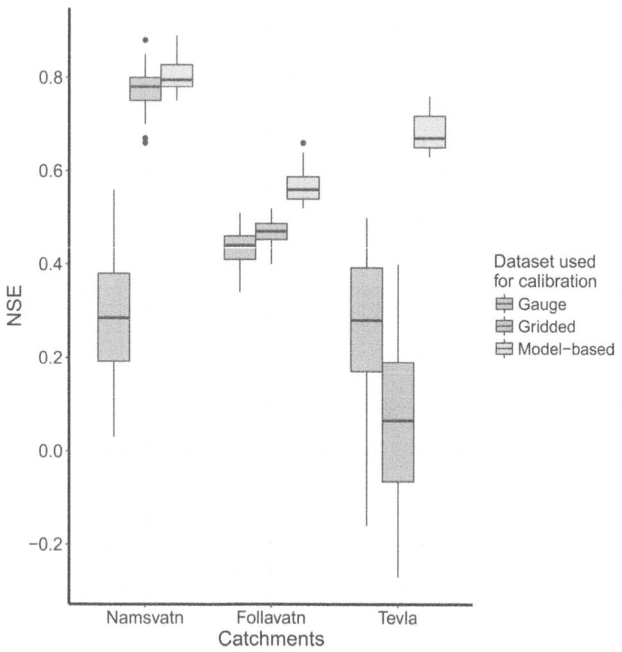

Figure 9 | Box plot of NSE computed for the simulated flow during the verification period (2014–2016) using model-based dataset as input but using three different sets of the 50 optimal parameters calibrated using the three forcing datasets (gauge, gridded and model-based).

DISCUSSION

This study investigated the potential of NWP model-based meteorological reanalysis as an alternative to traditional gauge observations for hydrological modelling. In this paper, we showed that the performance of the model-based data was nearly as good or even better than the gauge and observational gridded dataset.

For this assessment, we adopted an MC approach to model calibration (Seibert 1997; Seibert 2003; Steele-Dunne *et al.* 2008; Ledesma & Futter 2017). Even though the MC method with uniform random sampling (100,000 runs in this study) is time and resource consuming, the approach can map most of the feasible parameter combinations; hence, it provides a solid basis for investigating the uncertainty in the model parameter and the response. In this study, we found that the model parameter uncertainty and the uncertainty in the simulated flow using the model-based data as an input was comparable to or even lower than those associated with the gauge and gridded dataset.

Several studies (Te Linde *et al.* 2008; Lauri *et al.* 2014; Essou *et al.* 2016b; Roth & Lemann 2016) assessed the use of reanalysis dataset as an alternative input data for hydrological modelling. This study extends current work with the evaluation of model-based data as an alternative input to hydrological models used in inflow predictions for operational hydropower production planning. In addition, a few studies (Steele-Dunne *et al.* 2008) in the literature focussed on the investigation of model parameter uncertainty depending on different forcing datasets and its consequence in the model response as presented in this paper.

Essou *et al.* (2016b) tested the global and regional reanalysis dataset as an input to a hydrological model in 370 catchments in the United States comparing the output with observational gridded data. Their results showed that the regional reanalysis dataset, which is assimilated using ground-based precipitation observations, produced simulated river flow similar to observed flows. Even though the performance of the global reanalysis was also similar to observed flows, they found that performance was degraded by precipitation seasonality biases.

For well-regulated hydropower schemes, long-term predictions are generally more important than short-term forecasts. Vice versa, poorly regulated schemes are in higher need of short-term inflow forecasts. Such schemes risk flood-spill, when inflow exceeds through flow capacity. Among the three hydropower systems investigated in this study, Namsvatn consists of a large reservoir with high regulation capacity, while the other two catchments have a relatively small regulation capacity. Further, it can be noted that all catchments are snow-fed. Hence, long-term operational planning is required to predict the long-term volume and seasonal distribution based on snow storage in the catchments of all three hydropower systems. In this study, the model-based data simulated the flow volume as well as the observed; hence, it can be a potential alternative to gauge measurements for long-term operational hydropower planning. However, the comparison of model-based data with gauge observations showed that model-based data underestimate the daily extreme precipitation. Moreover, the HBV model that is based on model-based data failed to simulate some of the observed high peaks. This is probably of less importance for the hydropower systems, which consist of large reservoirs with high regulation capacity, than for systems with low regulation capacity.

NWP is an evolving field with the advancement of data science and computer technology (advanced data assimilation techniques). The quality of model-based data products will further improve in the future. Hence, the performance of hydrological model simulations using model-based data products can be foreseen to become better than today in the future.

The calibrated models will be forced by meteorological forecasts (1–10 days) to predict the inflow for short-term hydropower operational planning. We did a verification test on calibrated models using three different forcing datasets and found a clear advantage of using a dataset for calibration which is similar to the source of the dataset being used for the operational simulation. This result implies that the model-based data could be a better alternative for calibrating and updating hydrological models used for inflow forecasting when the forecasting dataset and the model-based dataset come from the same NWP model. Model-based data derived from the same NWP model, which is used to generate meteorological forecasts, will be provided on the same grids and derived using the same methods. Consequently, the data structure of the model-based historical data and the meteorological forecasts would be similar, and the model parameter uncertainty would be similar in the calibration period and in the forecast period. The use of the past records of meteorological forecasts to evaluate predicted flow from models calibrated using the three forcing datasets used in this study is recommended as a future study for this work.

CONCLUSIONS

The evaluation of model-based input data for hydrological modelling in this study showed that model-based precipitation and air temperature can be a potential alternative to those obtained from gauge measurements and observational gridded data.

The correlation between model-based and gauge data was varying among gauge locations, and the median value of correlation for daily precipitation was 0.8. However, the performance of model-based input data with daily timestep was nearly as good or even better than the gauge or gridded data for the model calibration. It was found that the model parameter uncertainty and simulation uncertainty associated with model-based data appeared as similar to gauge and gridded datasets. Further, the annual simulated flow volume using the model-based data as an input was nearly the same as the observed annual runoff volume.

These results indicate that model-based data can be a potential alternative input to the hydrological models used for inflow predictions for long-term operational hydropower planning. This could be very useful in remote catchments with few gauges and in areas where installing gauges is impossible. Further, results also imply that model-based data can be a promising data source for calibrating hydrological models used for short-term inflow predictions as meteorological forecasts would then have similar sources and similar data structure to the dataset used for the model calibration.

ACKNOWLEDGEMENTS

The authors thankfully acknowledge the Nord-Trøndelag Elektrisitetsverk (NTE) and Norwegian Meteorological Institute (met.no) for providing hydrometeorological data for this study. The authors thank Martí Garcia Cusí at the Norwegian University of Science and Technology (NTNU) for his initial work on the modelling. The authors gratefully acknowledge the Norwegian Research Council and Norconsult for funding this research work under the Industrial Ph.D. scheme (Project No.: 255852/O30).

CONFLICT OF INTEREST

The authors declare that there are no competing interests.

SUPPLEMENTARY MATERIAL

The Supplementary Material for this paper is available online at https://dx.doi.org/10.2166/nh.2020.076.

REFERENCES

Bergström, S. 1976 *Development and Application of a Conceptual Runoff Model for Scandinavian Catchments*, SMHI Rep. RHO 7. Swedish Meteorological and Hydrological Institute, Norrköping, Sweden.

Bergström, S. 1992 *The HBV Model: Its Structure and Applications*. Swedish Meteorological and Hydrological Institute, Norrköping, Sweden.

Berne, A. & Krajewski, W. F. 2013 Radar for hydrology: unfulfilled promise or unrecognized potential? *Advances in Water Resources* **51**, 357–366.

Beven, K. 2012 *Rainfall-Runoff Modelling: The Primer*, 2nd edn. John Wiley & Sons Ltd, Chichester, UK.

Beven, K. & Binley, A. 1992 The future of distributed models: model calibration and uncertainty prediction. *Hydrological Processes* **6**, 279–298.

Essou, G. R., Arsenault, R. & Brissette, F. P. 2016a Comparison of climate datasets for lumped hydrological modeling over the continental United States. *Journal of Hydrology* **537**, 334–345.

Essou, G. R., Sabarly, F., Lucas-Picher, P., Brissette, F. & Poulin, A. 2016b Can precipitation and temperature from meteorological reanalyses be used for hydrological modeling? *Journal of Hydrometeorology* **17**, 1929–1950.

Førland, E., Allerup, P., Dahlström, B., Elomaa, E., Jónsson, T., Madsen, H., Perälä, J., Rissanen, P., Vedin, H. & Vejen, F. 1996 *Manual for Operational Correction of Nordic Precipitation Data*. DNMI-Reports 24, p. 66.

Häggmark, L., Ivarsson, K.-I., Gollvik, S. & Olofsson, P.-O. 2000 MESAN, an operational mesoscale analysis system. *Tellus A: Dynamic Meteorology and Oceanography* **52**, 2–20.

Hasan, M. M., Sharma, A., Mariethoz, G., Johnson, F. & Seed, A. 2016 Improving radar rainfall estimation by merging point rainfall measurements within a model combination framework. *Advances in Water Resources* **97**, 205–218.

Haylock, M., Hofstra, N., Tank, A. K., Klok, E., Jones, P. & New, M. 2008 A European daily high-resolution gridded data set of surface temperature and precipitation for 1950–2006. *Journal of Geophysical Research: Atmospheres* **113**, D20119.

Hwang, Y., Clark, M., Rajagopalan, B. & Leavesley, G. 2012 Spatial interpolation schemes of daily precipitation for hydrologic modeling. *Stochastic Environmental Research and Risk Assessment* **26**, 295–320.

Killingtveit, Å. & Sælthun, N. R. 1995 *Hydrology*. Norwegian Institute of Technology, Division of Hydraulic Engineering, Trondheim, Norway.

Kirchner, J. W. 2009 Catchments as simple dynamical systems: catchment characterization, rainfall-runoff modeling, and doing hydrology backward. *Water Resources Research* **45**, W02429.

Kuczera, G. & Williams, B. J. 1992 Effect of rainfall errors on accuracy of design flood estimates. *Water Resources Research* **28**, 1145–1153.

Lauri, H., Räsänen, T. & Kummu, M. 2014 Using reanalysis and remotely sensed temperature and precipitation data for hydrological modeling in monsoon climate: Mekong River case study. *Journal of Hydrometeorology* **15**, 1532–1545.

Lawrence, D. & Haddeland, I. 2011 Uncertainty in hydrological modelling of climate change impacts in four Norwegian catchments. *Hydrology Research* **42**, 457–471.

Ledesma, J. L. & Futter, M. N. 2017 Gridded climate data products are an alternative to instrumental measurements as inputs to rainfall–runoff models. *Hydrological Processes* **31**, 3283–3293.

Lorenz, C. & Kunstmann, H. 2012 The hydrological cycle in three state-of-the-art reanalyses: intercomparison and performance analysis. *Journal of Hydrometeorology* **13**, 1397–1420.

Lussana, C., Ole, E. T. & Francesco, U. 2016 *seNorge v2.0: An Observational Gridded Dataset of Temperature for Norway*. Norwegian Meteorological Institute, Norway.

Lussana, C., Saloranta, T., Skaugen, T., Magnusson, J., Tveito, O. E. & Andersen, J. 2018 seNorge2 daily precipitation, an observational gridded dataset over Norway from 1957 to the present day. *Earth System Science Data* **10**, 235.

Nash, J. E. & Sutcliffe, J. V. 1970 River flow forecasting through conceptual models part I – a discussion of principles. *Journal of Hydrology* **10**, 282–290.

Oke, A., Frost, A. & Beesley, C. 2009 The use of TRMM satellite data as a predictor in the spatial interpolation of daily precipitation over Australia. In: *Proceedings of the 18th World IMACS/ MODSIM Congress*, Cairns, Australia, 13–17 July 2009.

Parker, W. S. 2016 Reanalyses and observations: what's the difference? *Bulletin of the American Meteorological Society* **97**, 1565–1572.

Photiadou, C., Weerts, A. & Van Den Hurk, B. 2011 Evaluation of two precipitation data sets for the Rhine River using streamflow simulations. *Hydrology and Earth System Sciences* **15**, 3355–3366.

Rinde, T. 1999 *PINEHBV – A Workbench for Hydrological Modelling*. Nordic Workshop on HBV and Similar Runoff Models. Finnish Environmental Institute, Hydrology and Water Management Division, Helsinki.

Roth, V. & Lemann, T. 2016 Comparing CFSR and conventional weather data for discharge and soil loss modelling with SWAT in small catchments in the Ethiopian Highlands. *Hydrology and Earth System Sciences* **20**, 921–934.

Sælthun, N. R. 1996 *The Nordic HBV Model*. Norwegian Water Resources and Energy Administration, Oslo, Norway, Publication No. 7, pp. 1–26.

Seibert, J. 1997 Estimation of parameter uncertainty in the HBV model: paper presented at the Nordic Hydrological Conference (Akureyri, Iceland – August 1996). *Hydrology Research* **28**, 247–262.

Seibert, J. 2003 Reliability of model predictions outside calibration conditions: paper presented at the Nordic Hydrological Conference (Røros, Norway, 4–7 August 2002). *Hydrology Research* **34**, 477–492.

Sivasubramaniam, K., Sharma, A. & Alfredsen, K. 2018 Estimating radar precipitation in cold climates: the role of air temperature within a non-parametric framework. *Hydrology and Earth System Sciences* **22**, 6533–6546.

Steele-Dunne, S., Lynch, P., Mcgrath, R., Semmler, T., Wang, S., Hanafin, J. & Nolan, P. 2008 The impacts of climate change on hydrology in Ireland. *Journal of Hydrology* **356**, 28–45.

Takahashi, K., Takaya, Y. & Kobayashi, S. 2010 Re-analyses as predictability tools. In: *ECMWF Seminar on Predictability in the European and Atlantic Regions*, Shinfield Park, Reading, 6–9 September 2010.

Talagrand, O. 1997 Assimilation of observations, an introduction. *Journal of the Meteorological Society of Japan* **75**, 191–209.

Taskinen, A. & Söderholm, K. 2016 *Operational Correction of Daily Precipitation Measurements in Finland. Boreal. Environ. Res.* **21**, 1–24.

Te Linde, A. H., Aerts, J. C. J. H., Hurkmans, R. T. W. L. & Eberle, M. 2008 Comparing model performance of two rainfall-runoff models in the Rhine basin using different atmospheric forcing data sets. *Hydrology and Earth System Sciences* **12**, 943–957.

Vaze, J., Post, D., Chiew, F., Perraud, J.-M., Teng, J. & Viney, N. 2011 Conceptual rainfall–runoff model performance with different spatial rainfall inputs. *Journal of Hydrometeorology* **12**, 1100–1112.

Villarini, G. & Krajewski, W. F. 2010 Review of the different sources of uncertainty in single polarization radar-based estimates of rainfall. *Surveys in Geophysics* **31**, 107–129.

Vu, M., Raghavan, S. V. & Liong, S. Y. 2012 SWAT use of gridded observations for simulating runoff – a Vietnam river basin study. *Hydrology and Earth System Sciences* **16**, 2801–2811.

Wagener, T., Wheater, H. S. & Gupta, H. V. 2004 *Rainfall-Runoff Modelling in Gauged and Ungauged Catchments*. Imperial College Press, London, UK.

Wolff, M. A., Isaksen, K., Petersen-Øverleir, A., Ødemark, K., Reitan, T. & Brækkan, R. 2015 Derivation of a new continuous adjustment function for correcting wind-induced loss of solid precipitation: results of a Norwegian field study. *Hydrology and Earth System Sciences* **19**, 951–967.

Yang, Y., Wang, G., Wang, L., Yu, J. & Xu, Z. 2014 Evaluation of gridded precipitation data for driving SWAT model in area upstream of three gorges reservoir. *PLoS ONE* **9**, e112725.

First received 28 May 2019; accepted in revised form 18 December 2019. Available online 22 January 2020

Evaluation of global forcing datasets for hydropower inflow simulation in Nepal

Bikas Chandra Bhattarai, John Faulkner Burkhart, Lena M. Tallaksen, Chong-Yu Xu IWA and Felix Nikolaus Matt

ABSTRACT

Discharge over the Narayani river catchment of Nepal was simulated using Statkraft's Hydrologic Forecasting Toolbox (Shyft) forced with observations and three global forcing datasets: (i) ERA-Interim (ERA-I), (ii) Water and Global Change (WATCH) Forcing Data ERA-I (WFDEI), and (iii) Coordinated Regional Climate Downscaling Experiment with the contributing institute Rossy Centre, Swedish Meteorological and Hydrological Institute (CORDEX-SMHI). Not only does this provide an opportunity to evaluate discharge variability and uncertainty resulting from different forcing data but also it demonstrates the capability and potential of using these global datasets in data-sparse regions. The fidelity of discharge simulation is the greatest when using observations combined with the WFDEI forcing dataset (hybrid datasets). These results demonstrate the successful application of global forcing datasets for regional catchment-scale modeling in remote regions. The results were also promising to provide insight of the interannual variability in discharge. This study showed that while large biases in precipitation can be reduced by applying a precipitation correction factor (p_corr_factor), the best result is obtained using bias-corrected forcing data as input, i.e. the WFDEI outperformed other forcing datasets. Accordingly, the WFDEI forcing dataset holds great potential for improving our understanding of the hydrology of data-sparse Himalayan regions and providing the potential for prediction. The use of CORDEX-SMHI- and ERA-I-derived data requires further validation and bias correction, particularly over the high mountain regions.

Key words | discharge, global forcing dataset, Himalaya, hydrological modeling, hydropower inflow simulation

Bikas Chandra Bhattarai (corresponding author)
John Faulkner Burkhart
Lena M. Tallaksen
Chong-Yu Xu IWA
Felix Nikolaus Matt
Department of Geosciences,
University of Oslo,
P.O. Box 1047, Oslo N-0316,
Norway
E-mail: *b.c.bhattarai@geo.uio.no*

INTRODUCTION

The Himalayan and adjacent Tibetan Plateau water supply is intricately linked to the livelihoods, economic and social, to millions of people (Bookhagen & Burbank 2010; Immerzeel *et al.* 2010; Remesan *et al.* 2019; Zhang *et al.* 2019). This region provides for drinking water, hydropower generation, agricultural demands, as well as water-powered grain mills and other agro-economic activities (Ménégoz *et al.* 2013). Large Asian rivers, such as the Indus, Sutlej, Ganges, and Brahmaputra, are widely acclaimed for their great cultural, spiritual, economic, and ecological significance (Kumar 2017). The Ganges River and its tributaries alone fulfill significant water demands for more than 250 million people. Originating in the Himalayas, the river travels over 2,500 km, aggregating water from its tributaries

doi: 10.2166/nh.2020.079

to become the third largest freshwater delta (Chowdhury & Ward 2004).

Due to the high precipitation rates (1,500–2,500 mm/year; Dahal & Hasegawa 2008) and steep elevation gradients, hydropower has significant potential in the region and, if well managed, could provide a resource for economic growth. Hydropower potential depends on climatic conditions, particularly on precipitation, evaporation, temperature, and snow/ice in the catchment (Edenhofer *et al.* 2011). However, the barriers for sustainable water resource management in this region are manifold. First, there are numerous socio-political issues related to trans-border management issues (Biswas 2011). Despite efforts, a comprehensive framework to guide the development and international cooperation has yet to be ratified (Biswas 2011). Second, portions of the region are heavily glaciated and climate change is exacting an immediate and tangible impact (Immerzeel *et al.* 2010; Dehecq *et al.* 2019).

In this paper, we focus on a third significant barrier: providing robust and reliable analysis of the water resource potential and impacts from climate change. Data scarcity and the complexity of the terrain creating large spatial gradients and variability in weather and climate make such analyses a challenging task. The extremely heterogeneous topography in the region presents a great challenge for the accurate measurement of meteorological variables, giving rise to data scarcity (Pellicciotti *et al.* 2012). Although there are some meteorological stations, they are not well distributed in space. To overcome these challenges, regional climate models and reanalysis data offer gridded datasets of many meteorological variables, although with a rather coarse spatial resolution (Guo & Su 2019). The Coordinated Regional Climatic Downscaling Experiment (CORDEX) (Giorgi *et al.* 2009) provides extensively used regional datasets for past and future climate. For a historical perspective, the reanalysis dataset ERA-Interim (ERA-I) from the European Center for Medium-Range Weather Forecasts (ECMWF) (Dee *et al.* 2011) is widely used (Li *et al.* 2013, 2018; Xu *et al.* 2016). Bharti & Singh (2015) reported that the ERA-I precipitation is largely overestimated over the Indian Himalayan region. To address the bias inherited in ERA-I, bias- and elevation correction-based monthly observations were carried out when creating the Water and Global Change Forcing Data ERA-I (WFDEI) datasets (Weedon *et al.* 2014; Kim *et al.* 2019).

Various factors make hydrological modeling in the region challenging, including a large spatial variability in hydrometeorological variables, steep gradients, marked seasonality driven by the Indian Monsoon, and contrasting moisture regimes between the high elevation Tibetan Plateau and regions in the vicinity of the Indian Ocean. Changes in the climatic condition in the region may lead to changes in regional water balance components impacting hydrologic regimes (Koch *et al.* 2011). Moreover, snow and glacier storage and melt play an important role for the river discharge generation (Radić & Hock 2014; Li *et al.* 2015, 2016). The effective assessment and implementation of hydropower projects and other water resource management projects depend on a thorough analysis of the discharge and hydrological storages (Bakken *et al.* 2013).

Hydrologic simulation of discharge and other water balance components (evapotranspiration, snow, and groundwater storage) (Bhattarai & Regmi 2015; Matt & Burkhart 2018; Li *et al.* 2019) is used for the analysis of available water resources both in the past and future analyses, but the quality of discharge simulation remains a challenge (Rochester 2010; Engeland *et al.* 2016; Kauffeldt *et al.* 2016). The choice of a suitable hydrological model and appropriate forcing data are critical for any analysis and will greatly affect the outcome (Kauffeldt *et al.* 2016). Currently, no standardized or community modeling framework within hydrology and prior studies of the water resources of the Himalayan region have been conducted using different types of hydrologic models ranging from simple conceptual models (e.g. Pradhananga *et al.* 2014; Bhattarai & Regmi 2015; Skaugen & Weltzien 2016; Bhattarai *et al.* 2018) to more advanced, distributed models (e.g. Pellicciotti *et al.* 2012; Jain *et al.* 2017). Hydrological modeling and water balance studies in Himalayan regions have taken a range of approaches and used different models to focus on topics such as glacier melt and retreat, water balance, flooding, and the impact of climate change (Bookhagen & Burbank 2010; Pellicciotti *et al.* 2012; Wortmann *et al.* 2014; Khanal *et al.* 2015). Recent studies have addressed the performance of large-scale forcing datasets, including the WFDEI and ERA-I, for discharge simulation in various regions across the world (e.g. Li *et al.* 2013, 2016; Weedon *et al.* 2014; Essou *et al.* 2016; Nkiaka *et al.* 2017), but, to date, the use of these global forcing datasets to simulate discharge in a Himalayan catchment is limited.

In this study, we use a distributed, conceptual hydrologic framework with an energy balance-based snow routine that has been demonstrated to perform well in the region (Xu *et al.* 2015; Hegdahl *et al.* 2016; Matt & Burkhart 2018). The main goal of this work is to evaluate the impact of four different forcing datasets on the simulated discharge and associated water balance assessment for the Narayani river catchment in Nepal. The forcing datasets include WFDEI, ERA-I, CORDEX, and a hybrid dataset referred to as Observed + WFDEI (all datasets are described in the Data and Methods section). Our specific objectives are to evaluate: (i) the discharge sensitivity to the forcing dataset for hydrologic modeling inflow, (ii) model performance using bias-corrected versus non-bias-corrected forcing as input to a (calibrated) hydrological model, and (iii) the sensitivity to forcing data for water balance assessment. Furthermore, we assess the overall ability of the Statkraft Hydrological Forecasting Toolbox (Shyft) (Burkhart *et al.* 2016) to simulate discharge in this complex and data-sparse regions.

STUDY AREA

The Narayani river catchment lies in the central part of Nepal. About 13% of the total area lies in China (Figure 1), and thus, it is a transboundary catchment. The main Narayani river gauging station is located in Narayanghat (27°42′30″N, 84°25′50″E) and is operated by the Department of Hydrology and Meteorology, Government of Nepal (DHM, GoN). The catchment area is 31,692 km^2 and is partly glacier covered (~8%; Omani *et al.* 2017) with elevation ranging from 175 m a.s.l. in the south to 8,148 m a.s.l. in the north. The hydrological regime is heavily influenced by the season. The seasons are defined as monsoon (June to September), post-monsoon (October to November), winter (December to February), and pre-monsoon (March to May) (Shrestha *et al.* 1999; Bhattarai *et al.* 2019). Almost 80% of total annual precipitation occurs during the monsoon period (Nayava 1974; Kripalani & Sontakke 1996). Tributaries to the Narayani river are either monsoon fed (those originating in middle and high mountain regions) or glacial and snow melt fed (those originating in a higher Himalayan

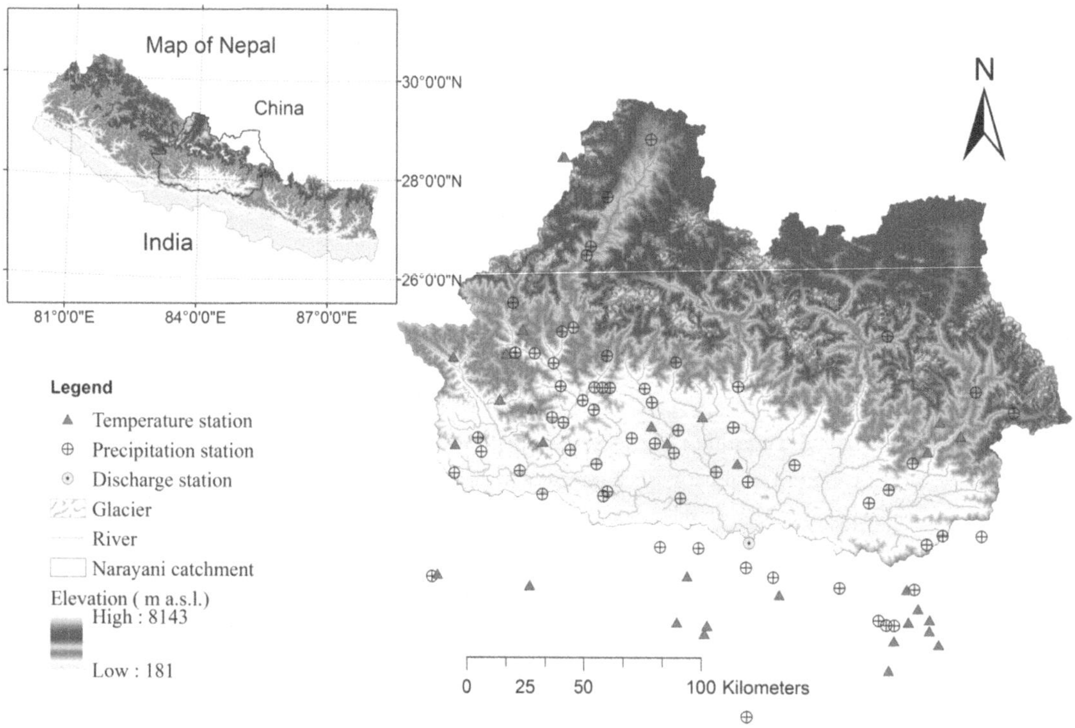

Figure 1 | Narayani river catchment with the location of the discharge station and meteorological stations (72 precipitation and 30 temperature) in the region (both within and outside the catchment boundary).

region). The Narayani river catchment lies also in the primary hydropower development region of Nepal, providing 44% of the total electric generation (Adhikari 2006).

DATA AND METHODS

Meteorological forcing data

Forcing variables used in the study include air temperature (T), precipitation (P), relative humidity (RH), wind speed (WS), and shortwave incoming solar radiation (S) at a daily resolution. Details of each dataset are described in the following sections.

Observed data

The DHM, GoN is responsible for collecting and disseminating hydrological and meteorological information for water resources, agriculture, energy, and other development activities in the country. Most of the measured data are based on the conventional manual observation. Measurements from the automatic weather station (AWS) are available only after the year 2000 (Karki 2010). The geographic location and installation date for each station are available from http://dhm.gov.np/meteorological-station/.

Observed temperature and precipitation station data which are located inside or in the vicinity of the catchment (Figure 1) were collected from the DHM. Data from the DHM were manually plotted for each station and screened for data quality. Clearly, erroneous departures from the historical pattern were removed manually from the datasets. The highest temperature station elevation is 3,870 m a.s.l., located in Chhoser (DHM st. no 633), and the highest precipitation station elevation is 3,705 m a.s.l., located in Mustang (DHM st. no 612).

To accommodate missing data, stations with less than 10 years of record or missing more than 15% of the observations were removed from the datasets. Ten precipitation stations (out of 82) did not meet this criterion and were removed from the datasets. Similarly, seven (out of 30) temperature stations were removed from datasets. This resulted in 72 precipitation stations and 23 temperature stations that were used for further analysis. Maximum numbers of stations with temperature and precipitation data were observed in the 2000–2009 period. Therefore, in this study, data for the period 2000–2009 were used.

All the stations are located below 4,000 m a.s.l., covering only 60% of the total catchment area (Figure 2(a)). Generally, temperature and precipitation data from higher elevations are missing, which add uncertainty to the current study. Daily mean temperature and mean annual precipitation for each station are plotted versus station elevation in Figure 2(b). Temperature shows as expected a consistent decrease with elevation, whereas precipitation shows a more mixed picture, with no clear trend below approximately 2,000 m a.s.l., but a notable decrease in precipitation above, i.e. from about 2,000 to 3,500 m a.s.l. Normally precipitation increases with elevation (Daly *et al.* 1994) in the mountainous region due to the orographic effect, but in the Himalayan region, an opposite pattern has been reported after a certain elevation level (e.g. Nayava 1980; Kansakar *et al.* 2004), in agreement with what is observed for the Narayani river catchment. From the analysis, station average daily temperature and mean annual precipitation for the period 2000–2009 were 10.7 °C and 1,292 mm/year, respectively.

Discharge observed at the Narayanghat station is available from the year 2000–2009 with no missing values, and this is also the period used for the hydrological model calibration and validation. Daily average discharge for the period 2000–2009 was 1,482.5 m^3/s.

Reanalysis and regional climate model data

A summary of the available time periods and the resolution of the forcing datasets are provided in Table 1. All datasets are well documented and are freely available.

ERA-I is a reanalysis global forcing dataset available from 1979, produced by the ECMWF. ERA-I temperature results from the assimilated surface temperature (Essou *et al.* 2016), while precipitation data are based on a reanalysis of precipitation fields generated with a meteorological model (Berrisford *et al.* 2011; Dee *et al.* 2011). The obtained precipitation data are not scaled using observation data. ERA-I is continuously updated once per month, with a delay of 2 months, and is freely available from http://apps.ecmwf.int/datasets/. Monthly mean meteorological variables

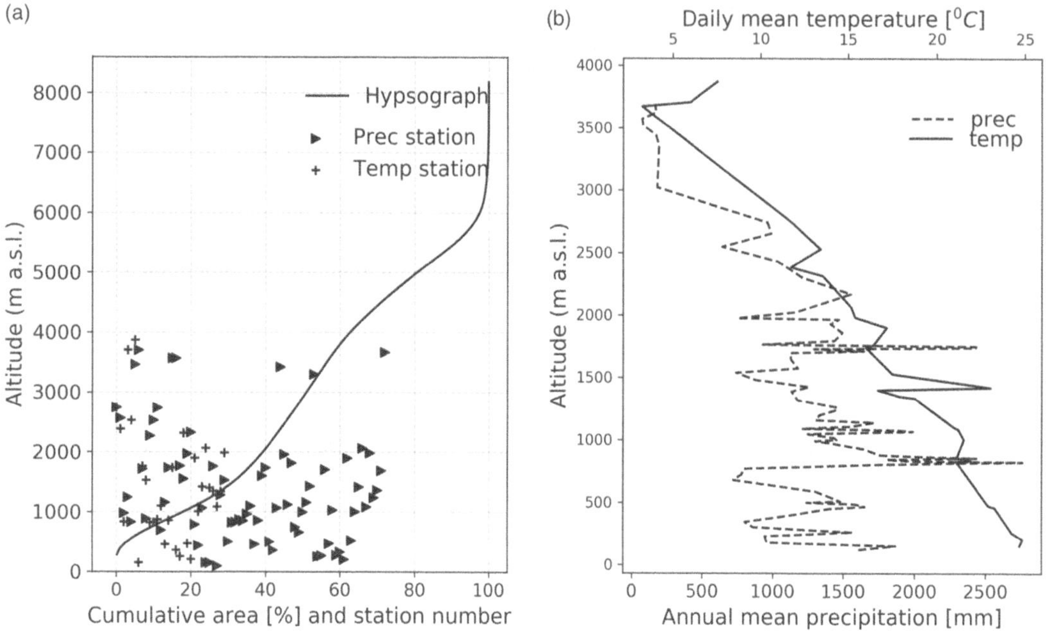

Figure 2 | (a) Station altitude (temperature and precipitation) plotted against the hypsographic curve for the Narayani river catchment and (b) station average daily mean temperature and annual mean precipitation (for the period 2000–2009) plotted against altitude. Station name, number, and data availability are provided in the Supplementary Data.

averaged over the Narayani river catchment using the Shyft interpolation routines (see the section 'Spatial-temporal interpolation of forcing data') are shown in Figure 3. The catchment average mean temperature for the period (2000–2009) was 7.6 °C, while catchment average annual precipitation was 4,660 mm/year.

The WFDEI is a global forcing dataset obtained by downscaling and bias-correcting ERA-I data (Weedon *et al.* 2014; Raimonet *et al.* 2017). The temporal and horizontal resolution of the dataset is shown in Table 1. WFDEI has two sets of rainfall generated by using either Climate Research Unit (CRU) or Global Precipitation Climatology Centre (GPCC) precipitation correction methods (Weedon *et al.* 2011). In this study, GPCC-corrected data were preferred to CRU because of their higher resolution and data quality (Weedon *et al.* 2014; Raimonet *et al.* 2017). The WFDEI dataset is freely available online from ftp.iiasa.ac.at. Catchment average mean daily temperature and annual precipitation (for the period 2000–2009) were found to be 7.9 °C and 1,764 mm/year, respectively.

CORDEX is a program sponsored by the World Climate Research Programme (WCRP), to produce an improved generation of regional climate change projections (Giorgi *et al.* 2009). CORDEX has two datasets, referred to Evaluation and Historical. Evaluation is run within reanalysis and is used to 'mimic' observations (i.e. represent real weather), whereas Historical is run within a climate model, and the results can only be used in a climatological sense. Daily datasets of the Evaluation product over the study area are available for the South Asia CORDEX

Table 1 | Summary of selected forcing datasets (*n* refers to the number of grid cells in the Narayani river catchment)

Forcing dataset	Data period	Spatial resolution (degrees)	Temporal resolution	*n*	Reference
ERA-I	1979–2015	0.75 × 0.75	3 h, 6 h	12	Dee *et al.* (2011)
WFDEI	1979–2016	0.5 × 0.5	Daily	20	Weedon *et al.* (2014)
CORDEX-SMHI	1980–2010	0.44 × 0.44	Daily	25	Sanjay *et al.* (2017)
Observed + WFDEI	1999–2010	–	Daily	–	

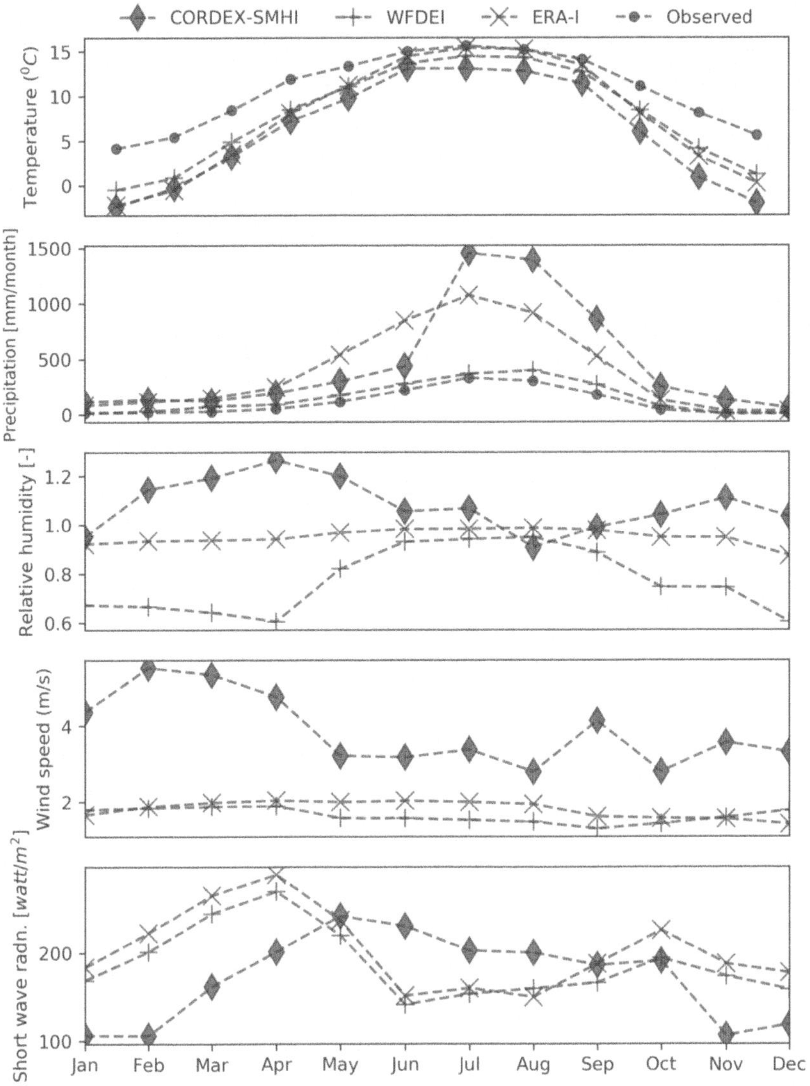

Figure 3 | Catchment average monthly mean Shyft-interpolated meteorological variables (for the period 2000–2009) for each forcing dataset.

domain (Region 6) from two institutional runs: Indian Institute of Tropical Meteorology (CORDEX-IITM) and Swedish Meteorological and Hydrological Institute (CORDEX-SMHI) (Sanjay *et al.* 2017). Based on a comparison of seasonal precipitation patterns over the study area, the CORDEX-SMHI was selected for this study. The horizontal resolution and available periods for the CORDEX-SMHI data are shown in Table 1. CORDEX-SMHI data can be freely downloaded on request from http://cccr.tropmet.res.in/home/index.jsp. Catchment average mean daily temperature and annual precipitation (for the period 2000–2009) are found to be 6.1 °C and 5,431 mm/year, respectively.

Topographical and land cover datasets

In this study, a digital elevation model (DEM) of 90 m spatial resolution from NASA's Shuttle Radar Topography Mission (NASA-SRTM) was used. The NASA-SRTM DEM is freely available for download from https://eros.usgs.gov/. The hydrology tool available in the software package, System for Automated Geoscientific Analysis (SAGA) (Conrad *et al.* 2015) under Qgis-2.18 (QGIS Development Team 2016), was used for automatic catchment delineation. Catchment delineation was based on the gauge point at Narayanghat and NASA-SRTM DEM. Catchment slope and aspect data for Shyft (see the Hydrological Model section) were generated during the catchment delineation subprocesses in Qgis. The calculated catchment domain was gridded into 2 km × 2 km cells for input to Shyft. The centroid of each vector grid cell was calculated using Qgis as further input to the model.

A land cover map (0.5° × 0.5°, resolution) providing forest, lake, glacier, and reservoir cover of the study area was extracted from the Moderate Resolution Imaging Spectroradiometer (MODIS) land cover data (Channan *et al.* 2014). MODIS land cover data are available free online at http://glcf.umd.edu/data/lc/. Land cover fractions of forest, lake, glacier, and reservoir cover area at the centroid point per each grid cell within the catchment were also calculated. The python code and algorithm for calculating land cover fraction per each vector grid cell at centroid are available online at https://github.com/felixmatt/shyft-gis.

Spatial interpolation of observed and gridded forcing data

Shyft is able to ingest daily spatially distributed meteorological data as input variables. Here, we interpolate both observations (station data) and gridded forcing data (WFDEI, ERA-I, and CORDEX-SMHI).

The gridded forcing data have a coarser resolution than the simulation domain in the hydrological modeling framework of Shyft (Table 1). Accordingly, Shyft provides interpolation routines used for downscaling the forcing datasets to the model grid cell scale. Data interpolation methods were selected based on prior studies by Sapkota (2016), Lombraña (2017), Matt *et al.* (2018), Matt & Burkhart (2018), and Teweldebrhan *et al.* (2018). Observed temperature and gridded data were extrapolated/interpolated by three-dimensional Kriging where elevation is the third dimension (Oliver & Webster 1990). Observed precipitation, gridded precipitation, global radiation, WS, and RH data were extrapolated/interpolated by inverse distance weighting (IDW) (Shepard 1968). A precipitation gradient of −0.07 mm/100 m (Table 2) and a temperature gradient of −0.6 °C/100 m are used during interpolation. The precipitation gradient was calculated from the observed precipitation data. The different interpolation methods along with their parameter values are presented in Table 2.

Table 2 | Parameter values used for IDW and Kriging interpolation

	IDW		Kriging	
	Parameter	**Value**	**Parameter**	**Value**
1	Maximum distance	400,000 m	Temperature gradients	−0.6 °C/100 m
2	Maximum number	10	Range	20,000 m
3	Distance measure factor	1	*z*-scale	20
4	Resolution	2 km × 2 km	Resolution	2 km × 2 km
5	–	–	Sill	25 °C
6	Precipitation gradient	−0.07 mm/100 m		

Hydrological model

The Shyft modeling framework was used to simulate daily discharge from the catchment. The modeling framework has three main hydrologic modeling routines (https://gitlab.com/shyft-os). These three models are different in the way to calculate evapotranspiration and snow estimation and melt.

In this study, the PT_GS_K model was used for discharge simulation. The PT_GS_K model uses the Priestley–Taylor (PT) method (Priestley & Taylor 1972) for estimating potential evapotranspiration, the Gamma snow routine (GS; described in the Gamma Snow section) for snowmelt, sub-grid snow distribution and mass balance calculation, and a simple storage–discharge function (K) for catchment response calculation (Lambert 1972; Kirchner 2009). Actual evapotranspiration (AE) was assumed to take place from the snow-free area and was estimated as a function of potential evapotranspiration and a scaling factor.

The catchment response function 'K' is based on the storage–discharge relationship concept described in Kirchner (2009) and represents the sensitivity of discharge to storage changes as given by the following equation. Generally, discharge is nonlinear and varies by many orders of magnitude, and it is recommended to use log-transformed discharge values to avoid numerical instability.

$$\frac{d(ln(Q))}{dt} = g(Q)\left(\frac{P-E}{Q} - 1\right) \tag{1}$$

where P, Q, and E are the rate of precipitation (input from either snow melt or rain), discharge, and AE (from the snow-free area), respectively, in units of depth per time.

The idea behind this method is that the discharge sensitivity to changes in storage, i.e. $g(Q)$, can be estimated from the time series of the discharge alone through fitting empirical functions to the data, such as the quadratic equation (Kirchner 2009), which is given by:

$$g(Q) = e^{c1+c2(ln(Q))+c3(ln(Q))^2} \tag{2}$$

where $c1$, $c2$, and $c3$ are the catchment-specific outlet parameters (hereafter called Kirchner parameters) obtained during the model calibration. There is no routing function in the model, but the Kirchner response function represents a delayed outflow from storage within the catchment.

Gamma snow

In Shyft, the Gamma snow routine, an energy balance approach (Equation (3)) for snow ablation, and the snow depletion curve following a gamma distribution are combined into a single routine. The energy balance approach in the Gamma snow routine is based on DeWalle & Rango (2008) and also briefly explained in Hegdahl *et al.* (2016). The net energy flux (ΔE) at the surface available for snow ablation is expressed as follows:

$$\Delta E = S.(1-\alpha) + L_{in} + L_{out} + H_{SE} + H_L + E_G \tag{3}$$

where S is the net shortwave radiation, L_{in} and L_{out} are the incoming and outgoing longwave radiations, H_{SE} and H_L represent sensible and latent heat fluxes, and E_G is the net ground heat flux calculated using a bulk-transfer approach. Two parameters defining the wind profile, intercept (wind constant), and slope (wind scale) are determined either by model calibration or as provided (Table 3). For a given time step (t), the snow albedo (α) at each cell depends on the minimum (α_{min}) and maximum albedo (α_{max}) as well as the albedo decay rate, temperature, and snowfall as described in Hegdahl *et al.* (2016):

$$\alpha_t = \begin{cases} \alpha_{min} + (\alpha_{t-1} - \alpha_{min}) \cdot \left(\dfrac{1}{2^{\frac{1}{\mathrm{FDR}}}}\right) & \text{if } T_a > 0^\circ\mathrm{C} \\ \alpha_{t-1} + (\alpha_{max} - \alpha_{min}) \cdot \left(\dfrac{1}{2\mathrm{SDR}}\right) & \text{if } T_a \leq 0^\circ\mathrm{C} \end{cases} \tag{4}$$

In Equation (4), FDR and SDR denote fast and slow snow cover decay rates, respectively. In this study, α_{max} and α_{min} are prescribed (refer Table 3).

Within the Gamma snow routine, precipitation falling in each cell is classified as solid or liquid depending on a threshold temperature (tx) and the actual cell temperature. Snow distribution within each cell is estimated by using a three-parameter gamma probability distribution. The third parameter in the gamma probability distribution represents the bare ground fraction in the cell. Finally, snowmelt depth (mm/day) is calculated by multiplying ΔE (available energy) with the latent heat of fusion for water.

A temperature index model which does not require glacier ice albedo was used to calculate glacier melt (see Hock 2003). The glacier reservoir was assumed to be infinite, and the glacier area was assumed to be constant throughout the simulation periods. Within a glacier-covered cell, glacier melt only happens from the snow-free fraction.

Parameters and calibration

Hydrological simulation from distributed models such as Shyft generally requires the estimation of model parameters through calibration with measured data (Madsen 2003). In the PT_GS_K routine, there are 14 parameters, out of which eight parameters (Table 3: top eight parameters) have been found to be significantly more sensitive than the rest (Teweldebrhan *et al.* 2018) and were selected for calibration in this study. The remaining six parameters were prescribed (Table 3: lower six parameters). The precipitation correction factor (p_corr_factor), which is used to correct bias, was also set as a calibration parameter.

Manual calibration can be time-consuming and subjective; therefore, an automatic calibration was carried out.

Table 3 | Model calibration parameters with upper and lower limits

Parameter	Description and unit	Parameter used in the submodel	Lower limit	Upper limit	Sources
*c*1	Outlet empirical coefficient 1 (–) Equation (4)	K	−8.0	0.0	Sapkota (2016); Lombraña (2017)
*c*2	Outlet empirical coefficient 2 (–) Equation (4)	K	−1.0	1.2	Sapkota (2016); Lombraña (2017)
*c*3	Outlet empirical coefficient 3 (–) Equation (4)	K	−0.15	−0.05	Sapkota (2016); Lombraña (2017)
wind scale	Slope in turbulent wind function (m/s)	GS	1.0	6.0	Sapkota (2016); Lombraña (2017)
tx	Temperature threshold rain/snow (°C)	GS	−3.0	2.0	Sapkota (2016); Lombraña (2017)
FDR	Fast albedo decay rate during melt (days)	GS	5.0	15.0	Hegdahl *et al.* (2016); Sapkota (2016)
SDR	Slow albedo decay rate during cold condition (days)	GS	20.0	40.0	Hegdahl *et al.* (2016)
p_corr_factor	Scaling factor for precipitation (–)	GS	0.4	2.0	Sapkota (2016)
Prescribed parameters					
ae scale factor	Scaling factor for AE (–)	AE	1.0	1.0	Sapkota (2016); Lombraña (2017)
Surface magnitude	Snow heat constant (mm SWE)	GS	30.0	30.0	Sapkota (2016)
wind constant	Intercept in turbulent wind function (–)	GS	1.0	1.0	Lombraña (2017)
min albedo	Minimum snow albedo used in snow routine	GS	0.6	0.6	Sapkota (2016)
max albedo	Maximum snow albedo used in snow routine	GS	0.9	0.9	Sapkota (2016)
max water	Frictional max water constant of snow (–)	GS	0.1	0.1	Hegdahl *et al.* (2016)

In the table, 'K' is the catchment response function; 'GS' is the Gamma snow; 'PT' is the Priestley and Taylor; and 'AE' is the actual evapotranspiration.

Upper and lower limits for each parameter are shown in Table 3. CREST v.2.1, the Shuffle Complex Evolution University of Arizona (SCE-UA) (Duan *et al.* 1992), is used as the kernel algorithm in the automatic calibration process. Typically, the procedure involves the selection of samples in the parameter space through the use of competitive evolution schemes, such as the simplex scheme, to reproduce better the observations. After several iterations, either due to convergence or when the maximum number of iteration is reached, the best set of parameter values based on the Nash–Sutcliffe coefficient of efficiency (NSE) is determined (Chu *et al.* 2010).

In this study, the model was calibrated and validated for each forcing dataset independently (hereafter named 'independent calibration mode'). The model calibration was based on observed discharge data from 2000 to 2004, and validation was based on data from 2005 to 2009.

Water balance estimation

A water balance analysis is a useful tool to describe the principal components of water in and out of a catchment (Rochester 2010), where the volume of water inflow should be balanced with water outflow, assuming no changes in storage ($\Delta S/\Delta t$). The following water balance components: mean annual precipitation, discharge, and AE were calculated. Water balance components were calculated using calendar years, so that the change in storage is mainly a difference in storage between first and last days of simulation.

The annual change in storage was calculated according to the following equation:

$$\frac{\Delta S}{\Delta t} = (P + G) - (Q + ET) \quad (5)$$

where P is precipitation, Q is the discharge, ET is actual evapotranspiration, G is the glacier melt, and the unit of measurement is mm/year. $\Delta S/\Delta t$ is the change in storage per time unit.

Model performance evaluation

To determine the agreements between observed and simulated discharges, NSE (Nash & Sutcliffe 1970), Kling–Gupta efficiency (KGE) (Gupta *et al.* 2009), square-root-transformed NSE (NSE_{sqrt}) (Seiller *et al.* 2012), benchmark series (G_{bench}) (Seibert 2001), and percentage volume difference (D_v) (Martinec *et al.* 1998) were used (Table 4). The NSE, NSE_{sqrt}, and KGE efficiency measures are used to access the predictive power of the hydrological model and can range from $-\infty$ to 1. An efficiency of 1 corresponds to a perfect match of simulated and observed discharges. To improve the estimation of the performance error, KGE considers three components: bias (α), variability (β), and linear correlation coefficient (r) to overcome the problems associated with NSE, i.e. NSE results in the underestimation of the streamflow variability and the runoff peaks (Gupta *et al.* 2009). NSE_{sqrt} gives more emphasis to the overall agreement between observed and simulated discharges (Seiller *et al.* 2012; Peña-Arancibia *et al.* 2015) as compared to NSE, so it is also included as an evaluation criterion. Percentage volume difference (D_v) gives the percentage bias between the simulated and observed series and can range from $-\infty$ to $+\infty$. D_v equals to zero indicates a perfect agreement between simulated and observed discharges. Since the observed discharge shows strong seasonal patterns, we also used a benchmark series (G_{bench}) for model evaluation. The monthly average discharge was used as a benchmark series. G_{bench} is negative if the model performance is poorer than the benchmark, zero if the model performs as well as the benchmark, and positive if the model is superior, with a highest value of one for a perfect fit. In this study, we aimed to achieve $NSE > 0.7$, $KGE > 0.7$, $NSE_{sqrt} > 0.7$, $G_{bench} > 0.5$, and D_v within $\pm 15\%$, during both the model calibration and validation periods.

Table 4 | Definitions of the evaluation criteria

Criteria	Mathematical expression	Description	Best value
D_v	$\frac{V_{obs} - V_{sim}}{V_{obs}} \times 100$	Volume difference in percentage	0
NSE	$1 - \frac{\sum_{i=1}^{n} (Q_{sim.i} - Q_{obs.i})^2}{\sum_{i=1}^{n} (Q_{obs.i} - \bar{Q}_{obs})^2}$	Nash–Sutcliffe efficiency	1
NSE_{sqrt}	$1 - \frac{\sum_{i=1}^{n} \left(\sqrt{Q_{sim.i}} - \sqrt{Q_{obs.i}}\right)^2}{\sum_{i=1}^{n} \left(\sqrt{Q_{obs.i}} - \sqrt{\bar{Q}_{obs}}\right)^2}$	Squared-root transformed Nash–Sutcliffe efficiency	1
KGE	$\sqrt{(r-1)^2 + (\alpha-1)^2 + (\beta-1)^2}$ Here, $\beta = \frac{\overline{Q_{sim}}}{\overline{Q_{obs}}}$; $\alpha = \frac{Q_{sim.std}}{Q_{obs.std}}$	Modified Kling–Gupta efficiency	1
G_{bench}	$\frac{\sum_{i=1}^{n} (Q_{obs.i} - Q_{sim.i})^2}{\sum_{i=1}^{n} (Q_{obs.i} - Q_{bench.i})^2}$	Goodness of fit with respect to the benchmark series	1

n: total number of days in the evaluation period; Q_{sim} and Q_{obs}: simulated and observed discharges; $Q_{sim.std}$ and $Q_{obs.std}$: standard deviation for simulated and observed discharges; $\overline{Q_{sim}}$ and $\overline{Q_{obs}}$: arithmetic mean for simulated and observed discharges; V_{sim} and V_{obs}: total discharge volume for simulated and observed discharges; Q_{bench}: monthly long-term average observed discharge.

RESULTS

Meteorological forcing data analysis

As discussed in the section 'Reanalysis and regional climate model data', the resolutions of gridded datasets are different. For the sake of comparison, Shyftinterpolated meteorological variables (for the period 2000–2009) for each forcing dataset were used. Temperature shows an overall agreement where all three datasets demonstrate similar seasonality and distinct cool or negative bias as compared to the observations during the winter and pre-monsoon seasons (Figure 3). The bias was the strongest in the CORDEX-SMHI data during the September–December period. Long-term spatially averaged temperature for July (Figure 4) shows that the CORDEX-SMHI was not able to capture higher daily average temperatures in the river valley. Daily average temperatures during July over the lower elevation region from ERA-I and CORDEX-SMHI were lower than the temperatures from the Observed and WFDEI interpolated datasets. What was most interesting is that negative catchment average temperatures, which were not at all captured by the interpolated observations (Figure 3), will have a significant impact on any simulation. We attribute the lack of negative catchment average temperatures to the lack of stations above 4,000 m a.s.l. combined with a too low-temperature lapse rate. This is a typical challenge when observations lack representativity to topography.

Precipitation varies notably among the forcing datasets. WFDEI precipitation agrees well with the observation (Figures 3 and 4). The strongest positive bias was observed for CORDEX-SMHI and ERA-I datasets (Figure 5). Long-term daily averaged precipitation for July–September shows the strongest deviation among forcing. The catchment precipitation for July was more than 150 mm/day in CORDEX-SMHI and ERA-I datasets, and the highest precipitation was found in the lower elevations. A seasonal comparison shows that the WFDEI was slightly higher than the observed precipitation during the monsoon season (Figure 6). Similar findings are reported for the Indus catchment by Dahri *et al.* (2016).

Monthly averaged wind speeds and incoming shortwave radiation from WFDEI and ERA-I were fairly similar but different for CORDEX-SMHI. Station

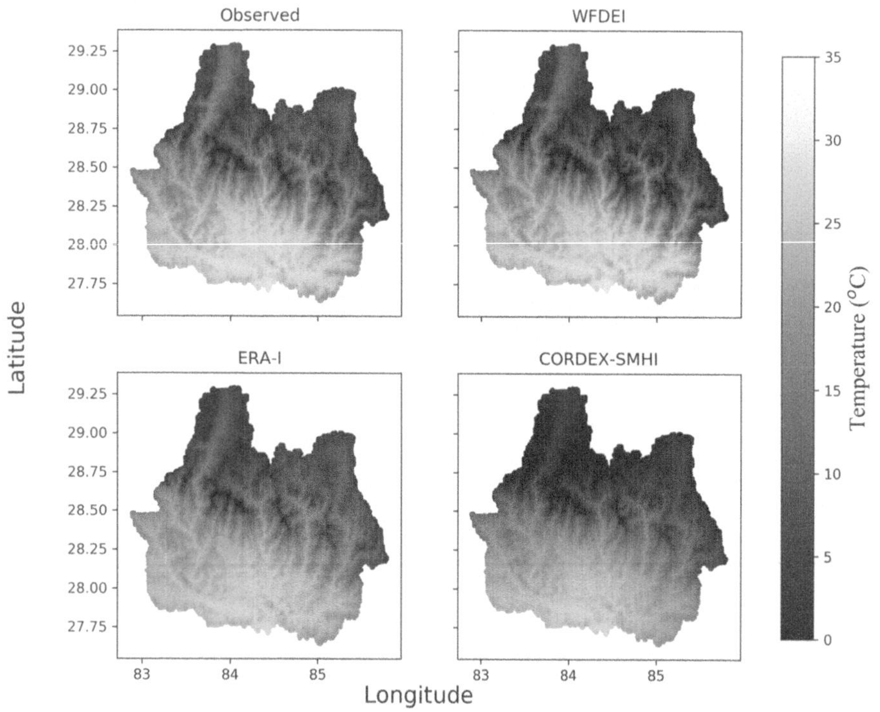

Figure 4 | Distributed mean daily Shyft-interpolated temperature (°C) for July over the Narayani river catchment for each forcing dataset.

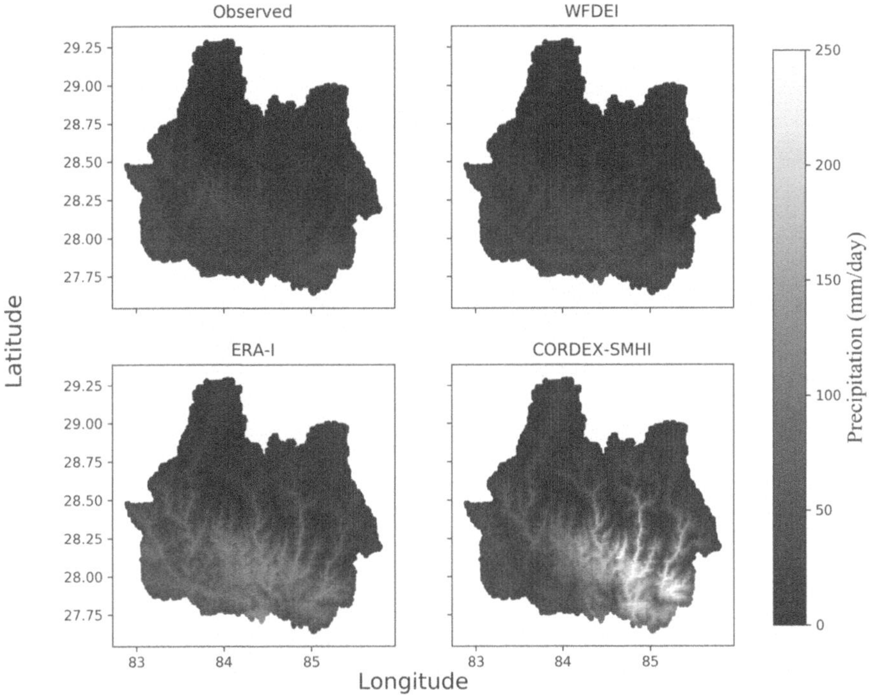

Figure 5 | Distributed daily mean Shyft-interpolated precipitation (mm) for July over the Narayani river catchment for each forcing dataset.

observations for RH, WS, and shortwave incoming global radiation were not available for the catchment, but previous global studies by Iizumi *et al.* (2014) and Weedon *et al.* (2014) showed that the WFDEI is more representative for the observations.

Model parameters

Shyft was calibrated for each forcing dataset, and the calibrated parameters are shown in Table 5. Normal ranges for all calibrated parameters are given in Table 3. The

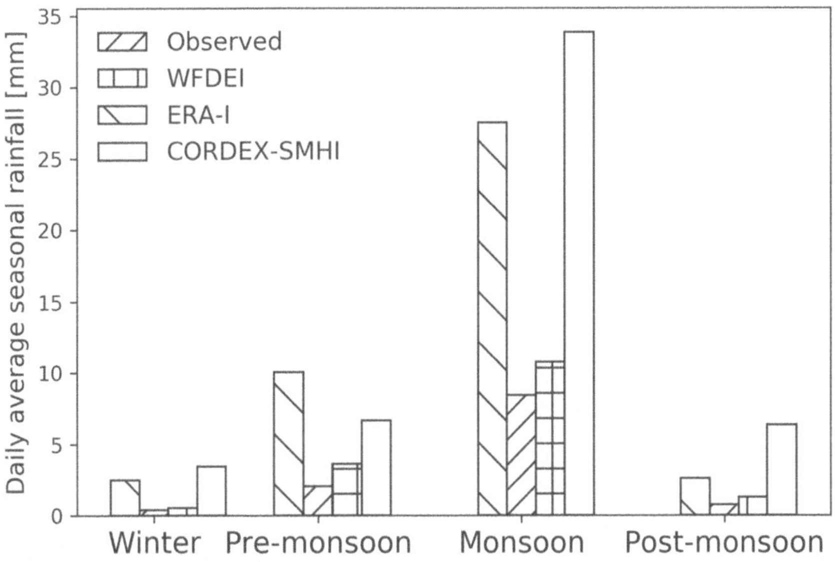

Figure 6 | Seasonal precipitations from different forcing datasets.

Table 5 | Estimated parameter values from the independent calibration mode for the four forcing datasets

Parameters	Unit	Observed + WFDEI	WFDEI	ERA-I	CORDEX-SMHI
*c*1	–	−5.51	−6.09	−6.64	−5.99
*c*2	–	0.44	0.25	0.14	−0.20
*c*3	–	−0.05	−0.05	−0.05	−0.07
tx	°C	−1.68	−1.38	−0.36	0.94
wind scale	m/s	2.01	4.70	4.84	3.40
FDR	days	11.59	12.03	9.94	10.75
SDR	days	39.98	32.66	29.04	31.55
p_corr_factor	–	1.34	1.09	0.41	0.4

highest range in values (greater than 0.15 variance in the normalized scale of −1 to 1) was seen for the parameters p_corr_factor (0.4–1.34), wind scale (2.01–4.84), threshold temperature (tx) (−1.68 to 0.94), and slow albedo decay rate (29.04–39.98). A relatively small variation in parameter values (less than 0.05 variance in the normalized scale of −1 to 1) was seen for the Kirchner coefficients (*c*1 (−6.64 to −5.51), *c*2 (0.14 to 0.44), and *c*3 (−0.05 to −0.07)) and fast albedo decay rate (9.94–12.03). The highest variance (0.27) in a normalized scale of −1 to 1 was found for p_corr_factor. The lowest and highest p_corr_factor values were observed for CORDEX-SMHI (0.4) and Observed + WFDEI (1.34) forcing dataset, respectively. The higher p_corr_factor for Observed + WFDEI is interesting and suggests an underprediction of precipitation in the catchment when gridded precipitation was based on Shyft-interpolated station observations. Unique to the CORDEX-SMHI dataset, the Gamma snow threshold temperature (tx) was found to be positive (0.94). Less sensitive parameters as reported by Teweldebrhan *et al.* (2018) were not calibrated but are listed with given values in Table 3.

Evaluation of discharge simulation using different forcing datasets

Figure 7 compares daily observed and simulated discharges using the different forcing datasets. Calibration and validation results in terms of error statistics are shown in Table 6. Simulations from Observed + WFDEI forcing give the best results in terms of error statistics during both calibration (NSE = 0.90, NSE_{sqrt} = 0.91, KGE = 0.94, G_{bench} = 0.43, and D_v = −0.91%) and validation (NSE = 0.90, NSE_{sqrt} = 0.93, KGE = 0.92, G_{bench} = 0.56, and D_v = −1.29%) periods. D_v in both calibration and validation was also best for Observed + WFDEI and found to be less than −2%. The second best performance was achieved for WFDEI (Table 6). For NSE, NSE_{sqrt}, and KGE, the ERA-I and WFDEI datasets were similar, but the volume differences during calibration were comparatively higher for ERA-I than WFDEI. Comparatively poorer model performance and higher volume differences (Table 6) were found

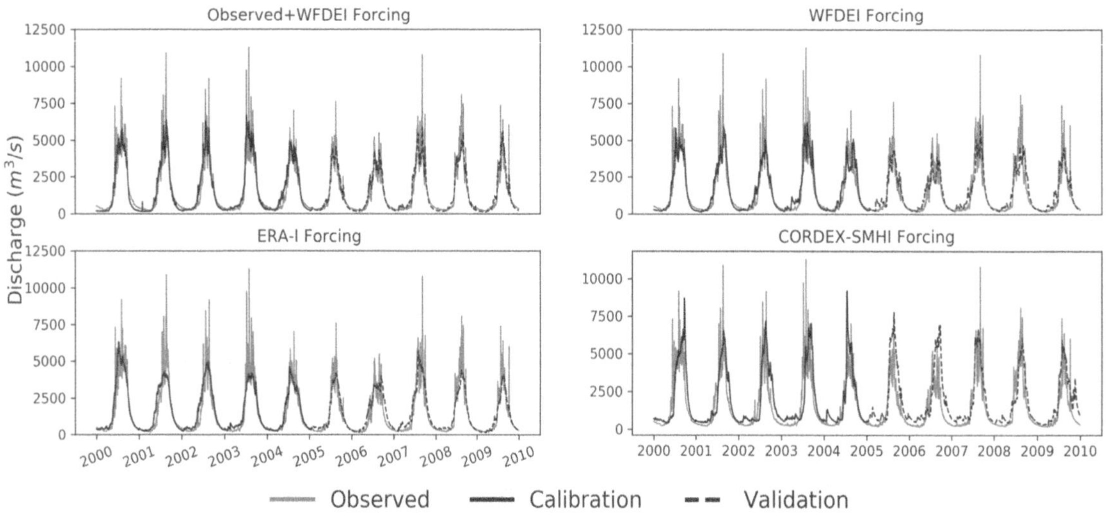

Figure 7 | Calibrated and simulated daily discharge for the Narayani river catchment for the period 2000–2009.

Table 6 | Performance statistics for different forcing datasets

	Observed + WFDEI		WFDEI		ERA-I		CORDEX-SMHI	
Error parameters	**Calibration**	**Validation**	**Calibration**	**Validation**	**Calibration**	**Validation**	**Calibration**	**Validation**
NSE	0.90	0.90	0.83	0.80	0.82	0.83	0.60	0.48
KGE	0.94	0.92	0.86	0.81	0.82	0.81	0.73	0.52
D_v	−0.91	−1.29	−1.72	−5.68	−4.01	−5.53	−19.9	−42.4
NSE_{sqrt}	0.91	0.93	0.88	0.84	0.88	0.87	0.70	0.57
G_{bench}	0.43	0.56	0.01	0.07	−0.03	0.19	−1.34	−1.47

for the model calibrated with the CORDEX-SMHI dataset. Goodness of fit with respect to the benchmark (G_{bench}) for ERA-I and CORDEX-SMHI during calibration periods was found negative, indicating poorer performance than the benchmark series. However, it should be noted the peak discharge was best simulated with CORDEX-SMHI data (Figure 7).

In Figure 8, the year 2004 was plotted to show in more details the representation of the annual cycle by the different forcing datasets. Overall, all datasets were able to simulate the cycle well; however, the CORDEX-SMHI simulation deviates somewhat from the rest and shows a higher peak flow early in the wet season.

Quantile–Quantile (QQ) plots for the different forcing datasets are shown in Figure 9. The QQ-plot for the calibration period is shown in Figure 9(a), for the validation period in Figure 9(b). To highlight low flows, Figures 9(c) and 9(d) show the QQ-plot on log-scale. The QQ-plot is a graphical technique to determine if two datasets come from the same population with a common distribution (Renard et al. 2010). Departures from the 1:1 reference line indicate discrepancies between the simulated and the observed discharge. Figure 9(a) shows that all simulations slightly overestimate discharge (with respect to the observed discharge) up to a value of around 4,000 m^3/s. Between 4,000 and 8,000 m^3/s, the CORDEX-SMHI simulations were higher than observed, whereas other simulations were lower than observed. After 8,000 m^3/s, CORDEX-SMHI simulations were able to capture peaks where others simulations were lower than observed. Although the

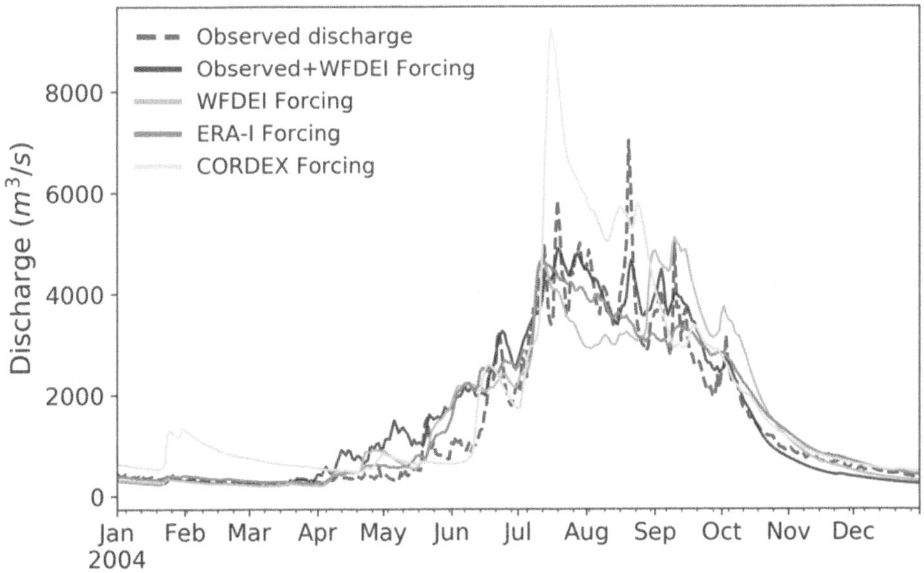

Figure 8 | Daily observed and simulated discharges from different forcing models for the year 2004.

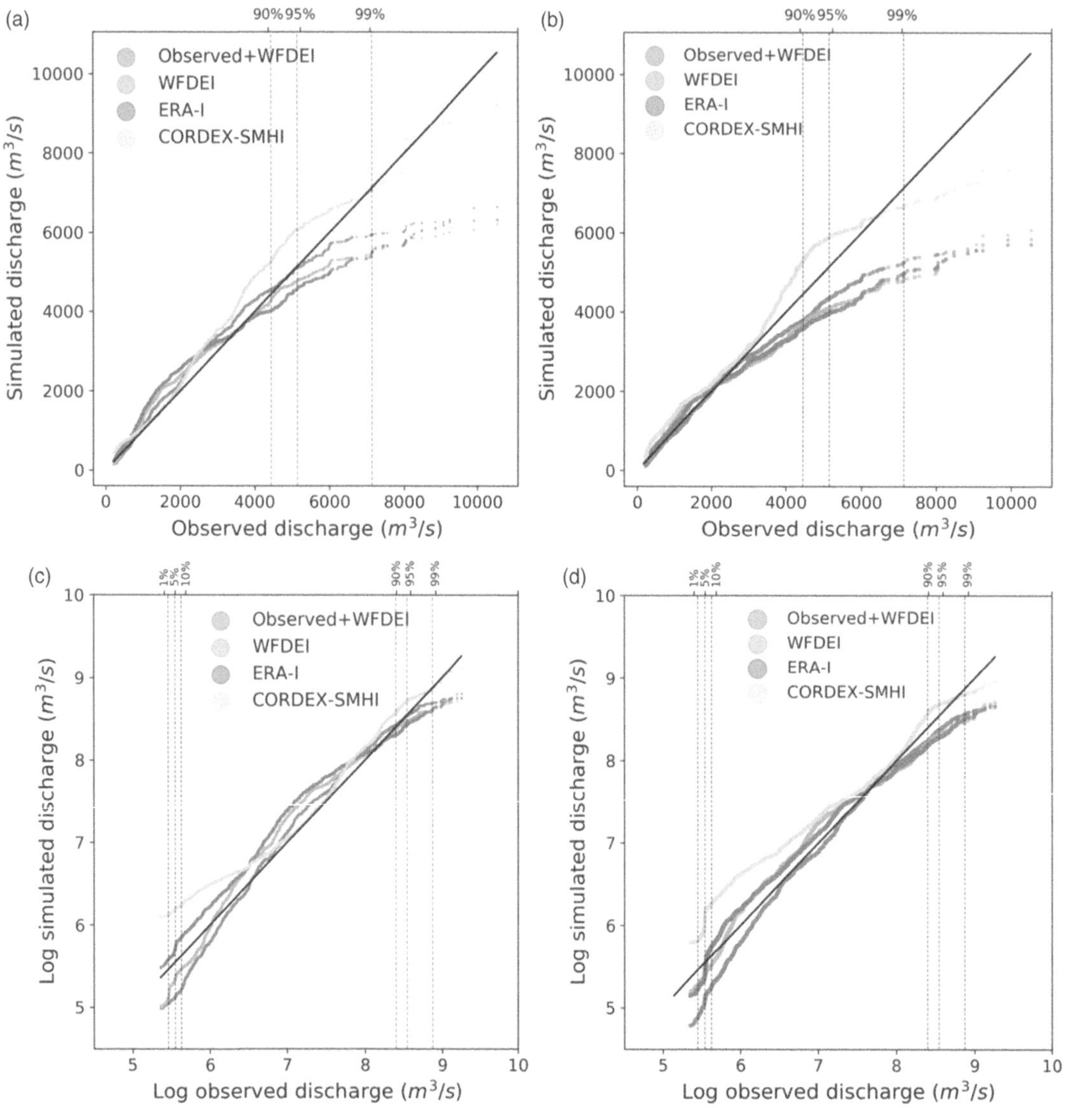

Figure 9 | QQ-plots of simulated versus observed discharge for the Narayani river catchment during model (a) calibration and (b) validation. Similarly, the QQ-plot for log-transformed observed and simulated discharges during model (c) calibration and (d) validation. The vertical lines denote the 1%, 5%, 10%, 90%, 95%, and 99% quantiles of the observed daily discharge.

Observed + WFDEI forcing dataset resulted in discharge simulations with the best performance, it was not able to capture the highest observed discharge (simulated discharge points after 95% quantile). The simulations from CORDEX-SMHI were able to capture relatively high peak-discharge events (above the 99% quantile); however, they have large deviations from the observations between 90% and 95% quantiles. For the low flows (i.e. 1–10% quantile), the simulation from the WFDEI forcing captures observation better than the Observed + WFDEI forcing (Figure 9(c)). However, low-flow simulation from ERA-I until 5% quantile was better than the rest.

Similarly, in the validation period (Figure 9(b)), simulated discharge and observed daily discharge for values less than 90% quantile from all forcing datasets fit well to the observations. Above 95% quantile, most forcing datasets fail to capture these highest discharges. Interestingly, the CORDEX-SMHI, which generally provides lower performance (Table 6), notable in the lower flow range, manages to capture the highest flows best. However, these simulations were mostly higher than the reference line, indicating a general overestimation during the full range of simulations. Similar to the calibration periods, low-flow simulation from WFDEI was also found closest to the reference (observations) during the validation periods (Figure 9(d)).

Water balance analysis

Figure 10 compares the catchment average (2000–2009) water balance components for the four different forcing datasets (the exact values are given in Table 7). Annual averages over the 10-year simulation period, 25.1%, 24.8%, 30.7%, and 17.8% of precipitation, were lost to evapotranspiration for Observed + WFDEI, WFDEI, ERA-I, and CORDEX-SMHI datasets, respectively. Among all forcing datasets, Observed + WFDEI shows lower runoff than the others. Higher runoff from CORDEX-SMHI forcing was observed, and it is associated with higher average precipitation during the monsoon period (Figure 3). The lowest (72.8 mm) total glacier melt during the simulation period was observed for the WFDEI dataset, while the highest (297.3 mm) was observed from ERA-I. The change in storage (Equation (5)) shows that 73.1, 32.7, 68.7, and 29.2 mm of water were surplus (a positive change in storage) in the catchment for Observed + WFDEI, CORDEX-SMHI, ERA-I, and WFDEI forcing data, respectively. The largest glacier melt contribution (19%) to total runoff was observed for ERA-I forcing, while the lowest was observed for WFDEI (4%). Different water balance components and their associated changes in storage are shown in Table 7.

Table 7 | Water balance components for different forcing (numbers in mm/year)

Water balance components	Observed + WFDEI	WFDEI	ERA-I	CORDEX-SMHI
Precipitation	1,771.1	1,965.7	1,901.4	2,234.1
Evapotranspiration	449.6	487.4	584.6	399.5
Glacier melt	242.2	72.8	297.3	132.5
Runoff	1,490.7	1,521.9	1,545.4	1,934.4
Change in storage (Equation (5))	73.0	29.2	68.7	32.7

DISCUSSION

Discussion on model parameters

As with any hydrologic model, the model used in this study is sensitive to the parameters; particularly the Kirchner coefficients, precipitation correction factor, and the threshold temperature (Teweldebrhan *et al.* 2018). While calibrating the model, different values of the parameters were obtained for different forcing datasets (Table 5). An examination of the model parameters revealed that the calibrated values of the precipitation correction factor, wind scale, threshold temperature, and slow albedo decay rates differ a lot

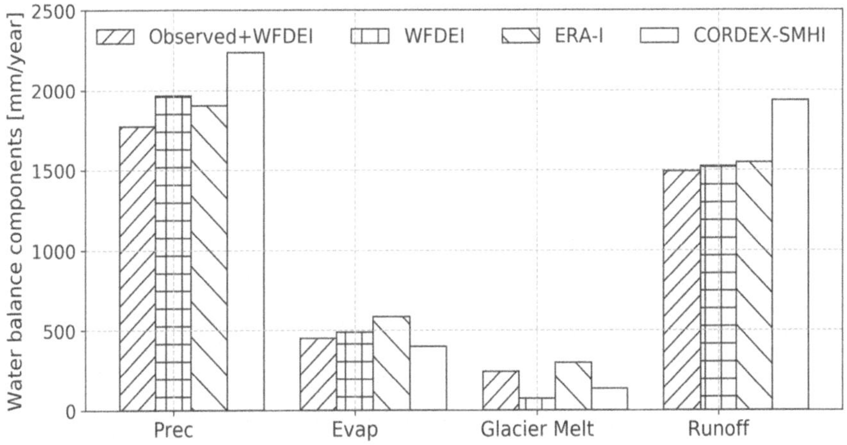

Figure 10 | Annual average (2000–2009) water balance components (Evap, actual evapotranspiration; Prec, precipitation calculated by using p_corr_factor) for different forcing datasets in the Narayani catchment.

between the forcing datasets. Larson & Peck (1974) and Lakew *et al.* (2017) pointed out that the model simulations are highly sensitive to the precipitation correction factor, which helps adjusting systematic bias embedded in the precipitation forcing. The precipitation correction factor may account for both under catch and lack of representative stations (Lakew *et al.* 2017) and is directly related to the bias in precipitation data. Prior studies by Engeland *et al.* (2016) and Teweldebrhan *et al.* (2018) suggested that the Kirchner parameters are highly sensitive since they have a large influence on a given simulation. We found that the variance for Kirchner response parameters was less than 0.05. An explanation is that these parameters are not dependent on the forcing datasets but rather on the physical catchment characteristics and thereby the concentration time of the catchment. Kirchner response parameters are dependent on the physical catchment characteristics and thereby the concentration time of the catchment and are therefore less sensitive to the forcing.

Potential factors controlling hydrological model efficiency during model calibration and validation

Different forcing datasets lead to different levels of performance efficiency in terms of discharge simulations (Table 6). Explaining these differences is not easy, as each forcing dataset has different characteristics such as the spatial resolution and the assimilation methods. Figures 3–5 present the catchment average monthly mean forcing variables for the different forcing datasets. From the figures, we can see that in general, the spatio-temporal distribution of precipitation from WFDEI was quite similar to the interpolated observations. The extent of higher precipitation than observation during pre-monsoon and monsoon seasons was larger for ERA-I and CORDEX-SMHI. Overestimation of precipitation during the monsoon season identified for CORDEX-SMHI in the Himalayan region (Ghimire *et al.* 2018) and for the ERA-I over the Indus catchment (Dahri *et al.* 2016). Moulin *et al.* (2009) and Zhu *et al.* (2014) indicate that a poor representativeness of precipitation over the catchment is a major source of uncertainty in discharge simulation.

Despite some of the biases in the forcing data, the performance of the model during calibration and validation showed that all datasets performed reasonably well. The good agreement between simulated and observed discharges during the model calibration and validation periods (Figure 7) is to a large extent due to the application of the precipitation correction factor. The multiplicative correction factor to the precipitation removes a substantial proportion bias in both precipitation and simulated discharges. It should be clarified here that such an adjustment of bias in discharge (through the input forcing correction) is an 'ad hoc' procedure to overcome the quality of forcing data. The 'p_corr_factor' shown in Table 5 indicates that the CORDEX-SMHI was reduced by a factor of 0.4, and ERA-I precipitation was reduced by a factor of 0.41 during the calibration period to minimize bias in discharge estimation. These numbers indicate that there were significant biases in these two forcing datasets and suggest that some kind of bias correction should be considered in any application. We also notice that the precipitation correction factor for WFDEI was near one. This is reasonable and encouraging, as this dataset is already bias-corrected (using observed precipitation). The precipitation correction factor for observed datasets was 1.34, indicating that the areal precipitation is underestimated. This underestimation is likely caused by a combination of undercatch of the precipitation gauges and the lack of representative observation stations. The undercatch depends on the precipitation gauge, wind, and precipitation phase (Mekonnen *et al.* 2015; Zhao *et al.* 2015), and the non-representativeness is caused by the limited number of observation stations at high altitudes. Better performance in discharge simulation by the application of a correction factor to precipitation was also shown by previous studies Larson & Peck (1974), Lawrence *et al.* (2009), and Lakew *et al.* (2017) and is a standard practice in the hydrologic analysis. The improvement of simulation from all forcing datasets versus prior results presented by Bhattarai *et al.* (2018) using the HBV model (Bergström 1995) in the same catchment is likely attributed to the distributed nature of Shyft.

Though we achieved high NSE, NSE_{sqrt}, and KGE using WFDEI, ERA-I, and Observed + WFDEI forcing datasets with reasonable D_v (Table 6), the simulations were still not able to capture the highest discharges as seen in the QQ-plots in Figure 9. As compared to high discharge, low discharge events were well captured by all forcing datasets.

Local simulated peak discharge during the monsoon season may be attributed to less representativeness of the input precipitation data. Particularly, the lack of ability to capture intense localized precipitation events could be the main reason for an inability to simulate high discharge concurrent with the observations. The CORDEX-SMHI simulation generated relatively higher discharge peaks as compared with the other datasets but has overall higher positive biases during the pre-monsoon season (Figure 3). The CORDEX-SMHI forcing datasets could provide an opportunity for the peak discharge analysis in the Himalayan catchment.

Discussion on the water balance analysis

The catchment water balance component analysis (Table 7) revealed notable differences among the forcing datasets. The discrepancies can, to some degree, be attributed to the uncertainty inherent in each forcing dataset. As mentioned by Nkiaka *et al.* (2017), different precipitation in each forcing dataset can strongly influence the optimized parameters that control the rates and threshold of hydrological processes in the catchment. Although measured evapotranspiration data are not available for comparison, the percentage of evapotranspiration estimates from WFDEI (24.79%) was found to be similar to the previous study by Sakai *et al.* (2004), Giertz *et al.* (2006), Sintondji *et al.* (2014), and Ragettli *et al.* (2015) in similar Himalayan catchments. The lowest evapotranspiration (399.5 mm/year) was found for the CORDEX-SMHI forcing dataset, which can be associated with high RH, and lower average shortwave radiation (Figure 3) than other forcing data. Lower evaporation (449.6 mm/year) and lower precipitation (1,771.1 mm/year) from the Observed + WFDEI than WFDEI may partly result from precipitation under-catch in the catchment.

Higher positive storage changes were found for the ERA-I and Observed + WFDEI forcing datasets as compared to CORDEX-SMHI and WFDEI. As compared to Observed + WFDEI forcing dataset, a lower storage change for ERA-I was observed and results from higher evapotranspiration in ERA-I. A change in storage was found smaller than the glacier melt for all forcing datasets. A smaller change in storage is due to the fact that it originates from snow that accumulates over the year and contributes to the glacier mass balance. Smaller changes in storage also indicate that the glacier mass balance was negative over the simulation period. Previous studies by Kulkarni (1992), Cogley *et al.* (2010), Bolch *et al.* (2012), Wagnon *et al.* (2013), and Gurung *et al.* (2016) also showed that the Himalayan glaciers are experiencing a higher retreat rate in recent decades. The highest glacier melt contribution was observed for ERA-I (19%), while the lowest glacier melt contribution is observed for WFDEI (4%). A study by Gupta *et al.* (2019) on the Marshyangdi River catchment (with 24% glacier cover area) shows that glacier melt contributes to 11.8% of the total discharge. A relatively lower percentage of snow and glacier melt contribution to the total discharge from the Modi River catchment (with 12% glacier cover area) is presented by Kayastha & Shrestha (2019). Similarly, a study by Nepal (2016) on the Dudhkoshi catchment (with 15% glacier cover area) shows that glacier melt contribution is 17% of the total discharge. A similar percentage of glacier melt contribution to the total discharge for Observed + WFDEI (16%) was observed for the Narayani river catchment, although the glacier cover area was only 8%. As suggested by Bonekamp *et al.* (2019), differences in glacier melt contribution to total discharge result from differences in meteorological forcing, as we also observed in our study.

Uncertainty in the model simulation and observation

Hydrological projections are subject to considerable uncertainty (Dobler *et al.* 2012) and are easily affected by various factors, including local and climatic conditions, optimized parameters (Shen *et al.* 2012), and the quality of forcing data (Teweldebrhan *et al.* 2018). Different factors also have a varying degree of impact on the discharge simulation. The challenge using a model with many model parameters is that we might get a good model fit but a less robust model for prediction and forecasting due to over-fitting. However, the uncertainty in the simulations will almost certainly be higher due to the increased uncertainty in the parameter values. In our study, the discharge was highly seasonal and particularly precipitation dependent. Inclusion of the precipitation correction factor as a calibrating parameter improves the predictability of the model at the expense of increases to the uncertainty in water balance components like evapotranspiration and snowmelt. However, only a single value of the precipitation correction factor (p_corr_factor) was applied. More complex precipitation correction could include seasonality and impact of orographic effects, though we feel this would result in greater uncertainty. Nonetheless, uncertainties remain.

First, uncertainties in the observed data result from the uneven spatial distribution and few monitoring meteorological stations, which are mainly located in lower elevation regions (Figure 2). Furthermore, many observations are manual. This poses a unique source of uncertainty in measurements, not necessarily greater or less from automated. Furthermore, there are numerous sources of errors associated in the establishment of the river stage and parameterization of rating curves. During model calibration, we assumed that observed discharge is correct. Some of deviations between model simulations and observations might actually be explained by errors in the observations and not in the model or the forcings, to assess the uncertainty in streamflow observations with required detailed knowledge about local river profiles and data used to establish the rating curve and it is outside the scope of this paper. Second, uncertainties might originate from the estimation of AE and other water balance components, which again is influenced by the precipitation estimates and other forcing data used for driving the model. And thirdly, there are the uncertainties in the model formulation itself. The model in this study assumed that the discharge in the river depends on the amount of water stored in the catchment, and we did not consider the impact of industrial and households' water consumption or any other regulation. Despite this, the simulation performance achieved by the model presented here is quite good, though the results are limited to the one catchment in Nepal. A further examination should evaluate whether the selected forcing datasets could be applied to simulated discharge at other mountain regions with longer time periods.

CONCLUSIONS

We aimed to identify the quality of discharge simulation for the Narayani catchment of Nepal based on different forcing datasets. The forcing datasets WFDEI, CORDEX-SMHI, ERA-I, and ground-based observation combined with WFDEI (i.e. Observed + WFDEI) were used to estimate the Naryani catchment discharge on a daily basis for the period of January 2000–December 2009. Because of the uniform coverage and data consistency, global forcing datasets from global and regional climate models were considered an important supplement to station data. In this study, the distributed Shyft hydrologic simulation platform was selected as a discharge simulation tool. The forcing data were interpolated using IDW for generating daily observed precipitation, while a gradient-based Kriging method was used for generating temperature fields matching the model resolution. In this study, Shyft was calibrated for the period of 2000–2004 and validation was done from 2005 to 2009.

Our analysis showed that large differences exist between different forcing datasets particularly in the amount of precipitation. Precipitation from ERA-I and CORDEX-SMHI was unrealistic leading to poor model performance. To improve the existing bias in precipitation particularly over the high mountain regions, further validation and algorithm improvements are required. With the application of a precipitation correction factor (p_corr_factor), a significant amount of bias in precipitation could be mitigated. Still, the performance of the bias-corrected forcing data, i.e. WFDEI, was better than the climate model datasets. Comparing model performance during model calibration and validation periods, relatively higher NSE, NSE_{sqrt}, and KGE with lower D_v were found for the Observed + WFDEI and WFDEI forcing datasets. The water balance analysis shows that higher evapotranspiration with large glacier melt is observed for the ERA-I forcing dataset, while the average evapotranspiration calculated from the WFDEI (24.79%) is similar to previous studies.

Therefore, based on the different results from the Shyft, we conclude that, in the data-poor Himalayan catchment, the WFDEI forcing dataset may be the best choice for water resource planning and hydropower inflow calculations. Discharge simulations resulting from the WFDEI forcing data were particularly promising for hydropower estimation and water resource assessment in data-scarce or ungauged regions. CORDEX-SMHI data captured higher peaks and may be suitable for the peak discharge analysis. However, to use CORDEX-SMHI and ERA-I in ungauged catchments for the water balance analysis, bias correction is required. Further analysis by implementing Shyft in the different Himalayan catchments for different forcing datasets is a recommended further step to assess the regional extensibility of the current results.

FUNDING

This work is a contribution to the Strategic Research Initiative 'Land Atmosphere Interaction in Cold Environments' (LATICE) of the University of Oslo and partially supported through the Norwegian Research Council's INDOR program under the Hydrologic sensitivity to Cryosphere-Aerosol interaction in Mountain Processes (HyCAMP) project (NRF No. 222195)

ACKNOWLEDGEMENT

We are thankful to the Department of Hydrology and Meteorology, Government of Nepal for providing observed hydrological and meteorological datasets. We also gratefully acknowledge Special Issue Editor Dr. Kolbjørn Engeland and anonymous reviewer for their comments on the research and manuscript.

SUPPLEMENTARY MATERIAL

The Supplementary Material for this paper is available online at https://dx.doi.org/10.2166/nh.2020.079.

REFERENCES

Adhikari, D. 2006 Hydropower development in Nepal. *NRB Economic Review* **18**, 70–94.

Bakken, T. H., Killingtveit, Å., Engeland, K., Alfredsen, K. & Harby, A. 2013 Water consumption from hydropower plants – review of published estimates and an assessment of the concept. *Hydrology and Earth System Sciences* **17** (10), 3983–4000.

Bergström, S. 1995 The HBV model. In: *Computer Models of Watershed Hydrology* (V. P. Singh, ed.). Water Resources Publications, Highlands Ranch, CO, USA.

Berrisford, P., Dee, D. P., Poli, P., Brugge, R., Fielding, K., Fuentes, M., Kbackslashr allberg, P. W., Kobayashi, S., Uppala, S. & Simmons, A. 2011 *The ERA-Interim Archive Version 2.0*. ECMWF, Shinfield Park, Reading, p. 23.

Bharti, V. & Singh, C. 2015 Evaluation of error in TRMM 3B42V7 precipitation estimates over the Himalayan region. *Journal of Geophysical Research: Atmospheres* **120** (24), 12458–12473.

Bhattarai, B. C. & Regmi, D. 2015 Impact of climate change on water resources in view of contribution of runoff components in stream flow: a case study from Langtang Basin, Nepal. *Journal of Meteorology and Hydrology* **9** (1), 74.

Bhattarai, S., Zhou, Y., Shakya, N. M. & Zhao, C. 2018 Hydrological modelling and climate change impact assessment using HBV light model: a case study of Narayani River Basin, Nepal. *Nature Environment and Pollution Technology* **17** (3), 691–702.

Bhattarai, B. C., Burkhart, J. F., Stordal, F. & Xu, C. Y. 2019 Aerosol optical depth over the Nepalese cryosphere derived from an empirical model. *Frontiers in Earth Science* **7**, 178.

Biswas, A. K. 2011 Cooperation or conflict in transboundary water management: case study of South Asia. *Hydrological Sciences Journal* **56** (4), 662–670.

Bolch, T., Kulkarni, A., Kääb, A., Huggel, C., Paul, F., Cogley, J. G., Frey, H., Kargel, J. S., Fujita, K., Scheel, M., Bajracharya, S. & Stoffel, M. 2012 The state and fate of Himalayan glaciers. *Science* **336**, 310–314.

Bonekamp, P. N., de Kok, R. J., Collier, E. & Immerzeel, W. W. 2019 Contrasting meteorological drivers of the glacier mass balance between the Karakoram and central Himalaya. *Frontiers in Earth Science* **7**, 107.

Bookhagen, B. & Burbank, D. W. 2010 Toward a complete Himalayan hydrological budget: spatiotemporal distribution of snowmelt and rainfall and their impact on river discharge. *Journal of Geophysical Research* **115** (F3), F03019.

Burkhart, J. F., Helset, S., Abdella, Y. S. & Lappegard, G. 2016 Operational Research: Evaluating Multimodel Implementations for 24/7 Runtime Environments. *American Geophysical Union, Fall General Assembly 2016, Abstract ID H51F*-1541.

Channan, S., Collins, K. & Emanuel, W. R. 2014 *Global Mosaics of the Standard MODIS Land Cover Type Data*, Vol. 30. University of Maryland and the Pacific Northwest National Laboratory, College Park, MD, USA.

Chowdhury, M. D. R. & Ward, N. 2004 Hydro-meteorological variability in the greater Ganges–Brahmaputra–Meghna basins. *International Journal of Climatology* **24** (12), 1495–1508.

Chu, W., Gao, X. & Sorooshian, S. 2010 Improving the shuffled complex evolution scheme for optimization of complex nonlinear hydrological systems: application to the calibration of the Sacramento soil-moisture accounting model. *Water Resources Research* **46** (9). https://doi.org/10.1029/2010WR009224.

Cogley, J. G., Kargel, J. S., Kaser, G., van der Veen, C. J., Paul, F., Cogley, J. G., Frey, H., Kargel, J. S., Fujita, K., Scheel, M., Bajracharya, S. & Stoffel, M. 2010 Tracking the source of glacier misinformation. *Science* **327** (5965), 522.

Conrad, O., Bechtel, B., Bock, M., Dietrich, H., Fischer, E., Gerlitz, L., Wehberg, J., Wichmann, V. & Böhner, J. 2015

System for automated geoscientific analyses (SAGA) v. 2.1.4. *Geoscientific Model Development Discussions* **8** (2), 2271–2312.

Dahal, R. K. & Hasegawa, S. 2008 Representative rainfall thresholds for landslides in the Nepal Himalaya. *Geomorphology* **100** (3–4), 429–443.

Dahri, Z. H., Ludwig, F., Moors, E., Ahmad, B., Khan, A. & Kabat, P. 2016 An appraisal of precipitation distribution in the high-altitude catchments of the Indus basin. *The Science of the Total Environment* **548–549**, 289–306.

Daly, C., Neilson, R. P. & Phillips, D. L. 1994 A statistical-topographic model for mapping climatological precipitation over mountainous terrain. *Journal of Applied Meteorology* **33** (2), 140–158.

Dee, D. P., Uppala, S. M., Simmons, A. J., Berrisford, P., Poli, P., Kobayashi, S., Andrae, U., Balmaseda, M. A., Balsamo, G., Bauer, P., Bechtold, P., Beljaars, A. C. M., van de Berg, L., Bidlot, J., Bormann, N., Delsol, C., Dragani, R., Fuentes, M., Geer, A. J., Haimberger, L., Healy, S. B., Hersbach, H., Hólm, E. V., Isaksen, L., Kållberg, P., Köhler, M., Matricardi, M., McNally, A. P., Monge-Sanz, B. M., Morcrette, J.-J., Park, B.-K., Peubey, C., de Rosnay, P., Tavolato, C., Thépaut, J.-N. & Vitart, F. 2011 The ERA-Interim reanalysis: configuration and performance of the data assimilation system. *Quarterly Journal of the Royal Meteorological Society* **137** (656), 553–597.

Dehecq, A., Gourmelen, N., Gardner, A. S., Brun, F., Goldberg, D., Nienow, P. W., Berthier, E., Vincent, C., Wagnon, P. & Trouvé, E. 2019 Twenty-first century glacier slowdown driven by mass loss in High Mountain Asia. *Nature Geoscience* **12** (1), 22–27.

DeWalle, D. R. & Rango, A. 2008 *Principles of Snow Hydrology*. Cambridge University Press, Cambridge.

Dobler, C., Hagemann, S., Wilby, R. L. & Stätter, J. 2012 Quantifying different sources of uncertainty in hydrological projections in an Alpine watershed. *Hydrology and Earth System Sciences* **16**, 4343–4360.

Duan, Q., Sorooshian, S. & Gupta, V. 1992 Effective and efficient global optimization for conceptual rainfall-runoff models. *Water Resources Research* **28** (4), 1015–1031.

Edenhofer, O., Pichs-Madruga, R., Sokona, Y., Seyboth, K., Kadner, S., Zwickel, T., Eickemeier, P., Hansen, G., Schlömer, S., von Stechow, C. & Matschoss, P. 2011 *Renewable Energy Sources and Climate Change Mitigation: Special Report of the Intergovernmental Panel on Climate Change*. Cambridge University Press, Cambridge.

Engeland, K., Steinsland, I., Johansen, S. S., Petersen-Øverleir, A. & Kolberg, S. 2016 Effects of uncertainties in hydrological modelling. A case study of a mountainous catchment in Southern Norway. *Journal of Hydrology* **536**, 147–160.

Essou, G. R. C., Sabarly, F., Lucas-Picher, P., Brissette, F. & Poulin, A. 2016 Can precipitation and temperature from meteorological reanalyses be used for hydrological modeling? *Journal of Hydrometeorology* **17** (7), 1929–1950.

Ghimire, S., Choudhary, A. & Dimri, A. P. 2018 Assessment of the performance of CORDEX-South Asia experiments for monsoonal precipitation over the Himalayan region during present climate: part I. *Climate Dynamics* **50** (7), 2311–2334.

Giertz, S., Jaeger, A., Schopp, M., Diekkrügerdiekkr¨diekkrüger, B. & Jaeger, A. 2006 An interdisciplinary scenario analysis to assess the water availability and water consumption in the Upper Ouémé catchment in Benin. *Advances in Geosciences* **9** (2006), 3–13.

Giorgi, F., Jones, C. & Asrar, G. R. 2009 Addressing climate information needs at the regional level: the CORDEX framework. *World Meteorological Organization (WMO) Bulletin* **58** (3), 175–195.

Guo, J. & Su, X. L. 2019 Parameter sensitivity analysis of SWAT model for streamflow simulation with multisource precipitation datasets. *Hydrology Research* **50** (3), 861–877.

Gupta, H. V., Kling, H., Yilmaz, K. K., Martinez, G. F. & Kling, H. 2009 Decomposition of the mean squared error and NSE performance criteria: implications for improving hydrological modelling. *Journal of Hydrology* **377**, 80–91.

Gupta, A., Kayastha, R. B., Ramanathan, A. L. & Dimri, A. P. 2019 Comparison of hydrological regime of glacierized Marshyangdi and Tamor river basins of Nepal. *Environmental Earth Sciences* **78** (14), 427.

Gurung, S., Bhattrai, B. C., Kayastha, R. B., Stumm, D., Joshi, S. & Mool, P. K. 2016 Study of annual mass balance (2011–2013) of Rikha Samba Glacier, Hidden Valley, Mustang, Nepal. *Sciences in Cold and Arid Regions* **8** (4), 0311–0318.

Hegdahl, T. J., Tallaksen, L. M., Engeland, K., Burkhart, J. F. & Xu, C.-Y. 2016 Discharge sensitivity to snowmelt parameterization: a case study for Upper Beas basin in Himachal Pradesh. *Hydrology Research* **47** (4), 683–700.

Hock, R. 2003 Temperature index melt modelling in mountain areas. *Journal of Hydrology* **282** (1–4), 104–115.

Iizumi, T., Okada, M. & Yokozawza, M. 2014 A meteorological forcing data set for global crop modeling: development, evaluation, and intercomparison. *Journal of Geophysical Research: Atmospheres* **119** (2), 363–384.

Immerzeel, W. W., van Beek, L. P. H. & Bierkens, M. F. P. 2010 Climate change will affect the Asian water towers. *Science* **328** (5984), 1382–1385.

Jain, K., Jain, S. K., Jain, N. & Xu, C.-Y. 2017 Hydrologic modeling of a Himalayan mountain basin by using the SWAT model. *Journal Hydrology and Earth System Sciences*. https://doi.org/10.5194/hess-2017-100

Kansakar, S. R., Hannah, D. M., Gerrard, J. & Rees, G. 2004 Spatial pattern in the precipitation regime of Nepal. *International Journal of Climatology* **24**, 1645–1659.

Karki, R. 2010 Status of automatic weather stations in Nepal and comparison of air temperature and precipitation data between automatic weather station and manual observation. In *IOM-104*. www.wmo.int/pages/prog/www/IMOP/publications/.

Kauffeldt, A., Wetterhall, F., Pappenberger, F., Salamon, P. & Thielen, J. 2016 Technical review of large-scale hydrological models for implementation in operational flood forecasting schemes on continental level. *Environmental Modelling & Software* **75**, 68–76.

Kayastha, R. B. & Shrestha, A. 2019 Snow and ice melt contribution in the daily discharge of Langtang and Modi Rivers, Nepal. In: *Environmental Change in the Himalayan Region* (S. Anup & T. Pankaj Thapa, eds). Springer, Cham, pp. 1–21.

Khanal, N. R., Mool, P. K., Shrestha, A. B., Rasul, G., Ghimire, P. K., Shrestha, R. B. & Joshi, S. P. 2015 A comprehensive approach and methods for glacial lake outburst flood risk assessment, with examples from Nepal and the transboundary area. *International Journal of Water Resources Development* **31** (2), 219–237.

Kim, D.-I., Kwon, H.-H. & Han, D. W. 2019 Bias correction of daily precipitation over South Korea from the long-term reanalysis using a composite Gamma-Pareto distribution approach. *Hydrology Research* **50** (4), 1138–1161.

Kirchner, J. W. 2009 Catchments as simple dynamical systems: catchment characterization, rainfall-runoff modeling, and doing hydrology backward. *Water Resources Research* **45** (2). https://doi.org/10.1029/2008WR006912

Koch, F., Prasch, M., Bach, H., Mauser, W., Appel, F. & Weber, M. 2011 How will hydroelectric power generation develop under climate change scenarios? a case study in the Upper Danube Basin. *Energies* **4** (10), 1508–1541.

Kripalani, R. H. & Sontakke, N. A. 1996 Rainfall variability over Bangladesh and Nepal: comparison and connections with features over India. *International Journal of Climatology* **16**, 689–703.

Kulkarni, A. V. 1992 Mass balance of Himalayan glaciers using AAR and ELA methods. *Journal of Glaciology* **38** (128), 101–104.

Kumar, D. 2017 River Ganges – historical, cultural and socioeconomic attributes. *Aquatic Ecosystem Health & Management* **20** (1–2), 8–20.

Lakew, H. B., Moges, S. A., Asfaw, D. H., Lakew, H. B., Moges, S. A. & Asfaw, D. H. 2017 Hydrological evaluation of satellite and reanalysis precipitation products in the Upper Blue Nile Basin: a case study of Gilgel Abbay. *Hydrology* **4** (3), 39.

Lambert, A. 1972 Catchment models based on ISO-functions. *Journal of the Institution of Water Engineers* **26**, 413–422.

Larson, L. W. & Peck, E. L. 1974 Accuracy of precipitation measurements for hydrologic modeling. *Water Resources Research* **10** (4), 857–863.

Lawrence, D., Haddeland, I. & Langsholt, E. 2009 Calibration of HBV hydrological models using PEST parameter estimation. *NVE Report*. Available from: www.nve.no.

Li, L., Ngongondo, C., Xu, C.-Y. & Gong, L. 2013 Comparison of the global TRMM and WFD precipitation datasets in driving a large-scale hydrological model in Southern Africa. *Hydrology Research* **44** (5), 770–788.

Li, H., Beldring, S., Xu, C.-Y., Huss, M. & Melvold, K. 2015 Integrating a glacier retreat model into a hydrological model – case studies on three glacierised catchments in Norway and Himalayan region. *Journal of Hydrology* **527**, 656–667.

Li, H., Xu, C.-Y., Beldring, S., Tallaksen, L. M. & Jain, S. K. 2016 Water resources under climate change in Himalayan basins. *Water Resources Management* **30** (2), 843–859. doi:10.1007/s11269-015-1194-5.

Li, H., Haugen, J. E. & Xu, C.-Y. 2018 Precipitation pattern in the Western Himalayas revealed by four datasets. *Hydrology and Earth System Sciences* **22**, 5097–5110.

Li, L., Shen, M. X., Hou, Y. K., Xu, C.-Y., Lutz, A. F., Chen, J., Jain, S. K., Li, J. J. & Chen, H. 2019 Twenty-first-century glacio-hydrological changes in the Himalayan headwater Beas River basin. *Hydrology and Earth System Sciences* **23**, 1483–1503.

Lombraña, J. U. 2017 *Evaluation of Snow Simulations in SHyFT*. Norwegian University of Science and Technology (NTNU), Trondheim.

Madsen, H. 2003 Parameter estimation in distributed hydrological catchment modelling using automatic calibration with multiple objectives. *Advances in Water Resources* **26** (2), 205–216.

Martinec, J., Rango, A., Roberts, R. & Baumgartner, M. F. 1998 *Snowmelt Runoff Model (SRM) User's Manual*. University of Berne, Department of Geography, Berne.

Matt, F. N. & Burkhart, J. F. 2018 Assessing satellite-derived radiative forcing from snow impurities through inverse hydrologic modeling. *Geophysical Research Letters* **45** (8), 3531–3541.

Matt, F. N., Burkhart, J. F. & Pietikäinen, J.-P. 2018 Modelling hydrologic impacts of light absorbing aerosol deposition on snow at the catchment scale. *Hydrology and Earth System Sciences* **22**, 179–201.

Mekonnen, G. B., Matula, S., Doležal, F. & Fišák, J. 2015 Adjustment to rainfall measurement undercatch with a tipping-bucket rain gauge using ground-level manual gauges. *Meteorology and Atmospheric Physics* **127** (3), 241–256.

Ménégoz, M., Gallée, H. & Jacobi, H. W. 2013 Precipitation and snow cover in the Himalaya: from reanalysis to regional climate simulations. *Hydrology and Earth System Sciences* **17**, 3921–3936.

Moulin, L., Gaume, E. & Obled, C. 2009 Uncertainties on mean areal precipitation: assessment and impact on streamflow simulations. *Hydrology and Earth System Sciences* **13**, 99–114.

Nash, J. E. & Sutcliffe, J. V. 1970 River flow forecasting through conceptual models part I – a discussion of principles. *Journal of Hydrology* **10**, 282–290.

Nayava, J. L. 1974 Heavy monsoon rainfall in Nepal. *Weather* **29** (12), 443–450.

Nayava, J. L. 1980 Rainfall in Nepal. *Himalayan Review* **12**, 1–18.

Nepal, S. 2016 Impacts of climate change on the hydrological regime of the Koshi river basin in the Himalayan region. *Journal of Hydro-environment Research* **10**, 76–89.

Nkiaka, E., Nawaz, N., Lovett, J., Nkiaka, E., Nawaz, N. R. & Lovett, J. C. 2017 Evaluating global reanalysis datasets as input for hydrological modelling in the Sudano-Sahel region. *Hydrology* **4** (1), 13.

Oliver, M. A. & Webster, R. 1990 Kriging: a method of interpolation for geographical information systems. *International Journal of Geographical Information System* **4** (3), 313–332.

Omani, N., Srinivasan, R., Karthikeyan, R., Smith, P., Omani, N., Srinivasan, R., Karthikeyan, R. & Smith, P. K. 2017 Hydrological modeling of highly glacierized basins (Andes, Alps, and Central Asia). *Water* **9** (2), 111.

Pellicciotti, F., Buergi, C., Immerzeel, W. W., Konz, M. & Shrestha, A. B. 2012 Challenges and uncertainties in hydrological modeling of remote Hindu Kush–Karakoram–Himalayan (HKH) Basins: suggestions for calibration strategies. *Mountain Research and Development* **32** (1), 39–50.

Peña-Arancibia, J. L., Zhang, Y., Pagendam, D. E., Viney, N. R., Lerat, J., van Dijk, A. I. J. M., Vaze, J. & Frost, A. J. 2015 Streamflow rating uncertainty: characterisation and impacts on model calibration and performance. *Environmental Modelling & Software* **63**, 32–44.

Pradhananga, N. S., Kayastha, R. B., Bhattarai, B. C., Adhikari, T. R., Pradhan, S. C., Devkota, L. P., Shrestha, A. B. & Mool, P. K. 2014 Estimation of discharge from Langtang River basin, Rasuwa, Nepal, using a glacio-hydrological model. *Annals of Glaciology* **55** (66), 223–230.

Priestley, C. H. B. & Taylor, R. J. 1972 On the assessment of surface heat flux and evaporation using large-scale parameters. *Monthly Weather Review* **100** (2), 81–92.

QGIS Development Team 2016 *QGIS Geographic Information System*. Open Source Geospatial Foundation. Retrieved from: http://qgis.osgeo.org.

Radić, V. & Hock, R. 2014 Glaciers in the earth's hydrological cycle: assessments of glacier mass and runoff changes on global and regional scales. *Surveys in Geophysics* **35** (3), 813–837.

Ragettli, S., Pellicciotti, F., Immerzeel, W. W., Miles, E. S., Petersen, L., Heynen, M., Shea, J. M., Stumm, D., Joshi, S. & Shrestha, A. 2015 Unraveling the hydrology of a Himalayan catchment through integration of high resolution *in situ* data and remote sensing with an advanced simulation model. *Advances in Water Resources* **78**, 94–111.

Raimonet, M., Oudin, L., Thieu, V., Silvestre, M., Vautard, R., Rabouille, C. & Moigne, P. L. E. 2017 Evaluation of gridded meteorological datasets for hydrological modeling. *Journal of Hydrometeorology* **18** (11), 3027–3041.

Remesan, R., Begam, S. & Holman, I. P. 2019 Effect of baseline snowpack assumptions in the HySIM model in predicting future hydrological behaviour of a Himalayan catchment. *Hydrology Research* **50** (2), 691–708.

Renard, B., Kavetski, D., Kuczera, G., Thyer, M. & Franks, S. W. 2010 Understanding predictive uncertainty in hydrologic modeling: the challenge of identifying input and structural errors. *Water Resources Research* **46**, 5.

Rochester, R. E. L. 2010 *Uncertainty in Hydrological Modelling: A Case Study in the Tern Catchment*. UCL, Shropshire, UK. Retrieved from: http://discovery.ucl.ac.uk/795428/ (May).

Sakai, A., Fujita, K. & Kubota, J. 2004 Evaporation and percolation effect on melting at debris-covered Lirung Glacier, Nepal Himalayas, 1996. *Bulletin of Glacier Research* **21**, 9–15.

Sanjay, J., Krishnan, R., Shrestha, A. B., Rajbhandari, R. & Ren, G. Y. 2017 Downscaled climate change projections for the Hindu Kush Himalayan region using CORDEX South Asia regional climate models. *Advances in Climate Change Research* **8** (3), 185–198.

Sapkota, A. 2016 *Regional Modelling of the Narayani Basin in Nepal*. Norwegian University of Science and Technology, Trondheim.

Seibert, J. 2001 On the need for benchmarks in hydrological modelling. *Hydrological Processes* **15** (6), 1063–1064.

Seiller, G., Anctil, F. & Perrin, C. 2012 Multimodel evaluation of twenty lumped hydrological models under contrasted climate conditions. *Hydrology and Earth System Sciences* **16** (4), 1171–1189.

Shen, Z. Y., Chen, L. & Chen, T. 2012 Analysis of parameter uncertainty in hydrological and sediment modeling using GLUE method: a case study of SWAT model applied to Three Gorges Reservoir Region, China. *Hydrology and Earth System Sciences* **16** (1), 121–132.

Shepard, D. 1968 A Two-Dimensional Interpolation Function for Irregularly-Spaced Data. Proceedings of the 1968 ACM National Conference, New York, 27–29 August 1968, pp. 517–524.

Shrestha, A. B., Wake, C. P., Mayewski, P. A., Dibb, J. E., Shrestha, A. B., Wake, C. P., Mayewski, P. A. & Dibb, J. E. 1999 Maximum temperature trends in the Himalaya and its vicinity: an analysis based on temperature records from Nepal for the period 1971–94. *Journal of Climate* **12** (9), 2775–2786.

Sintondji, L. O., Zokpodo, B., Ahouansou, D. M., Vissin, W. E. & Agbossou, K. E. 2014 Modelling the water balance of Ouémé catchment at the Savé outlet in Benin: contribution to the sustainable water resource management. *International Journal of Agriscience* **4**, 74–88.

Skaugen, T. & Weltzien, I. H. 2016 A model for the spatial distribution of snow water equivalent parameterized from the spatial variability of precipitation. *The Cryosphere* **10** (5), 1947–1963.

Teweldebrhan, A. T., Burkhart, J. F. & Schuler, T. V. 2018 Parameter uncertainty analysis for an operational hydrological model using residual-based and limits of acceptability approaches. *Hydrology and Earth System Sciences* **22**, 5021–5039.

Wagnon, P., Vincent, C., Arnaud, Y., Berthier, E., Vuillermoz, E., Gruber, S., Ménégoz, M., Gilbert, A., Dumont, M., Shea, J. M., Stumm, D. & Pokhrel, B. K. 2013 The cryosphere seasonal and annual mass balances of Mera and Pokalde glaciers (Nepal Himalaya) since 2007. *The Cryosphere* **7**, 1769–1786.

Weedon, G. P., Gomes, S., Viterbo, P., Shuttleworth, W. J., Blyth, E., Österle, H., Adam, J. C., Bellouin, N., Boucher, O. & Best, M. 2011 Creation of the WATCH forcing data and its use to assess global and regional reference crop evaporation over land during the twentieth century. *Journal of Hydrometeorology* **12** (5), 823–848.

Weedon, G. P., Balsamo, G., Bellouin, N., Gomes, S., Best, M. J. & Viterbo, P. 2014 The WFDEI meteorological forcing data set: WATCH forcing data methodology applied to ERA-Interim reanalysis data. *Water Resources Research* **50** (9), 7505–7514.

Wortmann, M., Krysanova, V., Kundzewicz, Z. W., Su, B. & Li, X. 2014 Assessing the influence of the Merzbacher Lake outburst floods on discharge using the hydrological model SWIM in the Aksu headwaters, Kyrgyzstan/NW China. *Hydrological Processes* **28** (26), 6337–6350.

Xu, H., Xu, C. Y., Sælthun, N. R., Zhou, B. & Xu, Y. 2015 Evaluation of reanalysis and satellite-based precipitation datasets in driving hydrological models in a humid region of Southern China. *Stochastic Environmental Research and Risk Assessment* **29** (8), 2003–2020.

Xu, H. L., Xu, C.-Y., Chen, S. D. & Chen, H. 2016 Similarity and difference of global reanalysis datasets (WFD and APHRODITE) in driving lumped and distributed hydrological models in a humid region of China. *Journal of Hydrology* **542**, 343–356.

Zhang, Y. G., Hao, Z. C., Xu, C.-Y. & Lai, X. D. 2019 Response of melt water and rainfall runoff to climate change and their roles in controlling streamflow changes of the two upstream basins over the Tibetan Plateau. *Hydrology Research*. doi: 10.2166/nh.2019.075.

Zhao, H., Yang, S., Wang, Z., Zhou, X., Luo, Y. & Wu, L. 2015 Evaluating the suitability of TRMM satellite rainfall data for hydrological simulation using a distributed hydrological model in the Weihe River catchment in China. *Journal of Geographical Sciences* **25** (2), 177–195.

Zhu, D., Xuan, Y. & Cluckie, I. 2014 Hydrological appraisal of operational weather radar rainfall estimates in the context of different modelling structures. *Hydrology and Earth System Sciences* **18** (1), 257–272.

First received 6 June 2019; accepted in revised form 14 January 2020. Available online 21 February 2020

Improving hydropower inflow forecasts by assimilating snow data

Jan Magnusson, Geir Nævdal, Felix Matt, John F. Burkhart and Adam Winstral

ABSTRACT

Accurate long-term inflow forecasts are essential for optimal planning of hydropower production. In snow-rich regions, where spring snowmelt is often the largest reservoir of water, inflow forecasts may be improved by assimilating snow observations to achieve more accurate initial states for the hydrological models prior to the prognosis. In this study, we test whether an ensemble Kalman based approach is useful for this purpose for a mountainous catchment in Norway. For 15 years, annual snow observations near peak accumulation at three locations were assimilated into a distributed hydrological model. After the update, the model was run for a 4-month forecasting period with inflows compared to a base case scenario that omitted the snow observations. The assimilation framework improved the forecasts in several years, and in two of the years, the improvement was very large compared to the base case simulation. At the same time, the filter did not degrade the forecasts largely, indicating that though the updating might slightly degrade performance in some years, it maintains the potential for large improvements in others. Thus, the framework proposed here is a viable method for improving snow-related deficiencies in the initial states, which translates to better forecasts.

Key words | data assimilation, hydrological modelling, inflow forecasting, snow modelling

Jan Magnusson (corresponding author)
Norwegian Water Resources and Energy Directorate (NVE),
Oslo,
Norway
E-mail: *jan.magnusson@gmail.com*

Geir Nævdal
NORCE Norwegian Research Centre AS,
Bergen,
Norway

Felix Matt
Statkraft AS,
Oslo,
Norway

Felix Matt
John F. Burkhart
Department of Geosciences,
University of Oslo,
P.O. Box 1047, Blindern N-0316, Oslo,
Norway

Adam Winstral
WSL Institute for Snow and Avalanche Research SLF,
Davos,
Switzerland

INTRODUCTION

To limit the increase in air temperature due to ongoing global warming, we need to make the best possible use of low-carbon power energy sources to reduce greenhouse gas emissions. For hydropower, reliable long-term inflow forecasts help production planners to optimally utilize the available energy. In areas with significant snow accumulations, we hypothesize that it is possible to improve such inflow forecasts by incorporating snow measurements into the prediction system. In snow-rich regions of Norway, hydropower companies routinely measure snow water equivalents (SWEs) at certain points in the catchment area near peak accumulation. Currently, the practice at Norwegian hydropower companies is to update forecasting models manually with these measurements. This procedure has major drawbacks such as limited reproducibility, unknown representativity of the measurements, lack of an objective and systematic approach for improving the forecasts and, finally, the method is also labor intensive. To overcome such shortcomings, we show how these measurements can be assimilated into a hydrologic forecasting model automatically using an objective method. Our main aim of this study is to test whether the

doi: 10.2166/nh.2020.025

proposed data assimilation method can improve the predictive skill of the forecasting system for a lead time of weeks to months.

In many hydropower dominated regions, and particularly the Nordics, the largest proportion of inflow to hydropower reservoirs on an annual basis often comes during the spring snowmelt period. We can predict this peak inflow to the reservoirs using either parametric snow cover models (e.g. Ohmura 2001) or energy-balance snow models based on physical principles (e.g. Anderson 1976), in combination with a hydrological model describing the water flow through the catchment. In combination with seasonal weather forecasts or by using climatological records, both types of snow models can provide inflow forecasts for lead times of several months. However, the parametric models, also called temperature-index models, are often employed in operational settings over the more data demanding energy-balance models since they only rely on air temperature and precipitation data alone. Uncertainty in the forecasted inflows depends on errors in the model structure, meteorological forcing data and initial conditions (Kuczera 1983; Beven & Binley 1992; Vrugt *et al.* 2005). For short-term forecasts, the uncertainty in the initial state variables can be reduced by updating the model with observed discharge using, for example, the Kalman filter (Fjeld & Aam 1981). For long-term predictions covering the snowmelt and summer season, the assimilation of snow states themselves may be of greater importance than the direct assimilation of runoff. In this study, we aim to minimize the total uncertainty in long-term runoff predictions by reducing the error in the initial snow states. Foremost, the simulated snow states at peak accumulation can be prone to large errors as potential forcing data errors accumulate throughout the snow accumulation season (Gragne *et al.* 2015). This can eventually lead to large errors in the predicted inflows during spring snowmelt. Thus, reducing the errors in the initial snow conditions may yield an improved long-term inflow forecast.

A large range of methods exist for updating snow models using observations. With direct insertion, the simulated states are adjusted to exactly match the observed values at the same location (e.g. Liston *et al.* 1999; Fletcher *et al.* 2012). However, this method will likely produce poor results if the errors in the measurement are large, and direct insertion cannot be used to update the model at locations lacking observations. The latter problem can be circumvented by applying some interpolation method, for example, by deriving spatially distributed grids of correction factors for the model parameters (e.g. Liston & Hiemstra 2008). In order to take the observation uncertainty into account, we can choose one of several methods originating from Bayes' theorem. Kolberg *et al.* (2006) and Kolberg & Gottschalk (2006) presented a method for assimilating snow-covered area information into the snow routine of a runoff model based on Bayes' theorem. Another approach is to use so-called optimal interpolation, which has been used for improving continental snow maps by incorporating snow depth observations (e.g. Brown *et al.* 2003; Barnett *et al.* 2005) or remote sensing data (e.g. Liu *et al.* 2013). Such methods can be further improved by using approaches that make use of the information provided by the state-space model. Two examples of such methods that have been applied for snow models are the particle filter (e.g. Leisenring & Moradkhani 2011; Magnusson *et al.* 2017) and the ensemble Kalman filter (EnKF) (e.g. Andreadis & Lettenmaier 2006; Slater & Clark 2006; De Lannoy *et al.* 2012; Magnusson *et al.* 2014). With the latter method, we can update snow models at unobserved locations and also take uncertainties in the forcing data and snow observations into account.

The above-mentioned studies have shown that model states can efficiently be improved by assimilating either ground or satellite-based snow observations. However, whether the assimilation of snow data before spring snowmelt also improves long-term predictions of inflows to, for example, hydropower reservoirs is still an open question. Thus, the aim of this study is to assess the value of snow observations for improving inflow forecasts covering the spring and summer season. We present an assimilation framework using the EnKF in combination with a temperature-index snowmelt model coupled to a hydrological model. The performance of the system was benchmarked against a base case simulation that did not utilize the snow observations. We also assessed whether eventual improvements in inflow forecasts depended on factors such as the average snow amounts in the catchment and if the number of assimilated snow observations influenced the quality of the predicted inflows.

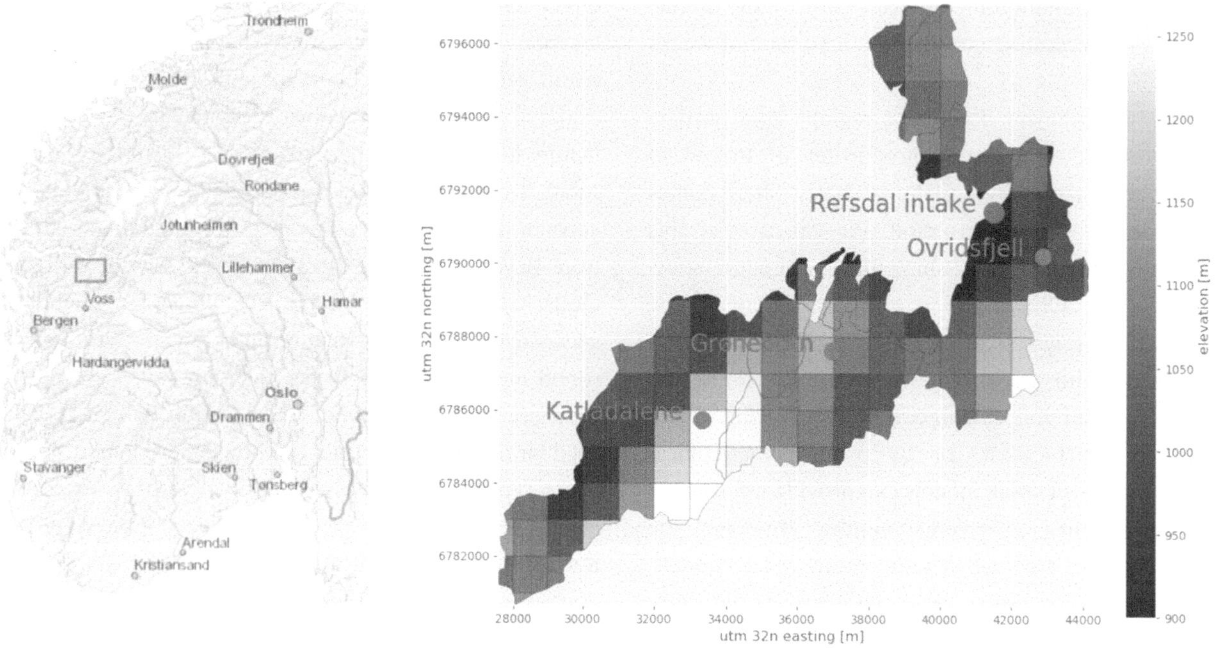

Figure 1 | The left panel shows the location of the Refsdal catchment in southern Norway, and the right panel displays the drainage basin for the catchment, including the three snow measurement locations (red dots) and the intake location of the Refsdal hydropower plant (green dot) where discharge is measured. The elevation of the grid cells used by the hydrological model is shown in the background of the right panel. Please refer to the online version of this paper to see this figure in colour: http://dx.doi.org/10.2166/nh.2020.025.

STUDY AREA AND DATA

Refsdal catchment

In this study, we used data covering the period from 2000 to 2015 for the Refsdal catchment (Figure 1). The watershed is located just south of the Sognefjord at the Norwegian west coast. The drainage basin covers an area of approximately 74 km^2 and altitudes range from 530 to 1,276 m.a.s.l. with an average elevation of 1,062 m.a.s.l. Hydropower reservoirs cover approximately 8% of the watershed area and forests cover a negligible part of the catchment (<1%). Thus, the catchment represents a typical mountainous landscape above the tree line at high latitudes.

For the study period, SWE has been measured at three locations (Katledalane – 1,280 m.a.s.l., Grønebotn – 1,060 m.a.s.l. and Ovridsfjell – 1,060 m.a.s.l.) throughout the watershed (see below for more information about those observations). Mean annual precipitation in the catchment equals approximately 2,200 mm, of which about 50% falls as snow.

Model forcing data

We used the seNorge Version 2 dataset of daily gridded average air temperature and precipitation for this study (Lussana & Tveito 2017). These gridded data have been generated for mainland Norway based on hundreds of observations statistically interpolated to a 1 km grid. The data records have been quality controlled, and detected errors have been corrected or removed using an automatic procedure. The precipitation data have, however, not been corrected for undercatch (Lussana *et al.* 2018), yet these data are inputs to operational hydrological models in Norway (Luijting *et al.* 2018) and have been used for research purposes (e.g. Huang *et al.* 2019). Thus, the precipitation grids likely underestimate the real-world precipitation, in some regions severely likely due to wind effects, low station coverage and poor representativity of the observation sites. For more details about the gridding procedure in seNorge for air temperature and precipitation and the quality of the grids, see Lussana *et al.* (2016) and Lussana *et al.* (2018), respectively.

Snow and inflow observations

Statkraft, a Norwegian hydropower company, conducts annual snow measurement campaigns at three locations inside the catchment. These measurement campaigns are typically conducted in early April. At each location, between 30 and 40 snow depth measurements are conducted in fixed intervals (10–25 m between each measurement, depending on topography) along straight transects at each site, and snow density is measured at a few points along the line. Each transect is positioned with the aim to provide a representative measurement for the area, and each campaign covering the three locations is undertaken within a couple of days. Per transect, an average SWE is calculated as follows:

$$\mathrm{SWE} = \frac{1}{n}\sum_{i=1}^{n} h_i * \frac{1}{m}\sum_{i=1}^{m} \rho_i$$

where h is the measured snow depth in meters at n locations along the route and ρ is the bulk snow density in kg/m^3 at m sites along the same snow transect. This SWE value is used herein to update the model predictions.

In addition, naturalized catchment discharge based on hydropower production records from the Refsdal power plant has been made available to us by Statkraft.

METHODS

Model description

In this study, we used the open-source hydrological toolbox Shyft (https://gitlab.com/shyft-os/shyft) for the data assimilation experiments. This framework has been developed to provide operational inflow forecasts for the hydropower industry. Shyft uses distributed modeling concepts and is optimized for the efficient simulation of the hydrologic processes relevant to the aforementioned purpose. The user can choose between both conceptual and more physically based modelling approaches for depicting various hydrological processes. This framework has been used for analyzing how parameter uncertainty in hydrological modelling influences reservoir inflow forecasts (Tweldebrahn *et al.* 2018).

In this study, the model was set up to run at a daily time step on a grid with a horizontal resolution of 1 km (see Figure 1). We simulated the snow cover development based on the methods applied in the Hydrologiska Byråns Vattenbalansavdeling (HBV) model (Lindström *et al.* 1997). This is a temperature-index based approach that also simulates and tracks the liquid water content of the snowpack as well as the snow distribution within each grid cell using a tiling approach. Potential evapotranspiration was computed using the Priestley–Taylor formula (Priestley & Taylor 1972). The subsurface response routine was based on the non-linear reservoir method presented by Kirchner (2009). In this study, routing of water between the individual grid cells was omitted since the catchment is small and we run the model on daily time steps.

For the simulations, we used the air temperature and precipitation data described above. A correction of precipitation amount is done through multiplying interpolated precipitation with a regional scaling parameter estimated during model calibration. This parameter should remove biases in the precipitation forcing that otherwise would severely degrade the simulations. Solar radiation, which is an input to the Priestley–Taylor algorithm and therefore affects the calculation of evapotranspiration, was set to a constant value since this part of the modelling chain has a minimal impact on the model performance during the snowmelt period for our study region. The model parameters were calibrated using the observed runoff for the period from 2009-9-1 to 2015-8-31. With the calibrated parameter values, the model shows a Nash–Sutcliffe efficiency (NSE) equal to 0.72 and a modified NSE of 0.65, for the whole study period. The modified NSE was computed using the climatology of observed discharge (monthly averages) as a benchmark instead of the average observed discharge following the methods outlined by Schaefli & Gupta (2007).

Data assimilation algorithm

For the data assimilation, we use a standard stochastic EnKF described briefly below. For further information about this filter, see e.g. Evensen (2009) and Vetra-Carvalho *et al.* (2018) and references therein. Let us begin by denoting the state vector of the system as s_k, where k represents the time. (In our case, the state vector contains the states of all the grid cells of the catchment.) The filter consists of a forecast step and an analysis step. In the forecast step, we propagate our model, which we denote by f, forward in time giving $s_k = f(s_{k-1}, \delta_{k-1})$, where δ_{k-1} represents a stochastic forcing term. In our case, the state vector consists of SWE in each grid cell of the model and the forcing term is the weather inputs consisting of air temperature and precipitation at each grid cell. An ensemble of forward models is run, each having its own state vector and stochastic forcing. The forward model for ensemble member i is denoted $s_k^i = f(s_{k-1}^i, \delta_{k-1}^i)$. When the measurements, which in our case are observations of a subset of the states, become available at a generic time k, we perform the analysis step. In this study, the measurements represent average SWE at three grid cells within the simulation domain. The elevation differences between the observations and the grid cells are lower than 100 m, and biases between the measurements and simulations arising due to this small altitude discrepancy were not accounted for in the data assimilation scheme. In the analysis step, the covariance matrix P_k of the ensemble of state vectors $[s_k^1, s_k^2, \ldots, s_k^N]$ is required. Here, N denotes the size of the ensemble. The covariance matrix P_k is defined as follows:

$$P_k = \frac{1}{N-1}\sum_{i=1}^{N}(s_k^i - \overline{s_k})(s_k^i - \overline{s_k})^T$$

where

$$\overline{s_k} = \frac{1}{N}\sum_{i=1}^{N} s_k^i$$

is the ensemble mean at time k. Let R denote the covariance matrix representing the measurement uncertainty and let the observed quantities y_k be related to the states as $y_k = Hs_k$ for a matrix H at time k. (In our case H will be a matrix consisting only of zeros and ones and is formed such that it selects those states that are observed.) Then, the Kalman gain matrix is given as $K = P_kH^T(HP_kH^T + R)^{-1}$ where superscript T denotes the matrix transpose. Now, ensemble member i $(i = 1, \ldots, N)$ is updated by the formula $s_k^{i,a} = s_k^i + K(y_k - (Hs_k^i + \epsilon_k^i))$ where ϵ_k^i are samples from a Gaussian distribution with mean zero and covariance matrix R. The superscript a is used to distinguish the posterior ensemble members from the prior ensemble members. The matrix R is the covariance matrix of the measurement uncertainties. In this study, we used 50 ensemble members since increasing the number of ensemble members above this value did not reveal any relevant improvements, supporting the use of the fewer ensemble members to reduce computational costs.

Forcing ensemble generation

In this study, the stochastic forcing term in the filter algorithm outlined above was generated by applying perturbations to the temperature and precipitation data presented above. We assume that the errors in air temperature display much smoother variations in space than precipitation and have therefore chosen the approach below to construct the perturbations on the forcing data. For air temperature, the seNorge data were perturbed by adding normally distributed noise constant over the whole catchment (i.e. all the cells have the same perturbation). The noise had zero mean and a standard deviation of 2.0 °C and was correlated in time with a correlation coefficient equal to 0.9 between consecutive days. For precipitation, the forcing data were perturbed using multiplicative noise drawn from a lognormal distribution. This perturbation was produced in the following steps. First, we generated random correlated fields with a decorrelation length equal to 5 km using a fast Fourier transform algorithm. For each grid cell, the random numbers were normally distributed with zero mean and unit standard deviation. Second, we transformed this normally distributed noise to a lognormal distribution using the methods and parameters described in Magnusson *et al.* (2017). Finally, the precipitation grids were perturbed using this noise. We also introduced a correlation in time between the precipitation grids with a correlation coefficient of 0.5 between consecutive days. For more details about the choices of method and parameters for generating the noise on the forcing data, see Magnusson *et al.* (2014, 2017). This method could further be refined by increasing the perturbations during cold and windy conditions when the precipitation gauges presumable measure precipitation with larger errors than during warm and calm conditions.

Description of experiments

Typically, long-term inflow forecasts are produced using either seasonal weather forecasts or some set of historical climatological data. In this study, the main aim is to assess whether improvements in the initial snow conditions also translates to more accurate long-term inflow forecasts during spring melt. Therefore, we isolate our analysis as much as possible on this part of the total uncertainty and try to minimize the effect of the remaining sources of errors.

As a base case, we first run the Shyft model using the seNorge weather data without any stochastic perturbations and for the whole period from 2000-9-1 to 2016-8-31. This simulation represents a typical reservoir inflow simulation for which historical data are available. We benchmark our data assimilation experiments against this simulation.

For the data assimilation experiments, we run the model with the stochastic forcings presented above until the snow observations become available during the first winter. Note that the snow measurements are typically performed in early April. At this point, we stop the model and update the SWE grids using the observations with the EnKF. We specify a measurement uncertainty for the SWE observations using a standard deviation equal to 50 mm. After the update, which typically occurs in early April, we run the ensemble for a period of 120 days without applying stochastic noise to the forcings. This period is usually long enough that most of the snowmelts in the catchment and factors such as the initial snow distribution does not affect the results largely. We then compare the simulated inflow of the base case and the data assimilation run over this period against the observed inflow. Finally, we repeat this procedure for each winter over the whole simulation period resulting in 15 inflow forecasts of 120 days available for further analysis. The advantage of this approach is that we can isolate the effect of changing the SWE field based on the measurements and thereby evaluate the effect of our data assimilation algorithm directly.

RESULTS

Figure 2 shows simulated and observed SWE for three representative winters: one winter with low amounts of snow (2009/2010, average measured SWE equals 521 mm), one with medium amounts (2003/2004, 881 mm) and one with high amounts (2004/2005, 1505 mm). During 2003/2004

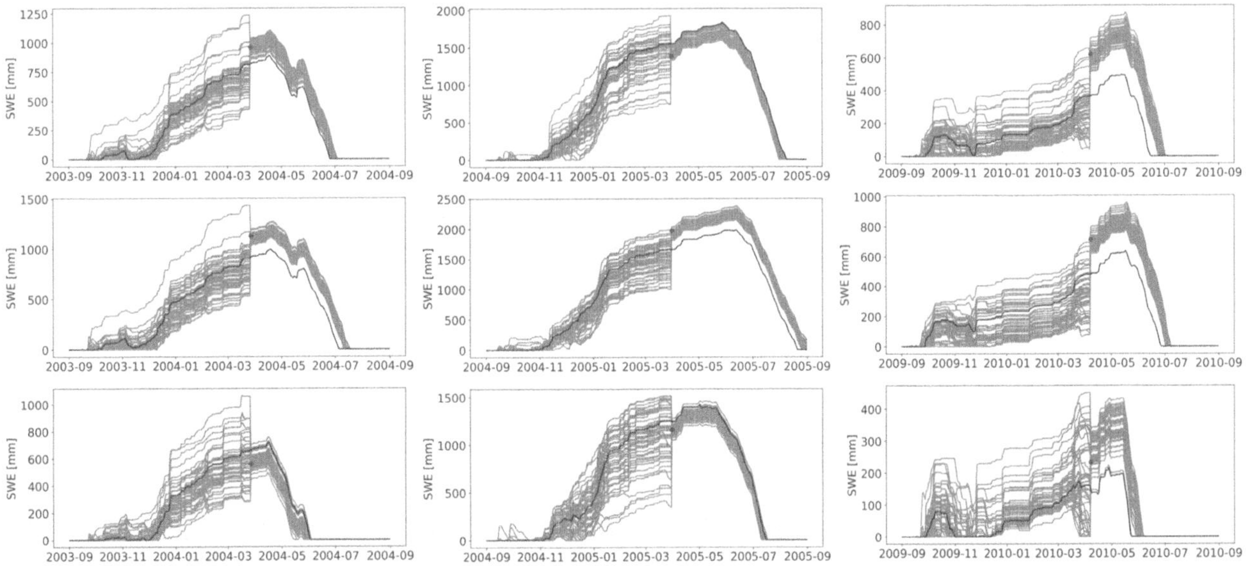

Figure 2 | Simulated and measured SWE for Grønebotn (top row), Katledalene (middle row) and Ovridsfjell (bottom row) for snow seasons 2003/2004 (left column), 2004/2005 (middle column) and 2009/2010 (right column). Ensemble members are shown in red, the base case in black and measurement as a blue dot. Note that the scales on the vertical axes differ between the panels. Please refer to the online version of this paper to see this figure in colour: http://dx.doi.org/10.2166/nh.2020.025.

and 2004/2005, the reference simulation matches the observations much better than in the drier winter (2009/2010), where the base run underestimates snow amounts at all three locations. The ensemble simulations show an increasing spread throughout the snow accumulation phase up to the point where the observations become available and encompass the observations in all years and locations displayed. The filter algorithm pushes the ensemble toward the observations and strongly reduces its spread, especially during the third winter. After the update, no perturbations have been added to the ensemble members to ensure better consistency with the base run during the 120-day long forecasting period and isolate the effect of the SWE update on the inflow forecasts. Therefore, the spread in the ensemble does not increase after the update.

As shown above, the filter algorithm adjusts the simulated SWE to better match the observations and improves the results compared to the base case (Figure 2). The question now arises whether this improvement in snow states also translates to better inflow forecasts. Figure 3 shows the forecasted cumulative discharge from the update time to the end of August for the same 3 years as displayed in Figure 2. In the beginning of the forecasting period, especially in 2005, both the base case and ensemble simulations tend to underestimate discharge, indicating that the model onsets snowmelt too late. Thus, the assimilation of snow measurements using the EnKF does not seem to improve the timing of simulated snowmelt. However, at the end of August, both the base run and ensemble simulations match the observed cumulative runoff rather well, in particular in 2005, even though of the large discrepancies in the start of the evaluation period. For 2004 and 2010, the assimilation of SWE improves the predicted discharge throughout the summer period. The improvement is particularly large in 2010 when the base case performs worst. This is also the year in which the base case underestimated the observed snow amounts most (compare with Figure 2). Thus, it seems as if the assimilation of snow data improves the results in years where the snow simulations deviate largely from the observations.

Figure 4 shows the performance of the inflow forecasts for all years in the study period. The forecasting period covers 120 days following the point in time where we updated the ensemble using the snow observations (see illustration of this forecasting period in Figure 3 highlighted by gray vertical dashed lines). For each year, we computed the percentage bias in total inflow over the forecasting

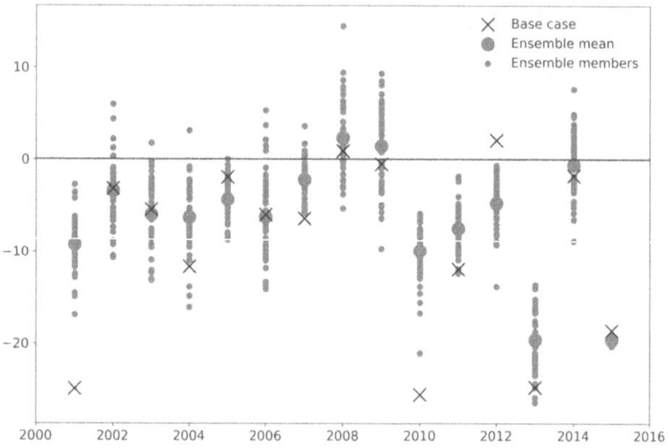

Figure 4 | Error in total inflow to the reservoir during the forecasting period of 120 days for each year shown as the percentage of observed inflow for the base case (black cross), ensemble mean (larger red dot) and individual ensemble members (small red dots). The zero line is marked in black. A positive value means an overprediction, whereas negative values indicate an underestimation by the simulation. Please refer to the online version of this paper to see this figure in colour: http://dx.doi.org/10.2166/nh.2020.025.

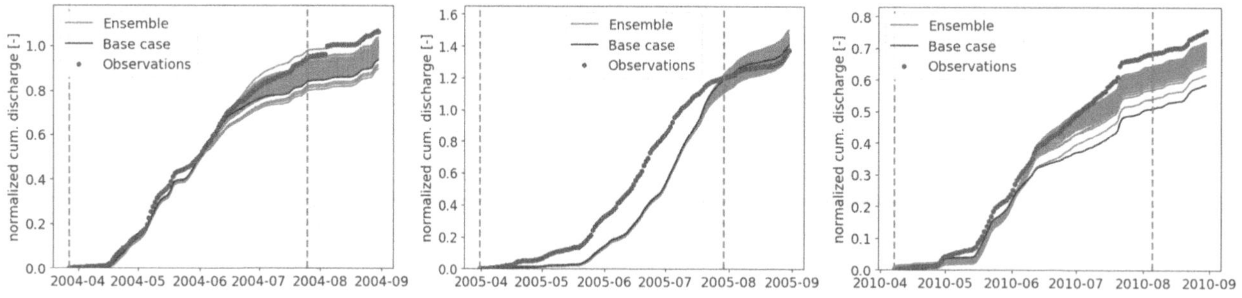

Figure 3 | Forecasted cumulative discharge from the update time to the end of August for 2004 (left), 2005 (middle) and 2010 (right). A dashed line is shown at the point in time where the SWE observations were assimilated, and another one 120 days afterwards. Note that the presented values have been normalized by the average observed cumulative discharge for the forecasting period using data from all available years.

period for both the ensemble run and the base case using the observed inflow. The ensemble means show a lower absolute bias than the base case in 7 years (2001, 2004, 2007, 2010, 2011, 2013 and 2014). In two of those years, the improvement by updating the model is large (2001 and 2010). However, for 2015, the data assimilation algorithm does not improve the inflow forecast even though the error is large. The improvement for 2013 is also small compared to the magnitude of the error. For 3 years, the base case simulation shows a slightly better performance than the ensemble mean (2005, 2009 and 2012). The ensemble mean is within an error of approximately 10% for all but two of the seasons (2013 and 2015), whereas the base case run has more years with an error higher than 10% (2001, 2004, 2010, 2011, 2013 and 2015). Thus, the data assimilation scheme effectively removes large errors in the inflow forecast observed in the base case run.

For the three representative winters displayed in Figure 2, the greatest improvement in simulated SWE due to the assimilated data occurred in the year with the lowest snow amounts (2009/2010). In this winter, the base case underestimated SWE substantially at all three measurement locations, whereas the assimilation run showed a much better match with the observations (see right panels in Figure 2). This reduced error in SWE also contributed to a large improvement in the inflow forecast compared to the base case (see right panels in Figures 3 and 4). Figure 5 shows the performance in the inflow forecasts against the average measured SWE at the three locations both for the base case and the ensemble run. For high snow amounts (SWE greater than 1,100 mm), the assimilation of snow observations does not seem to improve the results much compared to the base case. For medium snow amounts (SWE between 700 and 1,100 mm), the ensemble mean shows a somewhat better performance than the base run in 2 out of 3 years. Notably, in this study we got the largest benefit of assimilating snow observations that occur for years with low amounts of snow (SWE below 700 mm). For these years, the filter algorithm improves the simulated inflows compared to the base case in 3 out of 4 years, and in two of those years, the improvement is very large. Thus, these results indicate that the assimilation of snow observations is more important for years with low than high snow amounts in the study catchment.

In this study, snow observations were available at three different locations throughout the watershed (see Figure 1). In an additional set of experiments, we assimilated snow observations using all possible combinations of available locations to test how sensitive the inflow forecasts are to the constellation of assimilated observations. The mean absolute error in forecasted inflow for all these simulations is shown in Table 1. The smallest error is achieved when assimilating data from both Grønbotn and Katladalene, whereas updating the simulations with data from Ovridsfjell alone gave the worst results. Using exclusively Ovridsfjell data, assimilated model performance was worse than the

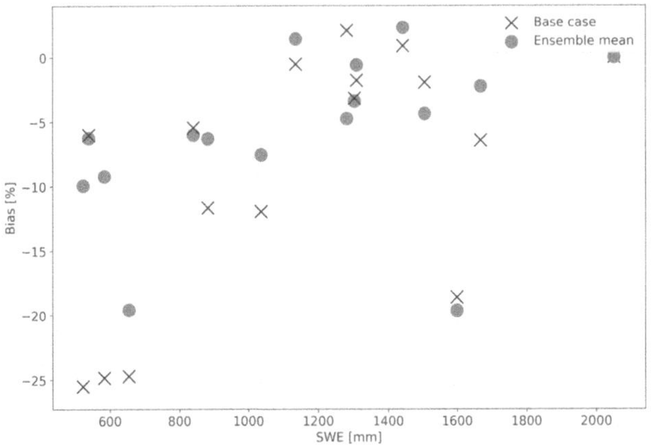

Figure 5 | Bias in forecasted inflows against the mean of the measured SWE at the three locations. The bias in forecasted inflow is given as a percentage of the observed inflow to the Refsdal hydropower plant. The assimilation of SWE tends to improve the inflow forecast for years with low snow amounts more than for snow-rich years.

Table 1 | Mean absolute error (MAE) in forecasted inflow depending on the number of observations available for assimilation

Grønebotn	Katladalene	Ovridsfjell	MAE (%)
			10
X			7
	X		6
		X	15
X	X		6
X		X	8
	X	X	8
X	X	X	7

The crosses mark for which locations the snow observations have been used in the updating algorithm. Thus, the top row is identical to the base case, and in the last row, all measurements have been used.

base case, indicating that incorporating these SWE observations degrades inflow forecasts. However, when using data from all three locations the data from Ovridsfjell does not seem to affect the performance severely, indicating that the procedure is robust when using data from several locations. Nevertheless, with the rather pragmatic approach presented here, it can be determined which snow observations should be used for assimilation and which observations can be dropped or relocated to a better position in the measurement setup.

DISCUSSION

Even though the base case simulation matches the observed inflow well in many years during the study period, this simulation overall underestimates measured inflow, and in a few years severely (cf. Figure 4), even though the model includes a precipitation correction parameter. This underestimation is likely to a large degree, related to the precipitation data we have used as model forcing. For most regions in Norway, these data seem to underestimate precipitation (Lussana *et al.* 2018). Most important, the gauges are prone to undercatch, particularly of solid precipitation, leading to an underestimation of actual precipitation (e.g. Sevruk 1996; Wolff *et al.* 2015). Though the model includes a precipitation correction factor, the simulations underestimate observed inflow. Other factors influencing the quality of the precipitation forcing data are the sparse density of measurement stations and potential misclassifications of precipitation phase (i.e. that a rainfall event was classified as snowfall or vice versa). For some years (e.g. 2001, 2010, 2013 and 2015), the effects mentioned above seem to influence the base case simulation severely, and result in a large underestimation of the observed inflow.

In several of the years where the base case run provides poor results, the ensemble simulations improve the results (Figure 4). We attribute this improvement to the more accurate snow states at the beginning of the forecasting period since the setup of the two simulations are otherwise identical for this period. Note that during the forecast, both runs are driven with the same unperturbed input data. Since we update the model using snow observations made at peak accumulation, the filter likely reduces errors related to the buildup phase of the snowpack, rather than processes related to ablation. Furthermore, we find that the assimilation of snow observations rarely degrades the inflow forecasts, and if so the difference with the base case is small. Thus, our setup of the filter algorithm seems successful and is a viable method for ensuring better inflow forecasts through more accurate initial states. The largest improvements in the forecasts are observed after winters with low amounts of snow (Figure 5). The updating using snow observations made near peak accumulation of the snow cover seems to reduce the errors in winters with the low amount of snow efficiently.

In addition to errors in the forcing data, poor results by the base case simulation may also be attributed to deficiencies in the snowmelt model. For some years, forecasted inflows occur with a delay compared to the observations (see, for example, 2005 in Figure 3). In fact, for all years with large snow accumulations (average measured SWE greater than 1,280 mm), the cumulative discharge forecasts lag behind observations at the onset of spring runoff. In this study, we have used a degree-day snowmelt model, including a bucket formulation for representing the routing of liquid water through the snowpack. This type of model does not capture all processes relevant for the onset of snowmelt and release of meltwater, and therefore introduces errors in the forecasts. However, in the case of seasonal inflow forecasting, the introduction of a more physically realistic model does not necessarily ensure better model performance. Such models require more variables as input, and those need to be of high quality in order to provide reliable results (e.g. Magnusson *et al.* 2015). In an operational setting, where the availability, representativeness and quality of the forcing data required by these physics-based solutions – whether provided by point observations or a weather forecasting model – add considerable uncertainties to the modeling chain (e.g. Raleigh *et al.* 2015). An alternative option for alleviating this problem might be to tune some of the parameters for the temperature-index snowmelt model using the EnKF during the ablation season by assimilating, for instance, the observed discharge data.

Utilizing the measurements from Grønebotn and/or Katladalene improved the forecasts, but the measurement from Ovridsfjell seemed to degrade the results. As there should not be any differences in how the measurements are obtained, the explanation is likely either found in the data assimilation part of the workflow, or because single snow observations may not be representative for larger areas. Ovridsfjell is the most eastern of the three measurement locations, and a potential east-west gradient in the local climate conditions may cause this measurement to give limited information about snow conditions for the western part of the catchment. To avoid any unreasonable influences of Ovridsfjell for areas the measurement is not representative for, one could use localization in the EnKF (see Vetra-Carvalho *et al.* (2018) and references therein for more information about localization). With localization, the effect of a measurement is down-weighted in regions where it may not be representative but still influence the simulations in more similar regions. A second possible explanation might be that there is a height difference between the measurement location and the average height of the grid cell assigned to that location. However, it is not easy to adjust the observations to the grid cell altitude since elevation-dependent gradients in snow amounts can be complex (Grünewald *et al.* 2014). Finally, the measurement setup was not originally designed for assimilation in a distributed hydrological model that also represents the subgrid variability of the snow distribution using a tiling approach. It is still an open research question of how to best design a snow measurement campaign for updating such a model, and whether to use observed snow distributions or average amounts within each grid cell.

There are a number of decisions we have made while testing our data assimilation approach that will influence the results, much of which could be improved upon with further research. One crucial decision is related to how we generate the ensemble of SWE grids, using perturbations on the forcing grids that is required for the EnKF (or other ensemble-based data assimilation methods). Obviously, there are uncertainties associated with the weather forcings, and to reflect this, we have added perturbations to the historical weather data available from the seNorge archive. Ideally, our forcings should represent the uncertainty in these data, and for the time being, we are generating a rather *ad hoc* perturbation to the weather data. In the future, these *ad hoc* perturbations could be replaced by high-resolution ensembles given by the latest weather forecasting models. Such ensembles exhibit physical consistency between the different variables due to the more realistic model used for generating the data and may also better reflect the uncertainty in the forcings. This may also improve the performance of the snow simulations, in particular when using an energy-balance approach, which is highly sensitive to biases in the inputs (Raleigh *et al.* 2015). However, the outputs from weather forecasting models likely need to be bias-corrected against ground observations to avoid large systematic errors before use in hydrological models. Furthermore, weather forecasting models also typically only provide data for the recent past making them unsuitable for the long-term analysis. Nevertheless, there exists a large potential for further testing and eventual improvement of the inflow forecasts by using better meteorological forecasts, snowmelt models and measurement setups.

CONCLUSIONS

In this study, we have assessed whether the assimilation of snow water equivalent (SWE) observations improves seasonal inflow forecasts for the snowmelt period. We assimilated snow observations, typically measured in the beginning of April, into a distributed conceptual hydrological model using an EnKF for 15 snowmelt seasons. The performance of the updating algorithm was tested by comparing simulated and observed inflows to a hydropower reservoir for 120-day forecast periods. The simulations with assimilated snow data were benchmarked against a base case that did not include the snow observations.

We find that the assimilation improves the seasonal inflow forecast for 7 years compared to the base case run, with a notable improvement for two of those years. For another 3 years, we observe a small decrease in model performance due to the assimilation. However, we do not see any larger degradations through the assimilation, indicating that though the updating might slightly degrade performance in some years, it maintains the potential for large improvements in others. In years where the assimilation does not improve inflow forecasts, this in fact may be related to factors other than inaccurate initial snow states, such as poor precipitation forcings during the forecasting period. In summary, the procedure tested here seems to work well for reducing forecast errors that are related to deficiencies in the modelled snow states.

The performance of the data assimilation scheme depended on which snow observations were used for updating the model. Data from one of the measurement locations tended to degrade the inflow forecasts. However, the sampling scheme for the snow observations available in this study was not originally designated for updating a gridded model using an ensemble updating technique. For such a purpose, it is still not clear how to best design a sampling procedure for snow, and further research is warranted in this direction.

ACKNOWLEDGEMENTS

The authors acknowledge Statkraft for providing access to data from the Refsdal catchment, and for providing financial support to Geir Nævdal and Jan Magnusson. The computations were performed on the Norwegian Academic Community Cloud (UH-IaaS available at http://www.uh-iaas.no/), using resources provided by the University of Bergen and the University of Oslo.

REFERENCES

Anderson, E. A. 1976 *A Point Energy and Mass Balance Model of a Snow Cover*. NOAA Technical Report NWS 19, Office of Hydrology, National Weather Service.

Andreadis, K. M. & Lettenmaier, D. P. 2006 Assimilating remotely sensed snow observations into a macroscale hydrology model. *Adv. Water Resour.* **29**, 872–886. https://doi.org/10.1016/j.advwatres.2005.08.004.

Barnett, T. P., Adam, J. C. & Lettenmaier, D. P. 2005 Potential impacts of a warming climate on water availability in snow-dominated regions. *Nature* **438**, 303–309. https://doi.org/10.1038/nature04141.

Beven, K. J. & Binley, A. M. 1992 The future of distributed models: model calibration and uncertainty in prediction. *Hydrol. Processes* **6**, 279–298. https://doi.org/10.1002/hyp.3360060305.

Brown, R. D., Brasnett, B. & Robinson, D. 2003 Gridded North American monthly snow depth and snow water equivalent for GCM evaluation. *Atmos.-Ocean* **41**, 1–14. https://doi.org/10.3137/ao.410101.

De Lannoy, G. J. M., Reichle, R. H., Arsenault, K. R., Houser, P. R., Kumar, S., Verhoest, N. E. C. & Pauwels, V. R. N. 2012 Multiscale assimilation of advanced microwave scanning radiometer – EOS snow water equivalent and moderate resolution imaging spectroradiometer snow cover fraction observations in northern Colorado: satellite-observed snow data assimilation. *Water Resour. Res.* **48**. https://doi.org/10.1029/2011WR010588.

Evensen, G. 2009 The ensemble Kalman filter for combined state and parameter estimation. *IEEE Control Syst.* **29**, 83–104. https://doi.org/10.1109/MCS.2009.932223.

Fjeld, M. & Aam, S. 1981 An implementation of estimation techniques to a hydrological model for prediction of runoff to a hydroelectric power station. *Model., Identif. Control* **2** (1), 37–56. https://doi.org/doi.10.4173/mic.1981.1.3.

Fletcher, S. J., Liston, G. E., Hiemstra, C. A. & Miller, S. D. 2012 Assimilating MODIS and AMSR-E snow observations in a snow evolution model. *J. Hydrometeorol.* **13**, 1475–1492. https://doi.org/10.1175/JHM-D-11-082.1.

Gragne, A. S., Sharma, A., Mehrotra, R. & Alfredsen, K. 2015 Improving real-time inflow forecasting into hydropower reservoirs through a complementary modelling framework. *Hydrol. Earth Syst. Sci.* **19**, 3695–3714. https://doi.org/10.5194/hess-19-3695-2015.

Grünewald, T., Bühler, Y. & Lehning, M. 2014 Elevation dependency of mountain snow depth. *Cryosphere* **8**, 2381–2394. https://doi.org/10.5194/tc-8-2381-2014.

Huang, S., Eisner, S., Magnusson, J., Lussana, C., Yang, X. & Beldring, S. 2019 Improvements of the spatially distributed hydrological modelling using the HBV model at 1 km resolution for Norway. *J. Hydrol.* https://doi.org/10.1016/j.jhydrol.2019.03.051.

Kirchner, J. W. 2009 Catchments as simple dynamical systems: catchment characterization, rainfall-runoff modeling, and doing hydrology backward. *Water Resour. Res.* **45**, W02429. doi:10.1029/2008WR006912.

Kolberg, S. & Gottschalk, L. 2006 Updating of snow depletion curve with remote sensing data. *Hydrol. Processes* **20** (11). https://doi.org/10.1002/hyp.6060.

Kolberg, S., Rue, H. & Gottschalk, L. 2006 A Bayesian spatial assimilation scheme for snow coverage observations in a gridded snow model. *Hydrol. Earth Syst. Sci.* **10**, 369–381. https://doi.org/10.5194/hess-10-369-2006.

Kuczera, G. 1983 Improved parameter inference in catchment models: 1. Evaluating parameter uncertainty. *Water Resour. Res.* **19** (5), 1151–1162. https://doi.org/10.1029/WR019i005p01151.

Leisenring, M. & Moradkhani, H. 2011 Snow water equivalent prediction using Bayesian data assimilation methods. *Stoch. Environ. Res. Risk Assess.* **25**, 253–270. https://doi.org/10.1007/s00477-010-0445-5.

Lindström, G., Johansson, B., Persson, M., Gardelin, M. & Bergström, S. 1997 Development and test of the distributed HBV-96 hydrological model. *J. Hydrol.* **201**, 272–288. https://doi.org/10.1016/S0022-1694(97)00041-3.

Liston, G. E. & Hiemstra, C. A. 2008 A simple data assimilation system for complex snow distributions (SnowAssim). *J. Hydrometeorol.* **9**, 989–1004. https://doi.org/10.1175/2008JHM871.1.

Liston, G. E., Pielke, R. A. & Greene, E. M. 1999 Improving first-order snow-related deficiencies in a regional climate model. *J. Geophys. Res. Atmos.* **104**, 19559–19567. https://doi.org/10.1029/1999JD900055.

Liu, Y., Peters-Lidard, C. D., Kumar, S., Foster, J. L., Shaw, M., Tian, Y. & Fall, G. M. 2013 Assimilating satellite-based snow depth and snow cover products for improving snow predictions in Alaska. *Adv. Water Resour.* **54**, 208–227. https://doi.org/10.1016/j.advwatres.2013.02.005.

Luijting, H., Vikhamar-Schuler, D., Aspelien, T., Bakketun, Å. & Homleid, M. 2018 Forcing the SURFEX/Crocus snow model with combined hourly meteorological forecasts and gridded observations in southern Norway. *Cryosphere* **12**, 2123–2145. https://doi.org/10.5194/tc-12-2123-2018.

Lussana, C. & Tveito, O. E. 2017 seNorge2 dataset [Data set]. Zenodo. Available at: https://doi.org/10.5281/zenodo.845733.

Lussana, C., Tveito, O. E. & Uboldi, F. 2016 seNorge v2.0, Temperature (No. 14/2016). Norwegian Meteorological Institute.

Lussana, C., Saloranta, T., Skaugen, T., Magnusson, J., Tveito, O. E. & Andersen, J. 2018 Senorge2 daily precipitation, an observational gridded dataset over Norway from 1957 to the present day. *Earth Syst. Sci. Data* **10**, 235–249. https://doi.org/10.5194/essd-10-235-2018.

Magnusson, J., Gustafsson, D., Hüsler, F. & Jonas, T. 2014 Assimilation of point SWE data into a distributed snow cover model comparing two contrasting methods. *Water Resour. Res.* **50**, 7816–7835. https://doi.org/10.1002/2014WR015302.

Magnusson, J., Wever, N., Essery, R., Helbig, N., Winstral, A. & Jonas, T. 2015 Evaluating snow models with varying process representations for hydrological applications: snow model evaluation. *Water Resour. Res.* **51**, 2707–2723. https://doi.org/10.1002/2014WR016498.

Magnusson, J., Winstral, A., Stordal, A. S., Essery, R. & Jonas, T. 2017 Improving physically based snow simulations by assimilating snow depths using the particle filter: snow assimilation with the particle filter. *Water Resour. Res.* **53**, 1125–1143. https://doi.org/10.1002/2016WR019092.

Ohmura, A. 2001 Physical basis for the temperature-based melt-index method. *J. Appl. Meteorol.* **40**, 753–761. https://doi.org/10.1175/1520-0450(2001)040<0753:PBFTTB > 2.0.CO;2.

Priestley, C. H. B. & Taylor, R. J. 1972 On the assessment of surface heat flux and evaporation using large-scale parameters. *Mon. Weather Rev.* **100**, 81–92. https://doi.org/10.1175/1520-0493(1972)100<0081:OTAOSH > 2.3.CO;2.

Raleigh, M. S., Lundquist, J. D. & Clark, M. P. 2015 Exploring the impact of forcing error characteristics on physically based snow simulations within a global sensitivity analysis framework. *Hydrol. Earth Syst. Sci.* **19**, 3153–3179. https://doi.org/10.5194/hess-19-3153-2015.

Schaefli, B. & Gupta, H. V. 2007 Do nash values have value? *Hydrol. Processes* **21**, 2075–2080. https://doi.org/10.1002/hyp.7072.

Sevruk, B. 1996 Adjustment of tipping-bucket precipitation gauge measurements. *Atmos. Res.* **42**, 237–246. https://doi.org/10.1016/0169-8095(95)00066-6.

Slater, A. G. & Clark, M. P. 2006 Snow data assimilation via an ensemble Kalman filter. *J. Hydrometeorol.* **7**, 478–493. https://doi.org/10.1175/JHM505.1.

Tweldebrahn, A. T., Burkhart, J. F. & Schuler, T. V. 2018 Parameter uncertainty analysis for an operational hydrological model using residual based and limits of acceptability approaches. *Hydrol. Earth Syst. Sci. Discuss.* 1–25. https://doi.org/10.5194/hess-2018-158.

Vetra-Carvalho, S., van Leeuwen, P. J., Nerger, L., Barth, A., Altaf, M. U., Brasseur, P., Kirchgessner, P. & Beckers, J.-M. 2018 State-of-the-art stochastic data assimilation methods for high-dimensional non-Gaussian problems. *Tellus Dyn. Meteorol. Oceanogr.* **70**, 1445364. https://doi.org/10.1080/16000870.2018.1445364.

Vrugt, J. A., Diks, C. G. H., Gupta, H. V., Bouten, W. & Verstraten, J. M. 2005 Improved treatment of uncertainty in hydrologic modeling: combining the strengths of global optimization and data assimilation. *Water Resour. Res.* **41**, W01017. https://doi.org/10.1029/2004WR003059.

Wolff, M. A., Isaksen, K., Petersen-Øverleir, A., Ødemark, K., Reitan, T. & Brækkan, R. 2015 Derivation of a new continuous adjustment function for correcting wind-induced loss of solid precipitation: results of a Norwegian field study. *Hydrol. Earth Syst. Sci.* **19**, 951–967. https://doi.org/10.5194/hess-19-951-2015.

First received 20 February 2019; accepted in revised form 14 January 2020. Available online 17 March 2020

Reproducing different types of changes in hydrological indicators with rainfall-runoff models

Carolina Massmann

ABSTRACT

Hydrological indicators support analyses about the impact of climate and anthropogenic changes on riverine ecosystems. As these studies often rely on hydrological models for estimating the future value of the indicators, it is important to investigate how well, and under which conditions, we can replicate changes in the indicators. This study looks at these questions by investigating the performance that can be achieved depending on the objective function for calibrating the model, the direction of the change in the indicator, the magnitude of this change and the properties of the catchments. The results indicate that, in general, indicators describing the magnitude of discharge (monthly and annual) can be adequately estimated with hydrological models, but that there are difficulties when estimating the characteristics of flow pulses, flow reversals and timing variables. For some of these indicators, it is not even possible to correctly estimate the direction of large changes. The analysis showed further that these problems cannot be resolved by adjusting the calibrated parameters, but that the model structure is unsuitable for modelling these indicators.

Key words | ecohydrology, hydrologic change, hydrological indicators, indices of hydrological alteration, model evaluation, rainfall-runoff model

Carolina Massmann
Institute for Hydrology and Water Management (HyWa),
University of Natural Resources and Life Sciences,
1190 Vienna,
Austria
and
Department of Civil Engineering,
University of Bristol,
Bristol BS8 1TR,
UK
E-mail: *carolina.massmann@boku.ac.at*

INTRODUCTION

The expected economic and population growth rates, coupled with the effect of climate change, raise fears that the number and severity of water-related conflicts might increase in the future. When dealing with competing uses of water, it is necessary that societies agree on the water quality and quantity allocated to fulfilling different needs and that they define, based on these agreements, clear targets for managing this resource (Chopra *et al.* 2005). The defined targets will depend, among other factors, on the current state and level of use of the aquatic resources as the closer the systems are to their natural state, the more likely that they will be managed for preserving ecosystems and their biodiversity (Acreman *et al.* 2014a). Accurate assessments of the current state of river ecosystems are thus important for supporting the definition of management objectives. Furthermore, such assessments constitute the basis for estimating how ecosystems could be affected by future interventions or changes (Acreman *et al.* 2014a), playing a prominent role in preparing for the future.

One important aspect these assessments need to consider is the health of river communities (i.e., plants, invertebrates and fish). However, due to the lack of available ecological data, such analyses are difficult to carry out (Acreman *et al.* 2014b). This is remedied by resorting to indicators describing the flow regime and using them as proxies for the ecological state by assuming that these indicators are

doi: 10.2166/nh.2020.073

the main determinants of the ecosystem properties (Poff *et al.* 1997). This allows for comparing the values of the indicators, for instance, at two points in time, and concluding that there is no impact on the river ecosystem if these values do not vary significantly (Acreman *et al.* 2014b). While it is recognized that this approach leaves out important factors influencing riverine communities (e.g., water temperature, sediment load), it is not unreasonable to do so as these factors tend to be correlated with river discharge (Colby 1956; Sinokrot & Gulliver 2000; Acreman *et al.* 2014b), which is regarded as the most important factor.

While the hydrological indicators can be directly calculated from discharge data, it is necessary to rely on data-based statistical approaches or hydrological models when discharge data are unavailable. Data-based approaches have been widely used for estimating long-term hydrological indicators by establishing links between catchment properties (e.g., soil properties or average precipitation) and the hydrological indicators using, for instance, regional regressions (Sanborn & Bledsoe 2006; Carlisle *et al.* 2010; Knight *et al.* 2012). Hydrological models, on the other hand, use climate time series as model inputs and provide discharge time series as output. The hydrologic indicators can then be calculated from the simulated hydrographs (Shrestha *et al.* 2014; Caldwell *et al.* 2015). While both approaches have their merits and disadvantages (see detailed account in Murphy *et al.* 2013), hydrological models tend to be better suited for carrying out more detailed analyses (e.g., annual variability) at more local scales (see Olsen *et al.* 2013), while statistical models might be preferred for obtaining a general picture or a first approximation of the long-term averages at more regional scales (see Carlisle *et al.* 2010). The present study ultimately aims at improving our abilities for estimating changes in the hydrological properties of streams and using this information for adapting to climate change. As this requires detailed local analyses, the study focuses on model-based rather than on data-based indicator estimates.

Previous studies investigating the factors influencing the quality of estimated hydrological indicators obtained with modelled discharge time series have focused primarily on the impact of different model calibration strategies and objective functions. This is not surprising since this is a practical question practitioners and researchers are often confronted with when working with more than one indicator as it has been observed that models tend to replicate the characteristics to which they are calibrated well, while they might have a quite bad performance for estimating other metrics (Murphy *et al.* 2013). Many efforts have, therefore, gone into finding a small set of objective functions that allow a close reproduction of several indicators. While this question has not yet been settled, the results indicate that considering the indicator of interest in the objective function has a positive impact on the quality of the estimated indicators. It is possible to consider single indicators by calibrating the model directly to the indicator of interest (Kiesel *et al.* 2017; Pool *et al.* 2017) or to combine indicators in the context of a multi-objective Pareto calibration (Hernandez-Suarez *et al.* 2018) or with objective functions that combine various indicators (Vis *et al.* 2015; Kiesel *et al.* 2017; Pool *et al.* 2017).

An aspect that has not been thoroughly considered in this context is our ability for replicating *changes* in hydrologic indicators. This is somewhat surprising, given that the assessment of changes in river ecosystems is one of the main applications of hydrologic indicators and there is a vast body of literature looking at the performance of rainfall-runoff models in periods with different climate properties to those observed during the calibration period (Coron *et al.* 2012; Seiller *et al.* 2012; Brigode *et al.* 2013; Patil & Stieglitz 2015; Dakhlaoui *et al.* 2017; Vormoor *et al.* 2018). Many of these studies were inspired by the work of Klemeš (1986), who advocated for assessing the usefulness of models for estimating hydrographs under changed climate conditions with differential split-sample tests. In such tests, the model is calibrated for a subset of climate conditions and then tested on periods with contrasting climate conditions. Most of these studies compare 'wetter' periods with contrasting 'dryer' periods as precipitation is the ultimate driver of discharge. If the analysis, however, does not focus on the annual or seasonal discharge, it might not be straightforward to identify contrasting periods *a priori* (Vormoor *et al.* 2018). This has implications when considering several indicators as it is expected that they will be sensitive to different climate characteristics, resulting in different sets of years characterizing contrasting climate properties depending on the indicator. In the only study addressing the ability of hydrological models for reproducing changes for a set of hydrological indicators that was identified, Shrestha *et al.* (2016) considered different 'cold' and 'warm' ENSO (El Niño Southern Oscillation) and PDO (Pacific Decadal Oscillation) phases for four Canadian subbasins. While this might not necessarily result in highly contrasting periods for each indicator, it is an approach that can be easily justified and which avoids the need for a sensitivity analysis for each indicator. This study follows another route as it investigates our ability for modelling changes in the IHAs irrespective of the changes in climate. This is achieved by analysing changes in performance between different periods with contrasting IHA values and not periods with contrasting climate conditions.

Considering (i) the importance of knowing how well we can reproduce changes in the IHAs and (ii) the lack of studies investigating this question systematically for a large number of catchments, the objectives of this study are:

- Investigating if models calibrated to achieve a good overall performance outperform models calibrated towards specific hydrological indicators when modelling different types of changes in the indicators.
- Investigating if the performance for estimating changes in the hydrologic indicators varies with the direction and magnitude of the considered changes.
- Identifying possible links between catchment properties and our ability for modelling changes in the indicators.

A strength of the analysis presented here is that it considers a large number of catchments exhibiting considerable variations in climate and catchment properties for obtaining more robust and representative results.

METHODOLOGY

Datasets and catchments

This study analyses 560 catchments located across the United States with no discharge gaps between 1981 and 2008. They are representative of a large spectrum in climate and catchment properties as shown by the ranges covered by the mean annual precipitation (between 304 mm and 3,250 mm), mean annual discharge (between 2 mm and 3,508 mm) and aridity index (between 0.21 and 4.65). These catchments belong to the GAGES II (Geospatial Attributes of Gages for Evaluating Streamflow, version II) dataset (Falcone 2011), which consists of gauges maintained by the USGS (United States Geological Survey) with at least 20 years of data or active in 2009. The gauges used in the present study are reference basins, which means they are among the least disturbed basins.

The climate data for the daily precipitation and the minimum and maximum temperatures were obtained from the Daymet 3 dataset (Thornton *et al.* 1997; Thornton *et al.* 2014) produced by the NASA Oak Ridge National Laboratory (ORNL). This dataset covers North America from 1980 to the present at a $1 \times 1\ \text{km}^2$ resolution. The daily discharge time series were downloaded from the USGS webpage.

The catchments were characterized with respect to seven variables describing the climate (Walsh seasonality, aridity index, maximum 1-day annual precipitation, percentage of precipitation falling as snow), subsurface properties (baseflow index (BFI)), topography (average slope) and the water balance (runoff coefficient). The climate properties were estimated with the Daymet 3 data. The approach described by Walsh & Lawler (1981) was used for calculating the seasonality index and the aridity index is defined as the quotient between precipitation and potential evapotranspiration. The BFI was taken from a gridded dataset provided by the USGS (Wolock 2003) and the average slope was estimated from a 10 m resolution digital elevation model obtained from the National Elevation Dataset (NED).

Hydrological model

The conceptual rainfall-runoff model used in this study is the most complex member of a set of eight models exhibiting a stepwise increase in complexity. The models were initially developed by Jothityangkoon *et al.* (2001), Atkinson *et al.* (2002) and Farmer *et al.* (2003) and later modified by Bai *et al.* (2009). The required model inputs consist of the daily precipitation and the maximum and minimum daily temperatures. Interception is modelled as a fraction of the precipitation input and described by the parameter *cei*. A day-degree-based snow accumulation and melt routine (described by the parameters *tt* and *dd*) implemented for 100 m elevation bands, computes then the liquid input to the soil bucket of size sb. The mechanisms available for water removal from this bucket are saturation overflow, subsurface flow (described by parameter ass), groundwater recharge (described by the coefficient k_d) and evapotranspiration. The water reaching the deep flow bucket is routed to the stream as defined by the baseflow recession coefficient *abf*. The evapotranspiration is modelled as a function of soil moisture at the current timestep and the potential evapotranspiration, estimated according to the Hargraeves–Samani approach (Hargraeves & Samani 1982). Unlike most conceptual models, this model uses a parameter representing the field capacity (fc) for distinguishing a saturated and an unsaturated zone in the soil bucket. These zones interact differently with the potential evapotranspiration and differ in their way of estimating the fraction of the catchment covered by vegetation (m) and the fraction of bare soil ($1 - m$). A sketch of the model structure can be found in Massmann (2019).

The hydrological model was run at a daily timestep between 1 October 1981 and 30 September 2008. A three-year warm-up period was considered for stabilizing the state variables, leaving the period between 1 October 1984 and 30 September 2008 (24 years) for the analysis. The model was calibrated under a Monte Carlo (MC) framework that relied on 40,000 model runs for each catchment. For each of these model runs, the parameter values were randomly sampled assuming a uniform distribution between the ranges shown in the Supplementary Material (Table S1).

Indices of hydrological alteration

For keeping the study manageable, it was necessary to select a subset of hydrological indicators. One option would have been to build on a recent study identifying the indicators correlating with the integrity of invertebrates and fish populations in different regions in the United States (Carlisle *et al.* 2017). However, since the correlated indicators vary depending on the region and between the invertebrate and fish communities, it would have been possible to focus only on one region. Instead, it was decided to select a subset of Richter's indices of hydrological alteration (IHAs) which characterize different flow conditions and components of the flow regime (Richter *et al.* 1996). Flow condition refers here to the magnitude of discharge which could be classified, for instance, into average, low and high discharges. The five components of the flow regime affecting the ecological processes in rivers are the magnitude of the discharge, the frequency, duration and timing of certain flow magnitudes and the rate of change of the discharge (Richter *et al.* 1996). This study considers 12 indicators which attempt to cover a wide range of flow conditions (average, high, very high, low, very low) and all five regime descriptors (Table 1).

Table 1 | Description of the indices of hydrological alteration used in the study

Name	Description	Unit	Regime property	Flow condition
Q Jan	Mean annual flow January (winter)	mm	Magnitude, timing	Average
Q Jul	Mean annual flow July (summer)	mm	Magnitude, timing	Average
Q95	95 exceedance percentile	mm	Magnitude, duration	Low extreme
Q5	5 exceedance percentile	mm	Magnitude, duration	High extreme
Q year	Mean annual discharge	mm	Magnitude	Average
dHFP	Duration of high flow pulses	days	Magnitude, duration	High
dLFP	Duration of low flow pulses	days	Magnitude, duration	Low
nHFP	Number of high flow pulses	–	Magnitude, frequency	High
nLFP	Number of low flow pulses	–	Magnitude, frequency	Low
Reversals	Number of reversals	–	Rate of change	Average
DOYx	Day of year maximum discharge	–	Timing	High extreme
DOYn	Day of year minimum discharge	–	Timing	Low extreme

While the information provided in Table 1 suffices for calculating most indicators, additional information might be necessary for understanding how the number of high flow pulses (HFPs) and low flow pulses (LFPs) was estimated. The approach used in this study consisted of taking the discharge time series for the 24 years considered in the analysis and estimating the 25th and the 75th flow percentiles. A pulse is then defined as an uninterrupted period during which the discharge is above the 75th percentile (HFP) or below the 25th percentile (LFP).

Study set-up

Estimation of changes in the hydrological indices

The steps that need to be followed for calculating the changes in the IHAs are (Figure 1):

- Calculate the annual value of the indicator using the observed discharge (IHA_o).
- Calculate the annual value of the indicator (IHA_{sim}) and the annual value of the Nash–Sutcliffe Efficiency (NSE) for each of the 40,000 simulated discharge timeseries.
- Sort the annual IHA_o values and separate them into four groups with increasing IHA_o value. This is done for analysing different types of changes (e.g., from very low to high indicator values and vice versa). Randomly select five years in each group and calculate the average IHA_o, IHA_{sim} and NSE value for these 5 years. For IHA_o, these averages are identified as $T_{o,g}$ with g standing for the group. For the IHA_{sim} and NSE, the 5-year averages are identified as $T_{h,g,c}$ and $T_{n,g,c,}$ respectively with c denoting the number of the MC run (i.e., ranging from 1 to 40,000). The usage of the average of five annual values, instead of one value accounting for the whole period, is inline of the use of the split KGE by Fowler *et al.* (2018), who found this results in consistently better performances during validation.
- Identify, for each group, the MC run within the highest NSE value and the lowest absolute IHA error ($abs(IHA_{sim}$- $IHA_o)$). Store these values in the vectors MCn and MCh, respectively, with MCn[g] and MCh[g] specifying the MC run with the highest NSE and lowest IHA error in group g, respectively.
- For all combination of the four groups (g1, g2, g3, g4) calculate the observed and simulated changes in the indicators. This results in 12 cases (i.e., g1 to g2, g1 to g3, g1 to g4, g2 to g1 and so on). The observed changes in IHAs are calculated as the difference in the observed IHA between both considered periods. For the modelled timeseries, two sets of changes are calculated. The first is based on the optimal Monte Carlo run with respect to the NSE, whereas the other set considers the optimal parameter set with respect to the IHA of interest.

Evaluation metrics

An essential aspect when assessing the performance of a model for reproducing changes is its ability to correctly reproduce the direction of change. This is assessed with the 'agreement in direction' (AiD) metric (Equation (1)):

$$AiD = \begin{cases} 1, \text{ if } (IHA_{sim,g2} - IHA_{sim,g1}) * (IHA_{o,g2} - IHA_{o,g1}) > 0 \\ 0, \text{ if } (IHA_{sim,g2} - IHA_{sim,g1}) * (IHA_{o,g2} - IHA_{o,g1}) \leq 0 \end{cases} \quad (1)$$

where $IHA_{sim,g1}$ and $IHA_{o,g2}$ stand for the simulated and observed IHA value in the groups specified by the subscripts g1 and g2. If the simulated and observed changes agree in their direction, the product will be positive and *a* will equal one. If the direction of change differs between the observed and simulated IHA, their product will be negative and AiD will equal zero.

It is further important to assess the magnitude of the error incurred when estimating changes in hydrological indicators. It can be distinguished between the nominal and the relative magnitude of the error. This differentiation is especially relevant when comparing catchments exhibiting large differences in discharge as errors in nominal units tend to be larger for catchments with higher discharge. On the other hand, the relative error (e.g., with respect to the mean annual discharge), tends to be larger for catchments with lower discharge (Murphy *et al.* 2013). This study considers both the nominal (NE) and the relative errors (RE),

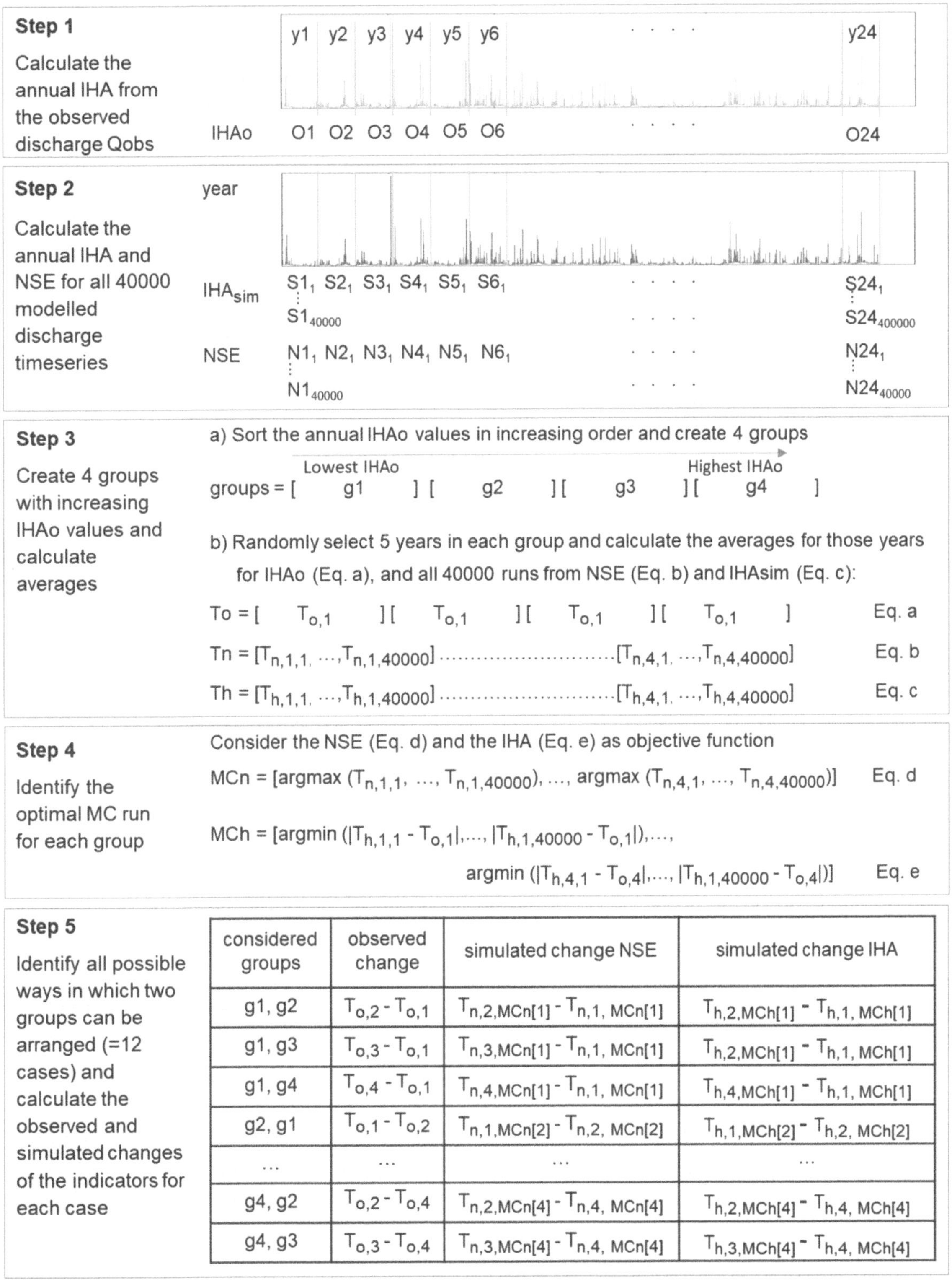

Figure 1 | Procedure for calculating the changes in the indicators for all 12 combinations of groups. $T_{o,i}$ stands for the observed IHA value in group g, while $T_{n,g}$ and $T_{h,g}$ stand for the 40,000 NSE and IHA values in group g. MCn[g] and MCh[g] identify the optimal MC run with respect to the NSE and IHA in period g.

described in Equations (2) and (3), respectively:

$$NE = |(IHA_{sim,g2} - IHA_{sim,g1}) - (IHA_{o,g2} - IHA_{o,g1})| \quad (2)$$

$$RE = \frac{|(IHA_{sim,g2} - IHA_{sim,g1}) - (IHA_{o,g2} - IHA_{o,g1})|}{IHA_{o,md}} \quad (3)$$

where $IHA_{o,md}$ stands for the mean value of the observed IHA values in both considered groups.

Data analysis

It is expected that both the ability for estimating the direction and the magnitude of changes might vary depending on the magnitude and direction of the considered change. For instance, it might be more difficult to get the direction of change right, the smaller the change is. The ability for correctly reproducing the magnitude of change might also vary depending on the type of change (i.e., if it corresponds to a wet-to-dry or a dry-to-wet situation and to a large or small change). The analysis was therefore carried out for different percentiles of observed IHA changes.

For shedding some light on the factors affecting the performance of hydrological models for estimating IHA changes, the relationship between some catchment descriptors (see section 'Datasets and catchments') and the performance metrics (AiD, NE and RE) was analysed using Pearson's correlation coefficient. This analysis was not carried out for the two variables describing the timing of events (DOYx and DOYn) as they are circular variables and thus require a different approach.

RESULTS AND DISCUSSION

How do models calibrated with different objective functions differ in their ability to reproduce IHAs?

The impact of the objective function on the calculated IHAs is shown in Figure 2, which depicts scatterplots relating the observed (*x*-axis) with the simulated IHAs (*y*-axis). The plots on the left were obtained with the Monte Carlo run achieving the highest NSE in each group (i.e., each plot has 560 (catchments) × 4 (groups) points), whereas the plots on the right side show the results when using the model run with the best value for the corresponding IHA.

The plots for the Q5 exceedance are similar to the plots for Q Jan, Q Jul, Q95 and Q year (Figure S1 in the Supplementary material), all of which describe the magnitude of the discharge (i.e., their units are mm). It can be seen that the IHAs calculated with the NSE calibrated model agree relatively well with the observed IHAs (left side plots). As the plots on the right side show the best IHA value that was achieved with the 40,000 Monte Carlo runs, it is not surprising that the observed and simulated values lie almost on the 1:1 line.

The patterns observed for the number of low pulses and reversals are similar to the patterns observed for the number of high pulses (Figure S1 in the Supplementary Material). While the plots on the right side agree better with the observed IHAs than the plots calibrated with the NSE, there is a considerable underestimation of the observed values even when the IHAs are used for calibrating the model. This underestimation is more pronounced for higher IHA values as the model is able to reproduce the values of the indicators at the lower end of the distribution to a certain extent, but there are large deficiencies when the IHAs exhibit higher values. For example, the model was unable to reach a value above 30 for the LFP indicator, while the observed values rise above 50. The fact that the hydrological model is unable to reproduce these features with any of the 400,000 MC runs indicates that the performance cannot be improved with more sophisticated calibration approaches, but that the problem lies in the model structure. The recent surge of interest in model structure-related issues has led to the development of modular hydrological models which enable researchers to compare and test different model structures (Clark *et al.* 2008; Fenicia *et al.* 2011) and to investigate the impact of different structural components on the hydrograph (Lane *et al.* 2019; Massmann 2020). The identification of an adequate model structure for a given catchment is, however, still an unresolved problem. The reason for this is that besides model structural uncertainty there are also measurement errors (e.g., in the discharge time-series) and uncertainties in the climate data which is used for forcing the models. Since model parameters can compensate for errors in the forcing, discharge measurements and model structure, it is difficult to isolate the effect of model structures on model performance (Gupta & Govindaraju 2019).

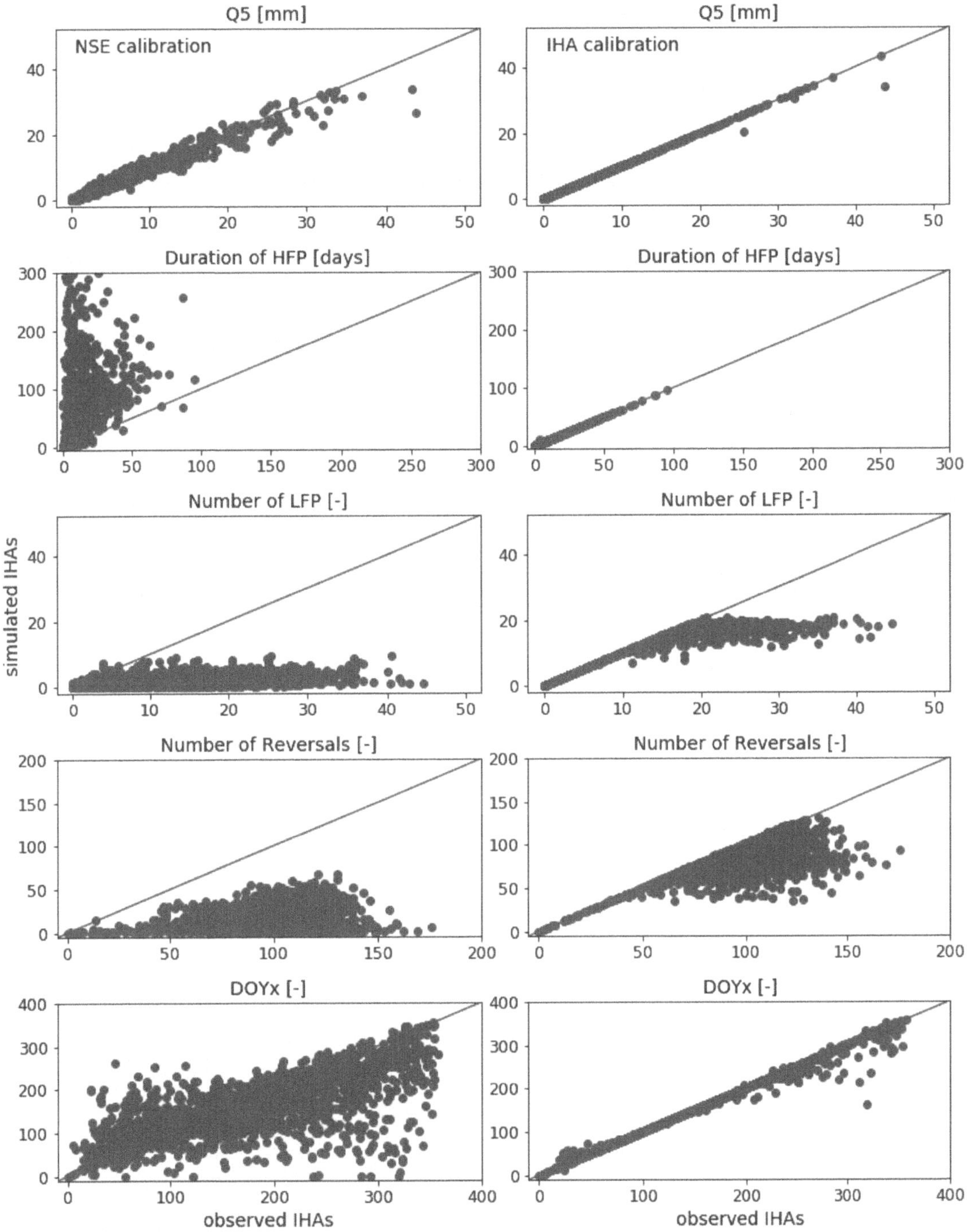

Figure 2 | Relationship between the observed and modelled IHAs considering all 4 groups and 560 catchments during calibration. The plots on the left consider the model calibrated with the NSE; the plots on the right show the results with the model calibrated with the corresponding IHA. Q5: 5th exceedance percentile; HFP: high flow pulses; LFP: low flow pulses; DOYx: day of year with minimum flow.

Due to the similarity of the plots for the indicators describing the duration of the HFPs and LFPs, only the results for the HFPs are presented in the manuscript. The model is able to correctly simulate this indicator as there is good agreement between the observed and simulated values on the right-hand side plot. However, it is observed that the optimal parameter set with respect to the NSE is not a good choice for modelling these indicators as it results in large overestimations of the duration of the pulses.

As the plots for the DOY of maximum and minimum discharge have similar patterns, the results are only shown for the day of maximum discharge. It is seen that the NSE is not able to reproduce this feature well, but there is no consistent over- or underestimation. The results reached with the IHA calibrated models agree in general with the observed IHA, but there is a large scatter. It is further interesting to note that the model calibrated with the IHA can model the day with minimum discharge better than the day with maximum discharge, but that this is the other way around when the NSE is used as objective function during calibration.

In summary, the hydrological model is able to simulate the indicators describing the magnitude of the flow and the timing of extreme events, but it is inadequate for modelling the number of pulses and reversals, which tend to be underestimated. These results agree well with previous studies. Shrestha *et al.* (2014) found, for instance, that the correlation between observed and modelled IHAs describing the magnitude tended to be above 0.5, while most IHAs related to flow pulses and reversals had values below 0.5.

The next sections investigate if our ability for correctly modelling the IHAs describing the magnitude and timing properties of hydrographs is reflected in good estimates of the changes in these IHAs. Analogously, we want to find out if the difficulties of hydrological models for reproducing the number of pulses and reversals result in low performances for modelling changes in these indicators as it might be possible that estimates of the IHAs obtained with the models are biased, but that their changes can be nevertheless modelled with some confidence.

Which objective function is most suitable for modelling IHA changes?

Table 2 allows for assessing the performance of the NSE and the IHAs as objective functions for calibrating the hydrological model used for estimating the IHAs. The table shows, for each IHA, the percentage of cases in which the model calibrated with the corresponding IHA has a better performance than the model calibrated with the NSE. The performance is evaluated with respect to two criteria: the

Table 2 | Percentage of cases in which the model calibrated with the corresponding IHA has a better performance for reproducing changes in the IHA than the model calibrated with the NSE

	Direction of change					Magnitude of change				
Type of change	—	—	− +	+ +	+ ++	—	—	− +	+ +	+ ++
Q Jan	13	19	39	28	53	36	29	30	41	45
Q Jul	15	22	35	37	52	27	37	51	61	51
Q95	24	37	30	46	38	45	44	69	60	60
Q5	55	27	46	30	38	42	34	41	36	33
Q year	71	15	51	29	100	45	44	40	44	33
nHFP	9	17	44	54	46	13	20	41	63	53
nLFP	50	60	45	28	34	68	65	57	84	86
dHFP	29	34	59	68	61	32	35	48	68	65
dLFP	52	55	43	39	48	81	85	91	93	84
Reversals	65	58	60	76	79	64	56	46	74	80
DOYx	60	56	51	63	57	56	57	54	59	55
DOYn	40	46	50	67	62	42	47	61	63	58

Cells with white and grey background indicate where the NSE and the IHA achieved the best performance, respectively. Cells in light grey colour show no significant difference (p-value <0.05) according to the z-score of proportions test. The results are shown for five change percentiles. The first column shows the results for the largest negative changes, the central column for the smallest (negative and positive) changes and the last column for large positive changes.

agreement in the direction and magnitude of the modelled and observed IHA changes. The results are shown for five different types of changes, each comprising 20% of the total cases. The first (second) column shows the results for very large (large) negative changes, the central column shows the result for small changes (both negative and positive) and the last two columns show the results for large and very large positive changes. It is seen, for example, that all values for the January discharge (Q Jan) are below 50. This means that the model calibrated with the IHA tends to have a better performance than the model calibrated with the NSE in less than 50% of the cases.

There is a tendency for the objective function that is best for modelling the direction of change to have also the best performance for modelling the magnitude of change. It is further seen that for the direction of change there is a tendency of having values closer to 50 in the central columns and more extreme values (either closer to zero or closer to 100) towards the edges. This indicates that the differences between the two objective functions are less pronounced for smaller than for larger changes. This relationship is, however, less evident for the magnitude of change.

Changes in the IHAs describing flow magnitudes (i.e., the first five indicators in Table 1) are generally better modelled with parameter sets obtained with the NSE (rather than with the IHA) as objective function. This indicates that the focus of the NSE on achieving a good overall performance in reproducing the hydrograph makes this a good objective function for modelling changes in the annual flows, monthly flows and flow percentiles. On the other hand, it suggests that single hydrological indicators might not constrain well enough all parameters to which the indicators are sensitive.

For the indicators describing the pulses and reversals, no clear trend could be observed. Finally, the indicators describing the day of the year in which the maximum and minimum values are observed, tend to be better modelled when the calibration is done with the IHA. As the DOY is calculated using the discharge of only one day in each year, it is important that the model reproduces adequately the relative magnitudes of the peaks and low period flows. This seems to be best achieved by focusing just on the days with extreme values rather than with a model calibrated to reproduce adequately overall model performance.

How well can we estimate the direction of IHA changes?

Figure 3 informs about the agreement in direction between the observed and simulated changes for the IHAs calculated using the NSE as objective function. It can be seen that the direction of change can be best reproduced for the Q year indicator. The ability for reproducing the direction of change increases with the magnitude of the change until changes of about 1,500 mm. From there on, the ability of the model for identifying the direction of changes starts to decline and it is observed, for the largest changes, that the model is only capable of getting the direction of changes right about 50% of the time.

The patterns for the HFPs and LFPs are similar, although the direction of change of the LFPs is less likely to be correctly reproduced than the direction of change of the HFPs. It is further interesting to note that for the number of pulses there is a clear increase in the ability for modelling the direction of changes as the magnitude of the changes increases, while this is not the case for the duration of the pulses, where the ability for reproducing the direction seems to be independent of the magnitude of the change.

The size of the considered change has only a small impact on the ability for reproducing the indicators describing the timing of extreme flows (DOYx and DOYn). In agreement with the results shown for the pulses, which showed that difficulties for reproducing the direction of change were larger for the indicators describing low flows, it is seen here that the day of minimum discharge has a lower agreement in direction than the day of maximum discharge. Moreover, as the agreement in direction of the DOYn varies around 0.5, it can be concluded that the model has no skill for modelling changes of this indicator.

The impact of the objective function used for defining the best parameter set is observed in Figure 4, which shows the results when using the IHA as objective function. The most striking difference between these plots at the ones presented in Figure 3 is their asymmetry, which indicates that the results vary depending on the type of change (i.e., from high to low indicator values or vice versa), something which is not observed when the models are calibrated with the NSE.

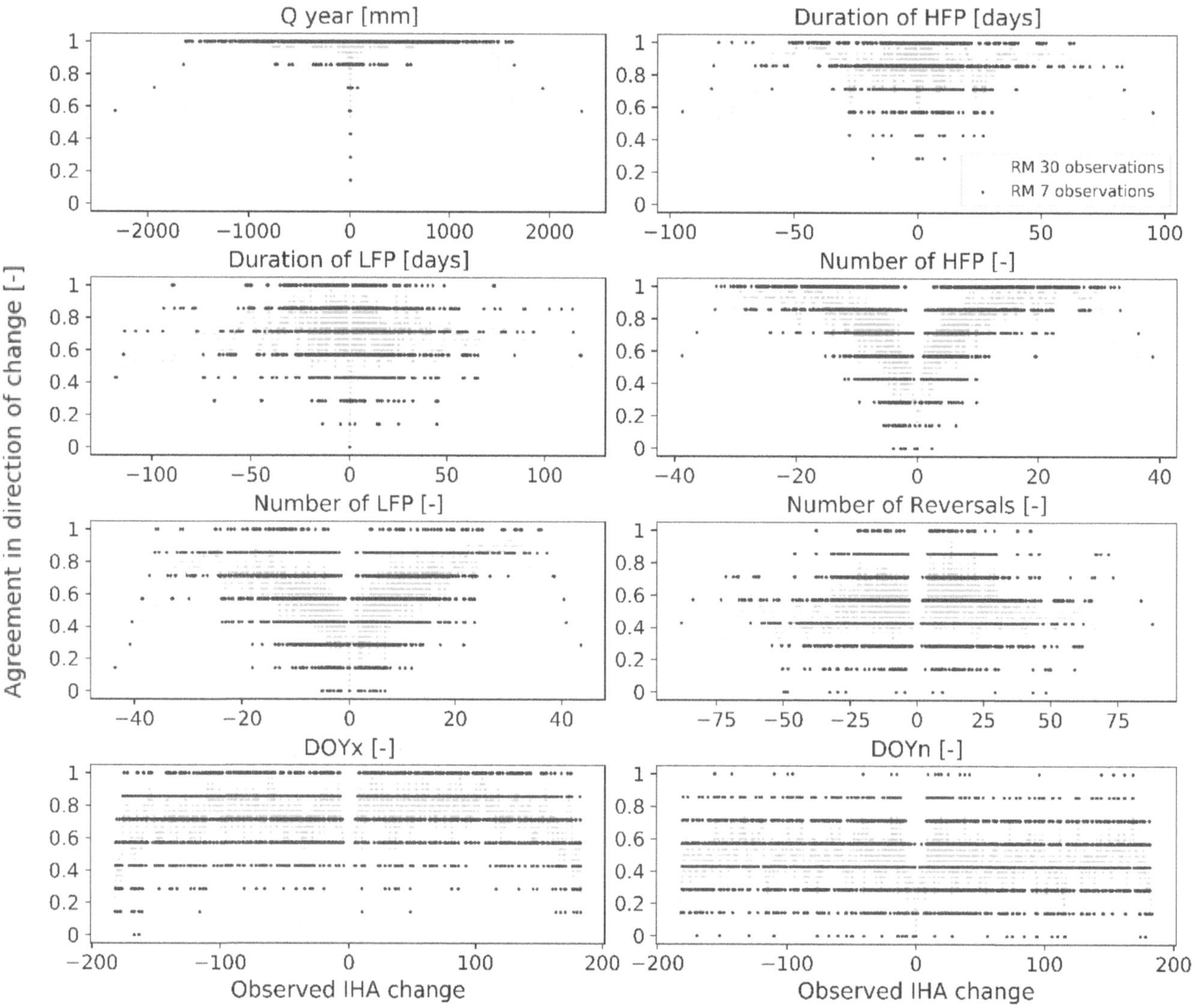

Figure 3 | Agreement in direction for all 12 cases in the 560 catchments using the optimal model run with respect to the NSE. The results are shown for the running mean of 30 (light colour) and 7 (dark colour) consecutive values. The *x*-axis shows the observed value of the corresponding change in IHA in the units of the IHA (specified in the title of each plot).

How well can we estimate the magnitude of IHA changes?

The first four rows of plots in Figure 5 show the absolute nominal error (NE) incurred when modelling changes in the IHAs with the model run optimizing the NSE. The plot for Q year is representative for all indicators describing the discharge magnitude. It shows an increase in the error as the modelled changes become larger. This increase is small for smaller changes but rises for more extreme changes. The error of the number of pulses and reversals shows an almost linear increase with little scatter as the magnitude of the change increases. The reason for this can be understood when looking at Figure 1, which shows only small differences in the simulated values of these IHAs regardless of the observed IHA values.

Besides an analysis of the NE, it is important to look at the relative error as it might be easier to grasp the relevance of an error when comparing its magnitude, for example, to the mean value of the indicator. Another advantage of using the relative error is that it facilitates comparisons of model performance between indicators. The last three rows of plots in Figure 5 show the relative error for some indicators. The relative error for the indicators describing

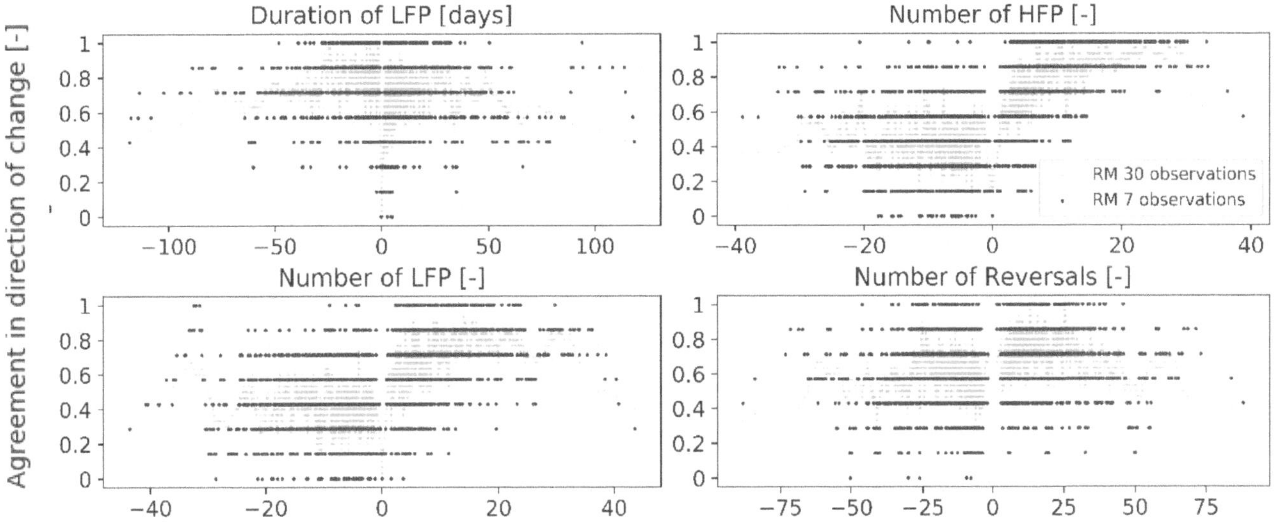

Figure 4 | Agreement in direction for all 12 cases in the 560 catchments using the optimal model run with respect to the IHA. The results are shown for the running mean of 30 (light colour) and 7 (dark colour) consecutive values. The *x*-axis shows the observed value of the corresponding change in IHA in the units of the IHA (specified in the title of each plot).

the discharge magnitude is high for small changes. For large changes, the errors are low and almost independent of the size of the modelled change. The REs for the number of pulses tend to increase as the modelled changes increase, which can again be explained by the little variability of the modelled IHA, even as the observed values increase (see Figure 1). The relative error of the timing IHAs is not considered as all catchments have the same 365 days and there is thus no need for normalizing the DOYx and DOYn.

Analysing the impact of the direction of change on the reproducibility of IHA changes

Sections 'How well can we estimate the direction of IHA changes?' and 'How well can we estimate the magnitude of IHA changes?' show how our ability for modelling IHA changes varies with the magnitude and direction of the considered IHA change. In most cases, it was possible to observe differences as a function of the magnitude of the change; the impact of the direction of change is, however, more subtle and requires additional analyses.

Table 3 shows the quotient between the mean absolute negative error and the mean absolute positive error for five change percentiles. For example, the percentiles considered for the largest changes were 0.1–1 (negative changes) and 99–99.9 (positive changes). The Q Jan indicator has an average error of 125.2 mm in the 0.1–1 percentile and an error of 63.5 mm in the 99–99.9 percentile (Table S4 in the Supplementary Material). The quotient of the negative and positive errors is thus 1.97 (Table 3), indicating that the error for negative changes is almost twice as large as the error for positive changes. Analogously, all other values larger than one in Table 3 indicate that the errors for negative changes are larger than the errors for positive changes, whereas this is the other way around when the values in Table 3 are smaller than one. Since negative changes are encountered when the indicator goes from high to lower values, values above one in Table 3 indicate that changes can be better modelled when they are from low to high indicator values.

For almost all considered indicators and change percentiles, the model is able to better model changes from low to high indicator values. These differences on performance as a function of the direction of the modelled change are more pronounced when the model is calibrated with respect to the IHA than to the IHAs. For the indicators describing flow magnitudes (i.e., Q year), this means that it is easier to model changes when models are calibrated in dryer periods and then run in wetter periods than the other way around. This is in agreement with previous studies (Vaze *et al.* 2010; Coron *et al.* 2012).

Besides analysing the impact of the direction of change on the mean error (Table 3), it was investigated how the

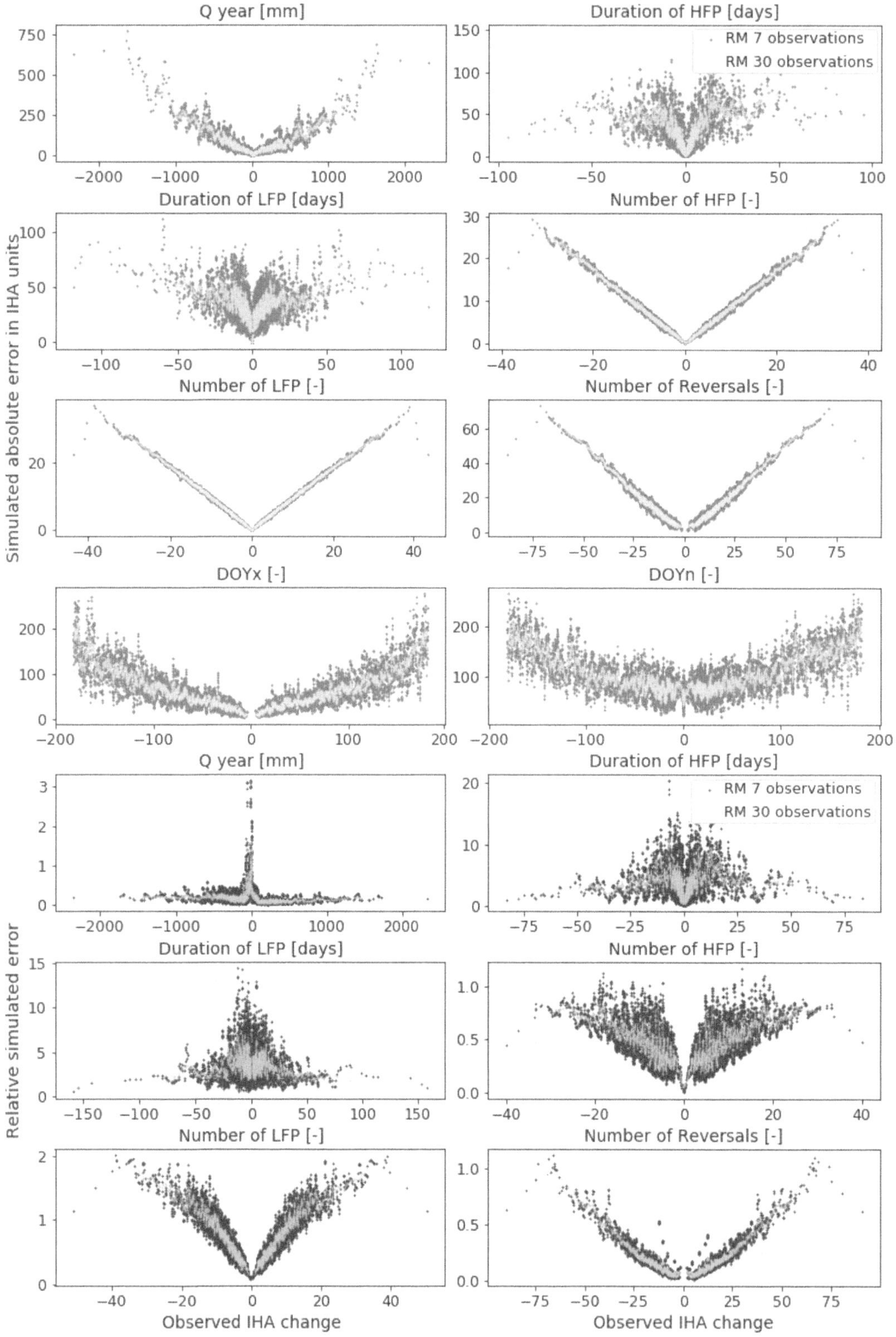

Figure 5 | Absolute nominal error (NE) in the first four rows and relative error (RE) in the last three rows for all 12 cases in the 560 catchments and using the optimal model run with respect to the NSE. The results are shown for the running mean of 30 (dark colour) and 7 (light colour) consecutive simulated errors. The *x*-axis shows the observed value of the corresponding change in IHA in the units of the IHA (specified in the title of each plot).

Table 3 | Quotient between mean absolute magnitude of the error for negative and their corresponding positive changes

	NSE calibration decreasing absolute size of the change →					**IHA calibration decreasing absolute size of the change →**				
Percentile quotient	0.1–1 / 99–99.9	1–15 / 85–99	15–25 / 75–85	25–35 / 65–75	35–45 / 55–65	0.1–1 / 99–99.9	1–15 / 85–99	15–25 / 75–85	25–35 / 65–75	35–45 / 55–65
Q Jan	1.97	1.51	1.23	1.15	1.01	2.57	1.63	1.23	1.27	1.94
Q Jul	0.85	1.32	1.27	1.14	1.10	2.65	1.88	1.94	1.74	1.75
Q95	0.92	1.12	1.11	1.02	1.09	1.26	1.50	1.35	1.43	1.71
Q5	1.34	1.74	1.41	1.29	1.19	1.77	1.23	1.15	1.23	1.25
Q year	1.12	1.78	1.34	1.25	1.19	1.49	1.12	1.16	1.20	1.28
nHFP	0.70	1.02	0.91	1.08	1.12	2.33	2.30	2.59	2.66	2.12
dHFP	1.05	1.03	1.03	1.15	0.92	1.22	1.34	1.44	1.48	1.50
nLFP	1.01	1.00	1.03	1.03	1.06	1.14	1.19	1.24	1.27	1.32
dLFP	1.01	1.01	1.01	1.04	1.00	1.04	1.10	1.14	1.20	1.19
Reversals	0.99	1.00	1.01	1.03	1.01	1.04	1.18	1.18	1.24	1.24

The results were estimated for five different change percentiles and for the NSE and IHA calibrated models.

direction of the change impacts our ability for reproducing the direction of change (Tables S2 and S3 in the Supplementary Material). There is almost no influence of the direction of change on our ability for identifying the direction of change if the NSE is used as objective function. The differences are more pronounced when the model is calibrated with the IHAs and when modelling the indicators describing the flow dynamics (i.e., pulses and reversals). It is interesting

Table 4 | Pearson's linear correlation coefficient between seven catchment descriptors (Wseas: Walsh's seasonality; D0max: maximum 1-day annual precipitation; slope: mean slope; aridity: quotient between precipitation and annual potential evapotranspiration; snow: percentage of precipitation falling as snow; BFI: baseflow index; RC: runoff coefficient) and the agreement in direction (AiD) and the nominal error (NE)

Variables		**Q Jan**	**Q Jul**	**Q95**	**Q5**	**Q year**	**nHFP**	**dHFP**	**nLFP**	**dLFP**	**Reversals**
AiD	Wseas	−0.17	−0.19	−0.19			−0.13	−0.18	−0.12		−0.21
	D0max	0.44	0.12	0.18			0.27	0.18	0.14		0.24
	Slope			0.14			−0.13				
	Aridity	−0.32	−0.31	−0.40	−0.14	−0.17	0.11	−0.13	−0.18	−0.13	−0.24
	Snow	−0.35	0.12				−0.45	−0.19			−0.15
	BFI	0.24	0.15	0.29		0.13	−0.23		0.15	0.12	0.18
	RC		0.25	0.26			−0.22				
NE	Wseas	0.28		−0.16	0.18	0.19	0.45	0.23	−0.47	−0.38	0.39
	D0max	0.44	0.13	0.32	0.37	0.39	−0.28	−0.18	0.39		−0.37
	Slope	0.42	0.26	0.32	0.32	0.37			−0.27	−0.15	
	Aridity	−0.47	−0.37	−0.49	−0.48	−0.40	0.70	0.13	−0.14		0.60
	Snow		0.39					0.20	−0.47		0.14
	BFI	0.70	0.49	0.56	0.64	0.63	−0.46	−0.11		−0.18	−0.43
	RC		0.18	0.41			−0.23		−0.20	0.25	

The cells without values had non-significant correlation coefficients (p-value = 0.05).

to see that changes in the HFPs are better modelled when they are from high to low flows, while changes in LFPs can be better modelled when they are from low to high values.

Identification of catchment descriptors influencing the quality of the IHA estimates

Table 4 shows the results of the correlation analysis between the catchment descriptors and two evaluation metrics (the agreement in direction and the NE). In general, there is a high agreement in the sign of the correlation coefficient observed for different IHAs. A higher agreement in the modelled direction of change is observed for wetter catchments (i.e., higher maximum 1-day precipitation and lower aridity), as well as for catchments with lower BFI. As previous studies have found that it is more difficult getting a good performance in more arid (van Esse *et al.* 2013; Weingartner *et al.* 2013) catchments and in catchments with higher BFI (Massmann 2020), these results indicate that the ability for getting the direction of the changes right is related to our ability of getting a good overall model performance.

The NE of the IHAs describing the magnitude of discharge has a negative correlation with aridity and a positive correlation with the maximum 1-day precipitation, indicating that dryer catchments, with lower discharges, tend to also have lower nominal errors. There is further a positive correlation with the BFI. The correlation for the other catchment descriptors is difficult to interpret and it might be more helpful to analyse the patterns visually. Figure 6 shows the spatial distribution of the average NE for each catchment and some representative IHAs. The IHAs describing the flow magnitude (Q year) tend to have low errors in the dryer central Great Plains and higher errors in the wet western coast. The number of reversals, on the other hand, has higher errors in dryer catchments and lower errors in the wetter ones. The error in the number of HFPs is low in the snowy Rocky Mountains area, while reaching high values in the eastern part characterized by a more homogeneous precipitation pattern across the year. The error in the number of LFPs is lowest in the western coast, which has a high seasonality in the precipitation and higher in the remaining United States. The errors in the duration of the pulses are highest in the central part and decrease towards the coasts. Finally, it is seen that changes in the day of the maximum discharge (DOYx) can be best modelled in areas with a large influence of snow and in the western coast, with clear seasonal precipitation patterns. The errors when modelling changes in the day on minimum discharge (DOYn) do not have a distinct spatial pattern.

Limitations of the study

Investigating the ability of hydrological models to model changes in IHAs is important in the context of climate change studies. This study sheds some light on the ability of hydrological models for modelling changes in IHAs, but the results cannot be used directly for inferring about the ability for modelling the impact of climate change as the link between IHAs and the climate dynamics (i.e., information about climate sensitivity) was not considered.

SUMMARY AND CONCLUSIONS

This study investigated the performance of a conceptual rainfall-runoff model for estimating changes in indicators of hydrological alteration.

- The results agreed with previous studies showing that indicators describing the magnitude of discharge are better modelled than indicators describing the number of pulses and reversals. An interesting finding in this context was that the difficulties for modelling the number of flow pulses and reversals cannot be attributed to calibration deficiencies, but that the model is unable to reach the observed values. This suggests that, instead of focusing on different calibration alternatives, we should investigate and address the causes of model failure. It would be further interesting to find out if this is a limitation of the specific rainfall-runoff model used in this study or a more general deficiency of the type of model (e.g., are distributed models able to achieve better results than lumped models?). Another question that could be investigated is if the quality of the IHAs estimates can be improved by redefining the indicators, for example, by constraining the number of considered pulses and reversals.

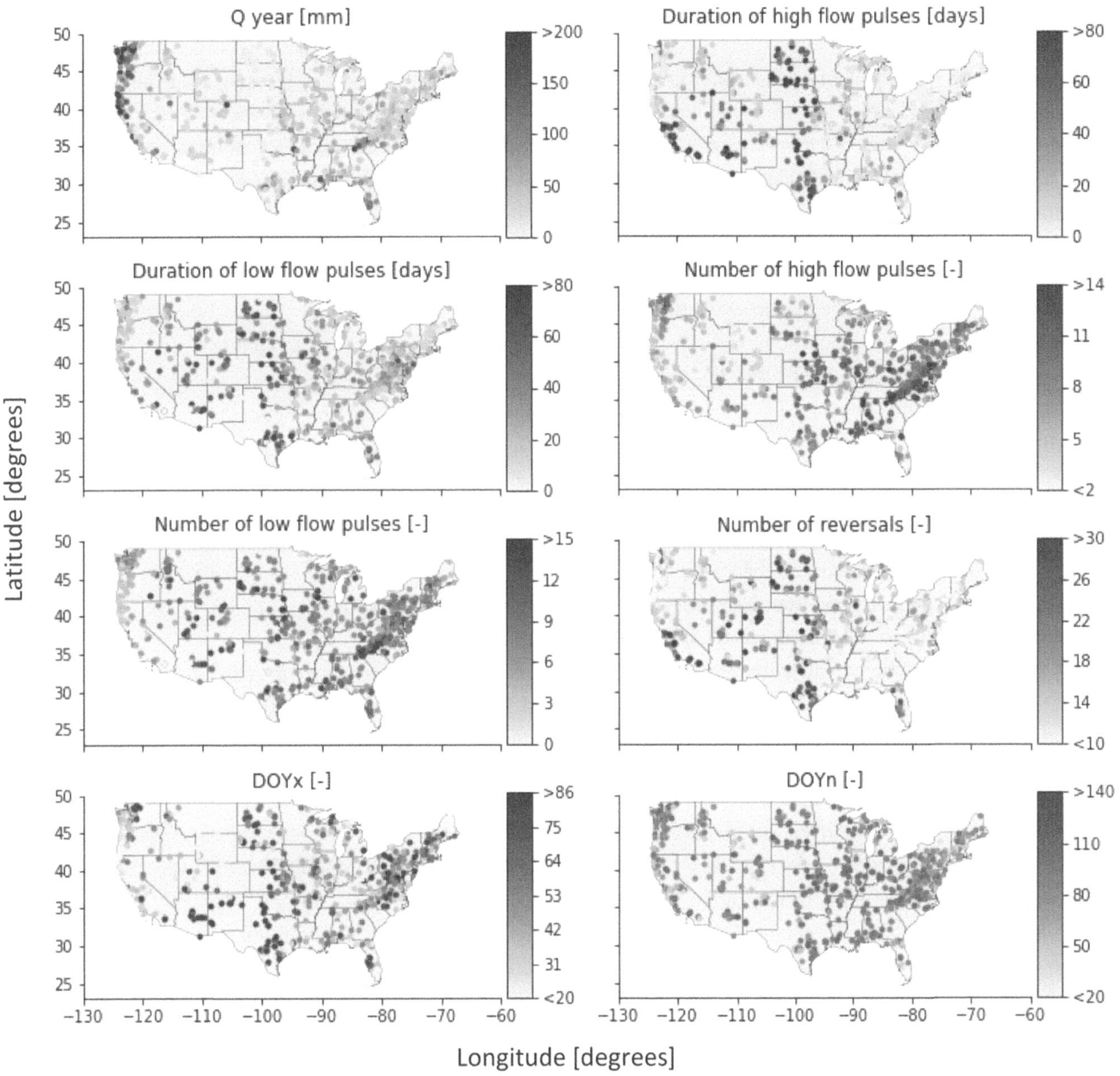

Figure 6 | Spatial distribution of the average nominal error (of all 12 cases) for each catchment and 8 representative IHAs.

- It was found that our ability for modelling changes in hydrological indicators is correlated with our ability for modelling the corresponding indicators.
- The quality of the estimates of IHA changes depends strongly on the considered indicator. The hydrological model was able to provide reliable estimates of the direction of the changes for the IHAs describing the discharge. While there are some difficulties in estimating the direction of the change when changes are small, the model achieves good performances as the magnitude of the changes increases. On the contrary, the model was not able to correctly estimate the direction of the changes for the duration of the pulses, the number of reversals and the day on minimum discharge, independently of the size of the considered change. For the number of pulses, the model had some skill for estimating the direction of

large changes, but the estimates for smaller changes were unreliable.

- The magnitude of changes might be adequately estimated across the entire range of variation of the changes for indicators describing the discharge magnitude. For pulses and reversals, on the other hand, there is a clear reduction in performance with increasing magnitude of the considered changes, decreasing the potential applicability of this approach for estimating the impacts of climate change on the discharge dynamics. These results do not point to a shortcoming of the models for general hydrology, but to limitations when used for modelling hydrologic indicators under change. The results highlight, furthermore, the need for developing alternative methodologies for understanding how climate change could affect riverine ecosystems.
- Regarding the influence of the objective function used for calibrating the hydrological model on the ability of reproducing changes in the IHAs, it was observed that the results vary depending on the considered IHA. For indicators describing discharge magnitude, the results tended to be better when the NSE was used as objective function. For flow pulses, reversals and timing variables, on the other hand, the magnitude of the error tended to be lower when the IHA was used for calibrating the model, while there seemed to be no clear tendency when estimating the direction of the changes.
- An analysis of differences in the quality of the IHA estimates depending on the direction of the changes showed that changes from low to high indicator values could be better modelled than changes from high to low values. This effect was more pronounced when the models were calibrated with respect to the IHAs. This has some implications when modelling the impacts of climate change as we will calibrate our models for the current climate and the use for estimating the discharge for dryer conditions. Since it was found that these changes (wet to dry) are more difficult to get right, it is important to avoid overestimating our ability for modelling dryer periods.
- An exception is found for the agreement in direction for the indicators describing HFPs and calibrated with the IHA, where the direction of changes could be best predicted for changes from high to low indicator values.
- An important strength of this study is that it relies on many catchments and that it considers different types of changes (with respect to the magnitude and direction of the changes). Such a stratification of the changes allows for additional insight. For example, we found that models are able to model adequately low IHA values of the indicators describing the discharge dynamics, but have problems in reproducing high values.

DATA AVAILABILITY STATEMENT

The primary data used in the study are publicly available from U.S. organizations. More detailed information about how it can be accessed is found in the text. Processed data are available from the author upon request.

ACKNOWLEDGEMENTS

Three anonymous reviewers provided numerous comments and suggestions that helped to greatly improve the manuscript. This research was funded by the Austrian Science Fund (FWF): project J-3689-N29.

SUPPLEMENTARY MATERIAL

The Supplementary Material for this paper is available online at https://dx.doi.org/10.2166/nh.2020.073.

REFERENCES

Acreman, M., Arthington, A. H., Colloff, M. J., Couch, C., Crossman, N. D., Dyer, F., Overton, I., Pollino, C. A., Stewardson, M. J. & Young, W. 2014a Environmental flows for natural, hybrid, and novel riverine ecosystems in a changing world. *Frontiers in Ecology and the Environment* **12** (8), 466–473. doi:10.1890/130134.

Acreman, M. C., Overton, I. C., King, J., Wood, P. J., Cowx, I. G., Dunbar, M. J., Kendy, E. & Young, W. J. 2014b The changing role of ecohydrological science in guiding environmental flows. *Hydrological Sciences Journal* **59**, 3–4, 433–450. doi: 10.1080/02626667.2014.886019.

Atkinson, S. E., Woods, R. A. & Sivapalan, M. 2002 Climate and landscape controls on water balance model complexity over changing timescales. *Water Resources Research* **38** (12). doi:10.1029/2002WR001487.

Bai, Y., Wagener, T. & Reed, P. 2009 A top-down framework for watershed model evaluation and selection under uncertainty. *Environmental Modelling and Software* **24**, 901–916. doi:10.1016/j.envsoft.2008.12.012.

Brigode, P., Oudin, L. & Perrin, C. 2013 Hydrological model parameter instability: a source of additional uncertainty in estimating the hydrological impacts of climate change? *Journal of Hydrology* **476**, 410–425. doi:10.1016/j.jhydrol.2012.11.012.

Caldwell, P. V., Kennen, J. G., Sun, G., Kiang, J. E., Butcher, J. B., Eddy, M. C., Hay, L. E., LaFontaine, J. H., Hain, E. F., Nelson, S. A. C. & McNulty, S. G. 2015 A comparison of hydrologic models for ecological flows and water availability. *Ecohydrology* **8**, 1525–1546. doi: 10.1002/eco.1602.

Carlisle, D. M., Falcone, J., Wolock, D. M., Meador, M. R. & Norris, R. H. 2010 Predicting the natural flow regime: models for assessing hydrological alteration in streams. *River Research and Applications* **26**, 118–136. doi:10.1002/rra.1247.

Carlisle, D. M., Grantham, T. E., Eng, K. & Wolock, D. M. 2017 Biological relevance of streamflow metrics: regional and national perspectives. *Freshwater Science* **36** (4), 927–940. doi:10.1086/694913.

Clark, M. P., Slater, A. G., Rupp, D. E., Woods, R. A., Vrugt, J. A., Gupta, H. V., Wagener, T. & Hay, L. E. 2008 Framework for understanding structural errors (FUSE): a modular framework to diagnose differences between hydrological models. *Water Resources Research* **44** (12). doi:10.1029/2007WR006735.

Colby, B. R. 1956 *Relationship of Sediment Discharge to Streamflow*. USGS Open-File Report 56-27, U.S. Dept. of the Interior, Geological Survey, Water Resources Division.

Coron, L., Andréassian, V., Perrin, C., Lerat, J., Vaze, J., Bourqui, M. & Hendrickx, F. 2012 Crash testing hydrological models in contrasted climate conditions: an experiment on 216 Australian catchments. *Water Resources Research* **48**, W05552. doi: 10.1029/2011WR011721.

Chopra, K., Leemans, R., Kumar, P. & Simons, H. (Eds.) 2005 *Ecosystems and Human Well-Being: Policy Responses*. Findings of the Condition and Trends Working Group of the Millennium Assessment. Island Press, p. 654.

Dakhlaoui, H., Ruelland, D., Tramblay, Y. & Bargaoui, Z. 2017 Evaluating the robustness of conceptual rainfall-runoff models under climate variability in northern Tunisia. *Journal of Hydrology* **550**, 201–217. doi:10.1016/j.jhydrol.2017.04.032.

Falcone, J. A. 2011 *GAGES-II: Geospatial Attributes of Gages for Evaluating Streamflow*. USGS Report. doi: 10.3133/70046617.

Farmer, D., Sivapalan, M. & Jothityangkoon, C. 2003 Climate, soil, and vegetation controls upon the variability of water balance in temperate and semiarid landscapes: downward approach to water balance analysis. *Water Resources Research* **39** (2), 1035. doi:10.1029/2001WR000328.

Fenicia, F., Kavetski, D. & Savenije, H. H. G. 2011 Elements of a flexible approach for conceptual hydrological modeling: 1. Motivation and theoretical development. *Water Resources Research* **47**, 11. doi:10.1029/2010WR010174.

Fowler, K., Peel, M., Western, A. & Zhang, L. 2018 Improved rainfall-runoff calibration for drying climate: choice of objective function. *Water Resources Research* **54** (5), 3392–3408. doi: 10.1029/2017WR022466.

Gupta, A. & Govindaraju, R. S. 2019 Propagation of structural uncertainty in watershed hydrologic models. *Journal of Hydrology* **575**, 66–81. doi:10.1016/j.jhydrol.2019.05.026.

Hargraeves, G. H. & Samani, Z. A. 1982 Estimating potential evapotranspiration. *Journal of the Irrigation and Drainage Division* **108** (3), 225–230.

Hernandez-Suarez, J. S., Nejadhashemi, A. P., Kropp, I. M., Abouali, M., Zhang, Z. & Deb, K. 2018 Evaluation of the impacts of hydrologic model calibration methods on predictability of ecologically-relevant hydrologic indices. *Journal of Hydrology* **564**, 758–772. doi:10.1016/j.jhydrol.2018.07.056.

Jothityangkoon, C., Sivapalan, M. & Farmer, D. L. 2001 Process controls on water balance variability in a large semi-arid catchment: downward approach to hydrologic model development. *Journal of Hydrology* **254** (1–4), 174–198. doi:10.1016/S0022-1694(01)00496-6.

Kiesel, J., Guse, B., Pfannerstill, M., Kakouei, K., Jähnig, S. C. & Fohrer, N. 2017 Improving hydrological model optimization for riverine species. *Ecological Indicators* **80**, 376–385. doi:10.1016/j.ecolind.2017.04.032.

Klemeš, V. 1986 Operational testing of hydrological simulation models. *Hydrological Sciences Journal* **31** (1), 13–24. doi:10.1080/02626668609491024.

Knight, R. R., Gain, W. S. & Wolfe, W. J. 2012 Modelling ecological flow regime: an example from the Tennessee and Cumberland River basins. *Ecohydrology* **5**, 613–627. doi:10.1002/eco.246.

Lane, R. A., Coxon, G., Freer, J. E., Wagener, T., Johnes, P. J., Bloomfield, J. P., Greene, S., Macleod, C. J. A. & Reaney, S. M. 2019 Benchmarking the predictive capability of hydrological models for river flow and flood peak predictions across over 1000 catchments in Great Britain. *Hydrology and Earth System Sciences* **23**, 4011–4032. doi:10.5194/hess-23-4011-2019.

Massmann, C. 2019 Modelling snowmelt in ungauged catchments. *Water* **11** (2), 301. doi:10.3390/w11020301.

Massmann, C. 2020 Identification of factors influencing hydrologic model performance using a top-down approach in a large number of U.S. catchments. *Hydrological Processes* **34**, 4–20. doi:10.1002/hyp.13566.

Murphy, J. C., Knight, R. R., Wolfe, W. J. & Gain, W. S. 2013 Predicting ecological flow regime at ungauged sites: a comparison of methods. *River Research and Applications* **29** (5), 660–669. doi:10.1002/rra.2570.

Olsen, M., Troldborg, L., Henriksen, H. J., Conallin, J., Refsgaard, J. C. & Boegh, E. 2013 Evaluation of a typical hydrological model in relation to environmental flows. *Journal of Hydrology* **507**, 52–62. doi: 10.1016/j.jhydrol.2013.10.022.

Patil, S. D. & Stieglitz, M. 2015 Comparing spatial and temporal transferability of hydrological model parameters. *Journal of Hydrology* **525**, 409–417. doi:10.1016/j.jhydrol.2015.04.003.

Poff, N. L., Allan, J. D., Bain, M. B., Karr, J. R., Prestegaard, K. L., Richter, B. D., Sparks, R. E. & Stromberg, J. C. 1997 The natural flow regime. *BioScience* **47** (11), 769–784. doi:10.2307/1313099.

Pool, S., Vis, M. J. P., Knight, R. R. & Seibert, J. 2017 Streamflow characteristics from modeled runoff time series – importance of calibration criteria selection. *Hydrology and Earth System Sciences* **21**, 5443–5457. doi:10.5194/hess-21-5443-2017.

Richter, B. D., Baumgartner, J. V., Powell, J. & Braun, D. P. 1996 A method for assessing hydrologic alteration within ecosystems. *Conservation Biology* **10**, 1163–1174. doi:10.1046/j.1523-1739.1996.10041163.x.

Sanborn, S. C. & Bledsoe, B. P. 2006 Predicting streamflow regime metrics for ungauged streams in Colorado, Washington, and Oregon. *Journal of Hydrology* **325** (1–4), 241–261. doi:10.1016/j.jhydrol.2005.10.018.

Seiller, G., Anctil, F. & Perrin, C. 2012 Multimodel evaluation of twenty lumped hydrological models under contrasted climate conditions. *Hydrology and Earth System Sciences* **16** (4), 1171–1189. doi:10.5194/hess-16-1171-2012.

Shrestha, R. R., Peters, D. L. & Schnorbus, M. A. 2014 Evaluating the ability of a hydrologic model to replicate hydro-ecologically relevant indicators. *Hydrological Processes* **28** (14), 4294–4310. doi: 10.1002/hyp.9997.

Shrestha, R. R., Schnorbus, M. A. & Peters, D. L. 2016 Assessment of a hydrologic model's reliability in simulating flow regime alterations in a changing climate. *Hydrological Processes* **30** (15), 2628–2643. doi: 10.1002/hyp.10812.

Sinokrot, B. A. & Gulliver, J. S. 2000 In-stream flow impact on river water temperatures. *Journal of Hydraulic Research* **38** (5), 339–349. doi:10.1080/00221680009498315.

Thornton, P. E., Running, S. W. & White, M. A. 1997 Generating surfaces of daily meteorological variables over large regions of complex terrain. *Journal of Hydrology* **190**, 214–251. doi:10.1016/S0022-1694(96)03128-9.

Thornton, P. E., Thornton, M. M., Mayer, B. W., Wilhelmi, N., Wei, Y., Devarakonda, R. & Cook, R. B. 2014 *Daymet: Daily Surface Weather Data on A 1-km Grid for North America, Version 2*. ORNL DAAC, Oak Ridge, Tennessee, USA (accessed June 2015). doi:10.3334/ORNLDAAC/1219.

van Esse, W. R., Perrin, C., Booij, M. J., Augustijn, D. C. M., Fenicia, F., Kavetski, D. & Lobligeois, F. 2013 The influence of conceptual model structure on model performance: a comparative study for 237 study catchments. *Hydrology and Earth System Sciences* **17**, 4227–4239. doi:10.5194/hess-17-4227-2013.

Vaze, J., Post, D. A., Chiew, F. H. S., Perraud, J.-M., Viney, N. R. & Teng, J. 2010 Climate non-stationarity – validity of calibrated rainfall–runoff models for use in climate change studies. *Journal of Hydrology* **394** (3–4), 447–457. doi:10.1016/j.jhydrol.2010.09.018.

Vis, M., Knight, R., Pool, S., Wolfe, W. & Seibert, J. 2015 Model calibration criteria for estimating ecological flow characteristics. *Water* **7**, 2358–2381. doi: 10.3390/w7052358.

Vormoor, K., Heistermann, M., Bronstert, A. & Lawrence, D. 2018 Hydrological model parameter (in)stability – 'crash testing' the HBV model under contrasting flood seasonality conditions. *Hydrological Sciences Journal* **63** (7), 991–1007. doi:10.1080/02626667.2018.1466056.

Walsh, R. P. D. & Lawler, D. M. 1981 Rainfall seasonality: description, spatial patterns and change through time. *Weather* **36**, 201–208. doi:10.1002/j.1477-8696.1981.tb05400.x.

Weingartner, R., Blöschl, G., Hannah, D. M., Marks, D. G., Parajka, J., Pearson, C. S., Rogger, M., Salinas, J. L., Sauquet, E., Srikanthan, R., Thompson, S. E. & Viglione, A. 2013 Prediction of seasonal runoff in ungauged basins. In: *Runoff Prediction in Ungauged Basins. Synthesis Across Processes, Places and Scales* (G. Blöschl, M. Sivapalan, T. Wagener, A. Viglione & H. Savenije, eds). Cambridge University Press, pp. 102–134.

Wolock, D. M. 2003 *Base-Flow Index Grid for the Conterminous United States*. USGS Report 2003-263. Available from: http://pubs.er.usgs.gov/publication/ofr03263

First received 16 May 2019; accepted in revised form 26 January 2020. Available online 30 March 2020

Uncertainty of annual runoff projections in Lithuanian rivers under a future climate

V. Akstinas, D. Jakimavičius, D. Meilutytė-Lukauskienė, J. Kriaučiūnienė and D. Šarauskienė

ABSTRACT

Uncertainties of runoff projections arise from different sources of origin, such as climate scenarios (RCPs), global climate models (GCMs) and statistical downscaling (SD) methods. Assessment of uncertainties related to the mentioned sources was carried out for selected rivers of Lithuania (Minija, Nevėžis and Šventoji). These rivers reflect conditions of different hydrological regions (western, central and southeastern). Using HBV software, hydrological models were created for river runoff projections in the near (2021–2040) and far (2081–2100) future. The runoff projections according to three RCP scenarios, three GCMs and three SD methods were created. In the Western hydrological region represented by the Minija River, the GCMs were the most dominant uncertainty source (41.0–44.5%) in the runoff projections. Meanwhile, uncertainties of runoff projections from central (Nevėžis River) and southeastern (Šventoji River) regions of Lithuania were related to SD methods and the range of uncertainties fluctuates from 39.4% to 60.9%. In western Lithuania, the main source of rivers' supply is precipitation, where projections highly depend on selected GCMs. The rivers from central and southeastern regions are more sensitive to the SD methods, which not always precisely adjust the meteorological variables from a large grid cell of GCM into catchment scale.

Key words | climate change, GCM, RCP, statistical downscaling, uncertainty analysis

V. Akstinas (corresponding author)
D. Jakimavičius
D. Meilutytė-Lukauskienė
J. Kriaučiūnienė
D. Šarauskienė
Lithuanian Energy Institute,
Breslaujos str. 3, LT-44403 Kaunas,
Lithuania
E-mail: *vytautas.akstinas@lei.lt*

INTRODUCTION

The accuracy of runoff projections highly depends on a wide range of factors related to climate change. Application of different climate scenarios and modelling tools for calculation of runoff projections increases the spread in the ensemble. When projecting river runoff, it is important to assess the uncertainties of selected tools and input data. Usually, the main sources of uncertainty are linked to global climate models (GCMs) and climate scenarios (RCPs). However, statistical downscaling (SD) methods can be regarded as an additional source of uncertainty as well. The GCM in combination with RCP provides the basis for investigation of future climate change. On the other hand, they are also the primary sources of systematic errors.

There are large biases comparing GCM output data with historical observations. Therefore, SD methods are used for the reduction of mentioned biases. Latif (2011) maintains that the primary uncertainty of projections is caused by the variability of natural hydro-meteorological processes. It is difficult to estimate such natural variability; hence, the assessment of uncertainties of GCMs is very important. The uncertainty interpretation as the range of runoff projection was successfully applied in several studies (Dobler *et al.* 2012; Bosshard *et al.* 2013). These studies constitute a solid basis for the exploration of uncertainties in runoff projections. The mentioned studies were conducted in a variety of locations using different climate and hydrological

doi: 10.2166/nh.2019.004

models and considered many hydrological parameters. This diversity of results provides in-depth analyses but makes it difficult to compare them as well. Therefore, the discussion about uncertainties related to climate change focused on climate sensitivity, which highly depends on the chosen GCM (Ahlström *et al.* 2013). GCMs are limited by the inherent simplifications of some processes in Earth's climate system. Accordingly, the model outputs involve different kinds of biases when comparing them to the observed climate variables (Räty *et al.* 2014) and the combinations of GCMs and RCP scenarios became the primary sources of climate projection uncertainty.

Addor *et al.* (2014) considered and systematically analysed a large number of uncertainty sources, which were estimated from simulations of future runoff projections in different Swiss catchments. This study revealed that GCMs and RCMs are usually the main sources of uncertainty and the uncertainty originating from the hydrological models dominated in the catchments, where the feeding sources of snow and ice melt is substantial. The relative contributions of four uncertainty sources (emissions scenarios, GCMs, local adjustment methods and HBV parameterisation) in hydrological projections of four catchments in Norway were discussed by Lawrence & Haddeland (2011). The results demonstrated that all of these sources can significantly contribute to the dispersion of projections of the mean annual flood. The importance of individual factors varied between catchments. It has been demonstrated that the selection of a GCM largely determines the variability in runoff projections. To identify long-term runoff changes, it is important to assess the uncertainties of the GCM in the historical period (Chen *et al.* 2017; Shen *et al.* 2018). Kundzewicz *et al.* (2018) proposed four measures for uncertainty reduction: increase of data reliability, reduction of uncertainties of GCMs, integration of regional climate models and hydrological models as well as solutions to optimise hydrological modelling.

There have been many studies where SD methods were applied for corrections of climate model outputs using observation data, where biases of data series of air temperature and precipitation for future periods are generally reduced (Hagemann *et al.* 2011; Räty *et al.* 2014). The application of SD methods helps to correct projections of meteorological variables. This way, the projections of surface runoff and river hydrological regime in impact assessment studies can be improved (Hagemann *et al.* 2011; Hundecha *et al.* 2016). Some studies also analysed the advantages and disadvantages of different SD approaches (Teutschbein & Seibert 2013; Maraun 2016). The mentioned scientific studies can be used for the selection of SD methods to improve GCM outputs for a fine temporal and spatial scale.

In Lithuania, uncertainty analysis of river runoff projections is not widely discussed. Kriaučiūnienė *et al.* (2013) assessed uncertainties of runoff projections made according to GCMs (ECHAM5 and HadCM3), SRES group emission scenarios (A2, A1B and B1) and calibration parameters of HBV model and established that the largest uncertainties were related to emission scenarios. Keršytė *et al.* (2015) evaluated the output of 24 GCMs of CMIP5 project which were simulated under all RCP climate scenarios. According to GCM outputs, some GCMs (GFDL-CM3, HadGEM2-ES and NorESM1-M) were selected as the best fit to reflect the local climate conditions of Lithuania. According to the mentioned scenarios and GCMs, the projections of runoff of the selected catchments of the Nemunas River basin were modelled (Stonevičius *et al.* 2017; Šarauskienė *et al.* 2018). Also, there are two European-scale studies, where alternative downscaling methods for bias corrections of precipitation and runoff projections in 11 river catchments from nine countries of Europe (including one in Lithuania – the Merkys River catchment) were chosen and discussed (Sunyer *et al.* 2015; Hundecha *et al.* 2016). However, these previously mentioned studies did not take into account the accuracy of runoff projections related to the potential uncertainty sources. The potential impact of SD methods on the correction of biases of GCM output has not been sufficiently investigated as well. Therefore, this research focuses on the evaluation of uncertainties of runoff projections according to climate scenarios, GCMs and SD methods in Lithuanian rivers from different hydrological regions in the near and far future. The evaluation of the accuracy of runoff projections will help to identify the uncertainty sources which have the most significant influence on the final results. Respectively, it will provide an opportunity to select more precise GCMs, climate scenarios and downscaling methods for accurate projections of annual runoff.

STUDY AREA AND DATA

The Nemunas River is a major Lithuanian river. The total length of the Nemunas is 937 km, while the river's basin area covers 98,200 km^2. Seventy-two per cent of Lithuanian territory falls within the Nemunas River basin. Lithuania falls within one climate zone. When the climate is homogeneous, the physico-geographical conditions have a larger influence on the formation of the rivers' runoff. Accordingly, the division into hydrological regions is done by the existing local physico-geographical conditions (relief, lithology, soils, land use, etc.), which differently transform precipitation into the surface and subsurface runoff.

Three river catchments (Minija – 2,942 km^2, Nevėžis – 6,140 km^2 and Šventoji – 6,888 km^2) were selected for this research. These catchments are from different hydrological regions of Lithuania (Western (LT-W), Central (LT-C) and Southeastern (LT-SE)) (Figure 1). The main source of runoff generation in western Lithuania is precipitation. The type of runoff generation in central Lithuania is mixed (snowmelt and rainfall). In southeastern Lithuania, the main feeding source is groundwater. Due to the previously mentioned physico-geographical factors and runoff generation patterns, the Lithuanian rivers from the same hydrological region have synchronic relations of the runoff.

The selected rivers are represented by the water gauging stations (WGS) of Kartena WGS (Minija River), Dasiūnai WGS (Nevėžis River) and Ukmergė WGS (Šventoji River). Runoff projections of the mentioned rivers were carried out for the near and far future. Nine meteorological stations (MSs) were selected for hydrological modelling of selected rivers (Figure 1). The weight of each MS was determined using the Thiessen polygon method for hydrological modelling in selected catchments. The reference periods of 1986–2005 was used for calibration (1986–1995) and validation (1996–2005). Therefore, the daily observations of the average air temperature (T, °C) and daily precipitation amount (P, mm) of

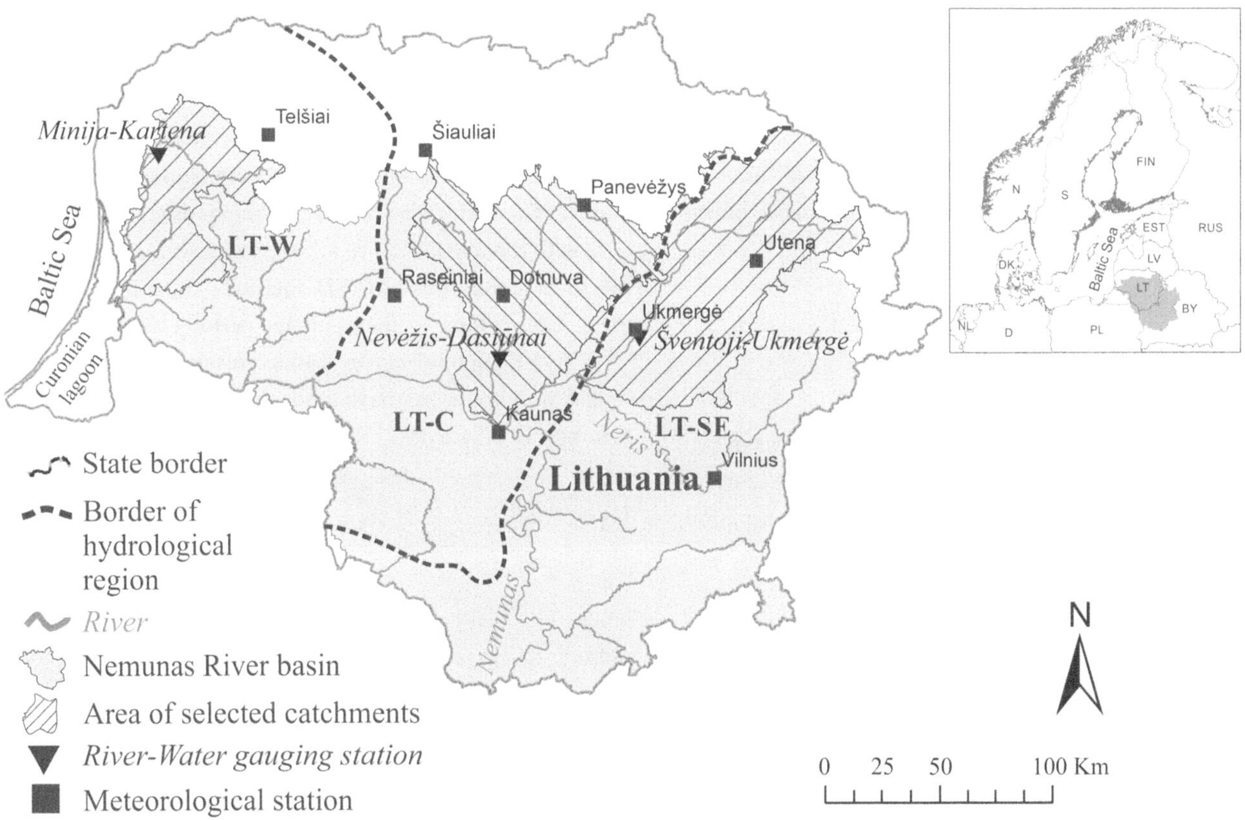

Figure 1 | The location of selected river catchments, water gauging stations, and MSs.

MSs, as well as daily discharge (Q, m^3/s) of selected WGSs, were used for this analysis.

The output (daily air temperature and daily precipitation amount) of three GCMs (GFDL-CM3, HadGEM2-ES and NorESM1-M) of the CMIP5 project generated by three RCP climate scenarios (RCP2.6, RCP4.5 and RCP8.5) were used for projecting the annual river runoff of the selected Lithuanian catchments in the 21st century. The raw data (air temperature and precipitation) of simulations of the reference period as well as projections (according to RCP2.6, RCP4.5 and RCP8.5) of the near future and far future have been taken from the NOAA (National Oceanic and Atmospheric Administration) GFDL (Geophysical Fluid Dynamics Laboratory) and WDCC (World Data Center for Climate) CERA data portals. Three GCMs (GFDL-CM3, NorESM1-M and HadGEM2-ES) from the mentioned databases with different spatial resolution were selected (Table 1). These GCMs were selected as the best fit (according to the median and range of selected meteorological variables of raw GCM output data) for climatic conditions of Lithuania (Keršytė *et al.* 2015) and reflect the uncertainty of the selected ensemble.

MSs located in different grid cells of the GCM were selected (Figure 2). Respectively, all historical observations of MSs (which coincide with particular grid cells) were used for SD methods to correct the systematic biases of GCM output in the reference period as well as biases of projections in the future.

METHODS

For evaluation of possible patterns and uncertainties of projections of river runoff in the near and far future according to observation data and available geographic information, hydrological models of the selected rivers were created (Figure 3). The general procedure used was as follows: the output data (T, P) of GCMs of GFDL-CM3, HadGEM2-ES and NorESM1-M according to RCP (2.6, 4.5 and 8.5) climate scenarios were adjusted to Lithuanian conditions by applying SD methods of bias correction (BC) with variable, change factor (CF) with variable and quantile mapping (QM). Applying the HBV software (Lindström *et al.* 1997), the resulting corrected data of T and P series were used to simulate projections of daily discharge in the near (2021–2040) and far future (2081–2100). The simulated values were compared to the values of the reference period (1986–2005) and the uncertainties of runoff projections were calculated according to the used uncertainty sources (RCPs, GCMs and SD methods). These steps in the procedure are described in detail in the following paragraphs.

In the periods of 2021–2040 (near future) and 2081–2100 (far future), projections of daily data of precipitation and temperature were adjusted using three different SD methods – BC, CF and QM. According to Sunyer *et al.* (2015), the selection of SD methods requires chosing the methods based on different underlying assumptions as well as including the change in mean and variance. Therefore, the well-known and widely applied SD methods were used in this research. Also, they can be flexibly used for adjustment of the several meteorological variables, such as precipitation and air temperature. The major idea of selected methods is to downscale data with low resolution to a fine spatial scale to reproduce the local conditions. All SD methods were trained with local observations for the reference period (1986–2005).

The BC method corrects the projected raw daily data of GCM outputs in mean and variance (Ho *et al.* 2012; Hawkins *et al.* 2013):

$$V_{BC}(t) = \overline{O_{REF}} + \frac{\sigma_{O,REF}}{\sigma_{V,REF}}(V_{RAW}(t) - \overline{V_{REF}}) \quad (1)$$

where V_{BC} is a corrected meteorological variable of GCM output, O_{REF} is observation in the historical reference period, V_{REF} is a meteorological variable of GCM output from the reference period, P_{RAW} is a meteorological variable of raw GCM output for the future period. The time mean is denoted by the bar above a symbol. Equation (1) was used to represent the relationship between distribution of O_{REF}

Table 1 | The spatial resolution of selected GCMs

No.	GCM	Abbreviation	Grid resolution	
			Longitude	Latitude
1.	GFDL-CM3	GFDL	1.5°	2°
2.	HadGEM2-ES	Had	1.875°	1.25°
3.	NorESM1-M	Nor	2.5°	1.895°

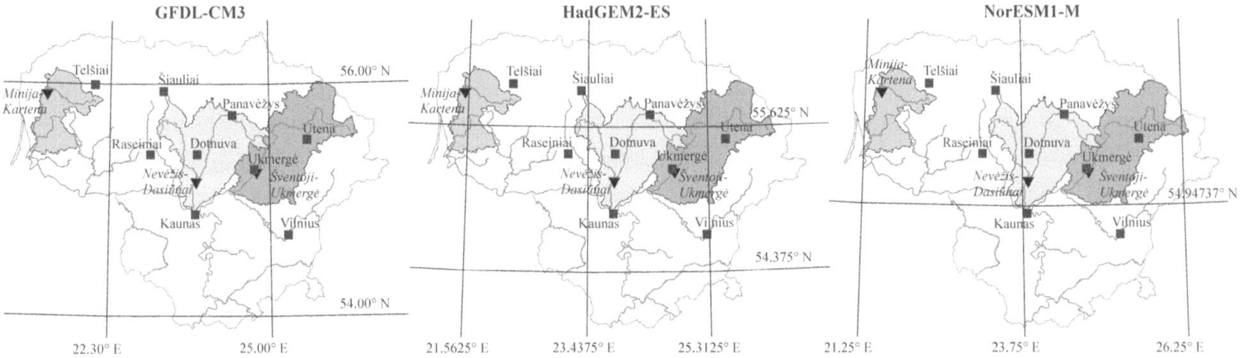

Figure 2 | Location of the MSs, selected river catchments, and spatial distribution of the grid cells of selected GCMs over Lithuania.

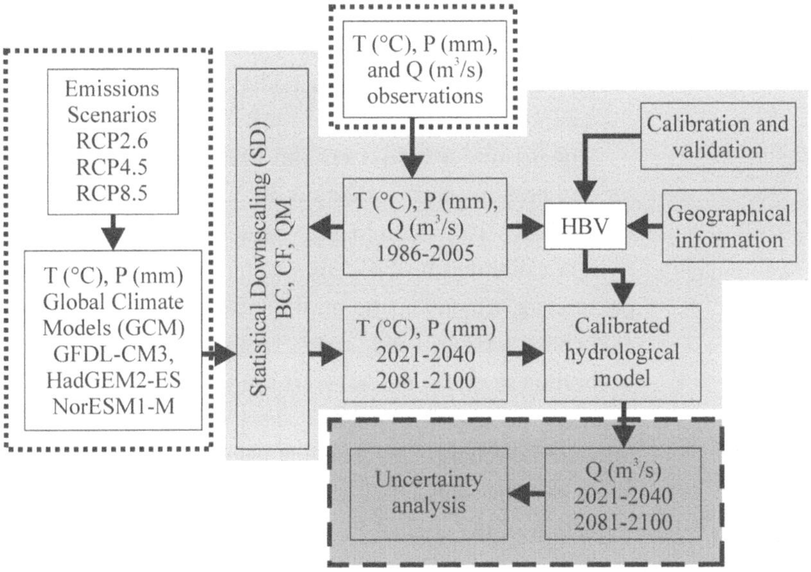

Figure 3 | The workflow of the uncertainty analysis of runoff projections related to selected uncertainty sources.

(observations in the reference period) and the distribution of V_{REF} (GCM simulations in the reference period), therefore $\sigma_{O,REF}$ and $\sigma_{V,REF}$ are standard deviations of daily observations and meteorological variable of GCM output in the reference period, respectively.

The CF method adjusts the observed variables according to the differences between projected variables of GCM output and simulated GCM output from the reference period. It is described by the following equation (Ho *et al.* 2012; Hawkins *et al.* 2013):

$$V_{CF}(t) = \overline{V_{RAW}} + \frac{\sigma_{V,RAW}}{\sigma_{V,REF}}\left(O_{REF}(t) - \overline{V_{REF}}\right) \quad (2)$$

which was used to represent the relationship between the distribution of V_{RAW} (GCM projection in the future) and the distribution of V_{REF} (GCM simulations in the reference period), therefore $\sigma_{V,RAW}$ and $\sigma_{V,REF}$ are the standard deviation of GCM output of the future projections and deviation of GCM output in the reference period, respectively.

The QM method (Gudmundsson *et al.* 2012) is based on the concept of transformation of the selected variable:

$$\begin{aligned} V_{Obs} &= h(V_{GCMREF}) \\ &= ECDF_{Obs-1}(ECDF_{GCMREF}(V_{GCMRAW})) \end{aligned} \quad (3)$$

where V_{Obs} is observed meteorological variable, $V_{GCM\ REF}$ is GCM output for the reference period, $V_{GCM\ RAW}$ is a meteorological variable, which is projected by GCM for the future period. $ECDF_{Obs-1}$ is an empirical cumulative distribution function for the observed period and $ECDF_{GCM\ REF}$ is empirical cumulative distribution function for the GCM reference period. First, all the probabilities in $ECDF_{Obs-1}$ and $ECDF_{GCM\ REF}$ are calculated at a fixed interval of 0.01. Then, h in each interval is estimated as the relative difference between the two different ECDFs. Interpolation between the fixed values is based on a monotonic tricubic spline interpolation. The correction of the number of wet days was done using the empirical probability of non-zero values in V_{Obs}. After that, all GCM values below this threshold were set to zero (Sunyer *et al.* 2015). The method was implemented by *Python* software.

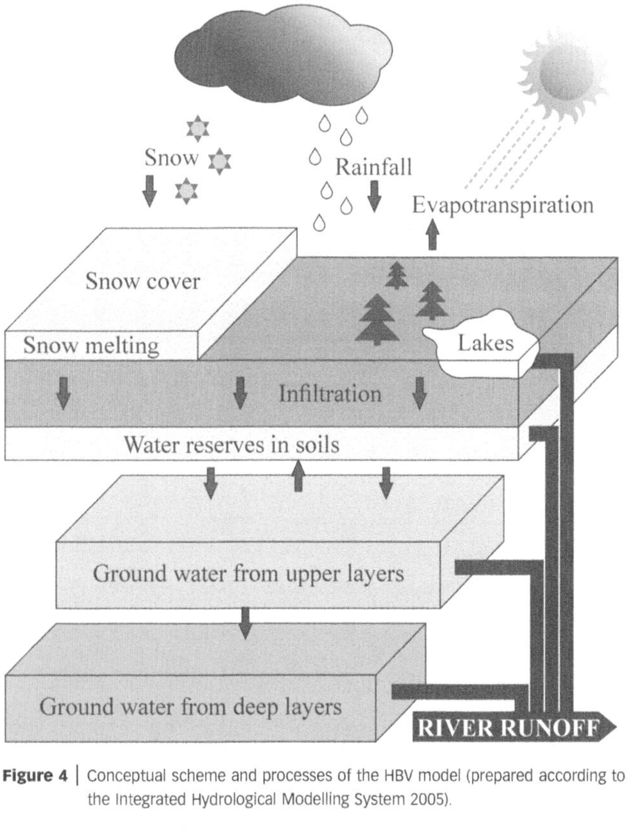

Figure 4 | Conceptual scheme and processes of the HBV model (prepared according to the Integrated Hydrological Modelling System 2005).

The HBV (*Hydrologiska Byråns Vattenbalansavdelning*) hydrological model created by SMHI (Swedish Meteorological and Hydrological Institute) is a rainfall-runoff model and describes hydrological processes in a river catchment scale (https://www.smhi.se/en/research/research-departments/hydrology/hbv-1.90007). The HBV model evaluates and calculates how in the river basin district, the atmospheric precipitation is transformed into river runoff taking into account temperature, evaporation, infiltration, accumulation in natural water bodies and the influence of the basin relief (Figure 4).

The periods of 1986–1995 and 1996–2005 were selected for calibration and validation of hydrological models, respectively. The hydrological model of each simulated river is calibrated in five stages using 16 basic calibration parameters, which depend on the local physical geographical characteristics and river basin attributes. During calibration steps, the models were evaluated by observed discharges, i.e., how simulated discharges coincide with measured discharges by changing values of calibration parameters. The simulated discharges of calibration and validation of the created hydrological models and the average rates (observed and simulated) of discharge for the used periods are presented in Table 2 as well as deviations (%) of simulated discharge from the observed values. The highest R^2 was obtained in the Minija River for calibration (0.88) and validation (0.83). Also, Table 2 shows the comparison of discharges where differences between measured

Table 2 | Statistics of calibration and validation of created hydrological models and deviation (%) of simulated discharge from the observed values

	Calibration			Validation		
		Average Q (m^3/s)			Average Q (m^3/s)	
River-WGS	R^2	Observed	Simulated (ΔQ, %)	R^2	Observed	Simulated (ΔQ, %.)
Minija-Kartena	0.88	17.7	18.4 (4.0)	0.83	16.8	16.6 (−1.2)
Nevėžis-Dasiūnai	0.86	38.9	34.6 (−11.1)	0.77	29.0	33.7 (16.2)
Šventoji-Ukmergė	0.75	46.5	44.5 (−4.3)	0.68	41.8	43.9 (5.0)

and simulated values are not high. The smallest deviation was in the Minija River (up to 4%) and the largest in the Nevėžis River (up to 16%). According to various studies, these discrepancies are small because in individual cases deviation errors of discharge measurement can reach 35% (Neff & Nicholas 2005). Taking into account the results of the calibration and validation of the model (Table 2) and the long data series used for these procedures, the created models are well prepared for projections of river runoff according to different climate scenarios in the future.

The evaluation of uncertainties associated with selected sources of uncertainties is necessary for projecting annual runoff changes in the future. In this study, the uncertainties of annual runoff projections consist of uncertainty sources as follows: climate scenarios (RCPs), GCMs and SD methods. In Lithuania, Kriaučiūnienė *et al.* (2013) evaluated the uncertainties of runoff projections using other sources of uncertainty (GCMs, SRES group climate scenarios and calibration parameters of HBV). Therefore, the uncertainty analysis of this research is based on a similar methodology. All possible combinations of uncertainty sources were made for evaluating the three sources of uncertainty (A, B, C) when each of them consists of three components (A_1, A_2, A_3, B_1, …, C_3). Accordingly, the 27 unique projections of runoff for each of the three catchments were created. The variable A represents the analysed source of uncertainty, while B and C are the remaining two sources of uncertainty. The same combinations of components (B_1, B_2, B_3, …, C_3) help to identify the uncertainties of A components (A_1, A_2, A_3). The uncertainties of source A were calculated by combining the same combinations of components B and C. The maximum value minus minimum value was estimated from the horizontal selections of A_1, A_2 and A_3 and the arithmetic average of the above-mentioned difference was calculated. The calculation of the contribution of each source to the spread in outcomes is based on the uncertainty caused by the three sources of uncertainty and calculates the percentage from other sources of uncertainty based on the average in difference.

RESULTS AND DISCUSSION

All deviations of annual runoff projections of the near and far future were calculated from their simulations in the reference period according to the same combinations of GCM and applied SD methods. Depending on different GCMs and SD methods, the projections of RCP scenarios fluctuated in a wide range. The deviations of annual runoff projections of the rivers of Minija, Nevėžis and Šventoji in the near and far future are shown in Figure 5. The projected annual runoff according to selected RCPs decreased on average from 13.3% in the near future to 33.9% in the far future compared to the reference period. In the near future, the lowest changes in river runoff were projected by RCP4.5 scenario, while the largest deviations and their variations were obtained according to RCP2.6 scenario. Meanwhile, the differences between RCPs increased in the far future because, on average, the RCP2.6 scenario projected the smallest decrease of river runoff but the largest range of possible projections. The most dramatic changes (up to a 47.2% decrease) of river runoff were projected by RCP8.5 in the far future.

The projections of river runoff determined by different GCMs showed similar patterns of deviations between the selected rivers and periods. The largest decrease of annual runoff was obtained applying the output of the Had climate model in both analysed periods, while the projections of the Nor model were the closest to the reference period. The projections with the highest range of deviations were obtained according to the GFDL model, especially in the far future.

The effect of SD methods on the projections of annual runoff was significant in the near and far future as well. The projections based on the BC and CF methods showed similar deviations in runoff projections. According to the mentioned methods, the average decrease of runoff consisted of 11.3% and 9.7% in the near future, and 18.5% and 18.7% in the far future, respectively. In all analysed rivers, the smallest average deviation of runoff projections from the reference period was obtained using the QM method. The deviations varied from −4.4% in the near future to −5.5% in the far future. However, the QM method provided the largest range of projected changes in the rivers Minija and Šventoji.

The obtained results of this research coincide with European study, where 15 combinations of RCM/GCM and eight different SD methods were used (Hundecha *et al.* 2016), because the decrease of annual runoff in the river catchments of Nevėžis and Šventoji were determined. In the

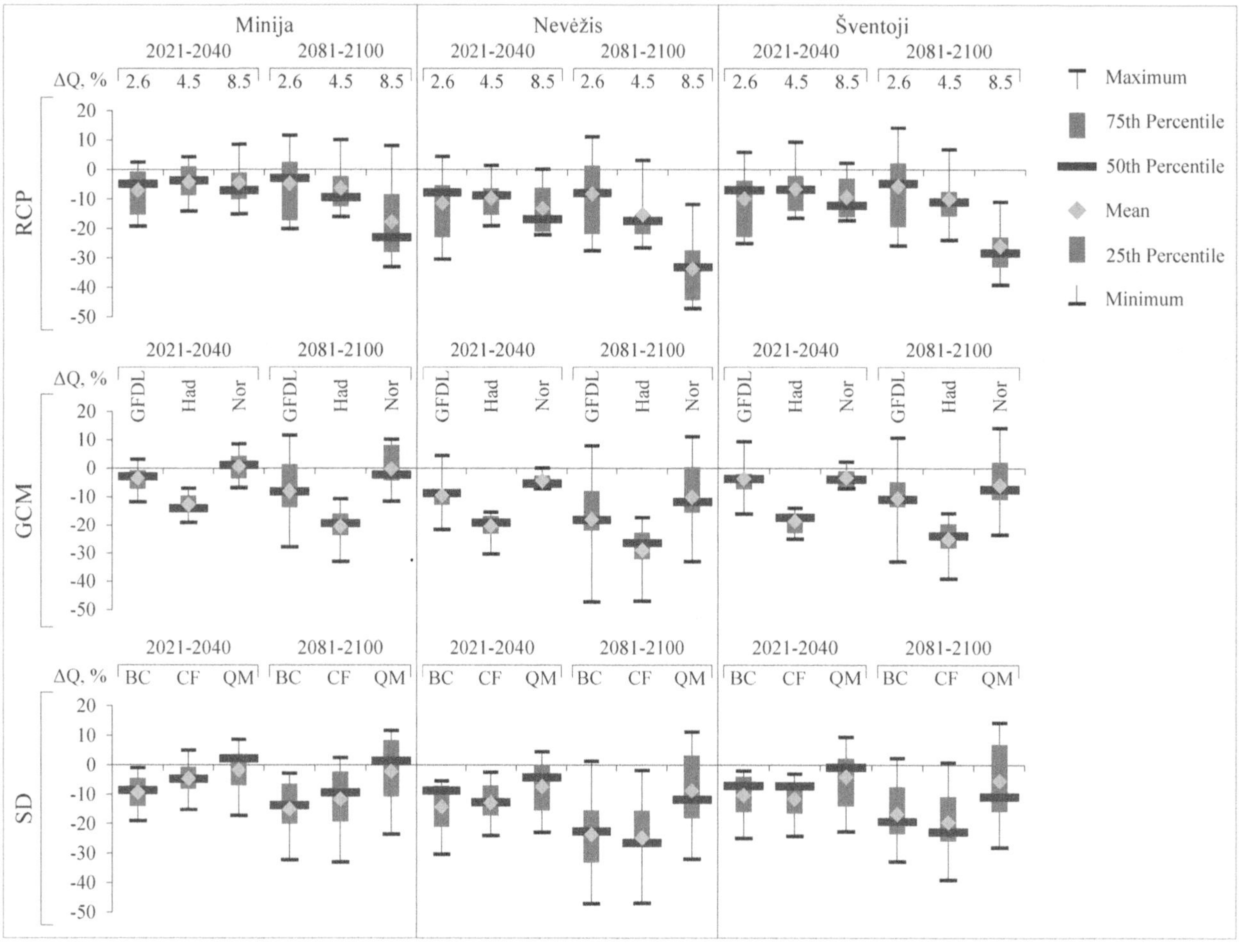

Figure 5 | Deviation of annual runoff projections from the simulations of the reference period in selected rivers according to RCP, GCM, and SD in the near and far future.

study of Hundecha *et al.* (2016), the general tendency of extreme flow projections tend to decrease in catchments where runoff generation from snowmelt is dominant; one of the selected case studies was from Lithuania – the Merkys River catchment, which falls within hydrological regions of southeastern Lithuania, where the Šventoji River is also located. Therefore, the established tendencies of runoff projections in the rivers (Šventoji and Nevėžis) of snowmelt-driven floods are linked to decrease as well as in the study of Hundecha *et al.* (2016).

The projection and assessment of tendencies of future river runoff are important for the identification of possible uncertainties regarding the selection of projection sources (RCP, GCM and SD). In this research, various combinations of these three sources provide wide range projections of river runoff. A high number of peaks (above reference period values) in the annual runoff hydrograph of the Minija River (Kartena WGS) (Figure 6) formed due to rainfall, which is the main feeding source of rivers in the western region of Lithuania, i.e., river runoff has a rapid reaction to heavy rainfall. The range of runoff projections in the Minija River varies depending on the season and GCM (Figure 6). However, the largest peaks occur in the winter season. Meanwhile, models of GFDL and Nor provide a wide variation of projections for the autumn season. In the 21st century, the mentioned GCMs provide lower projections (below reference period) of runoff for the summer season, while the values of discharge projections of the Had model range from average to very low.

The runoff of Nevėžis River (Dasiūnai WGS) has a very sensitive response to different climate scenarios and SD methods. Accordingly, runoff projections for different

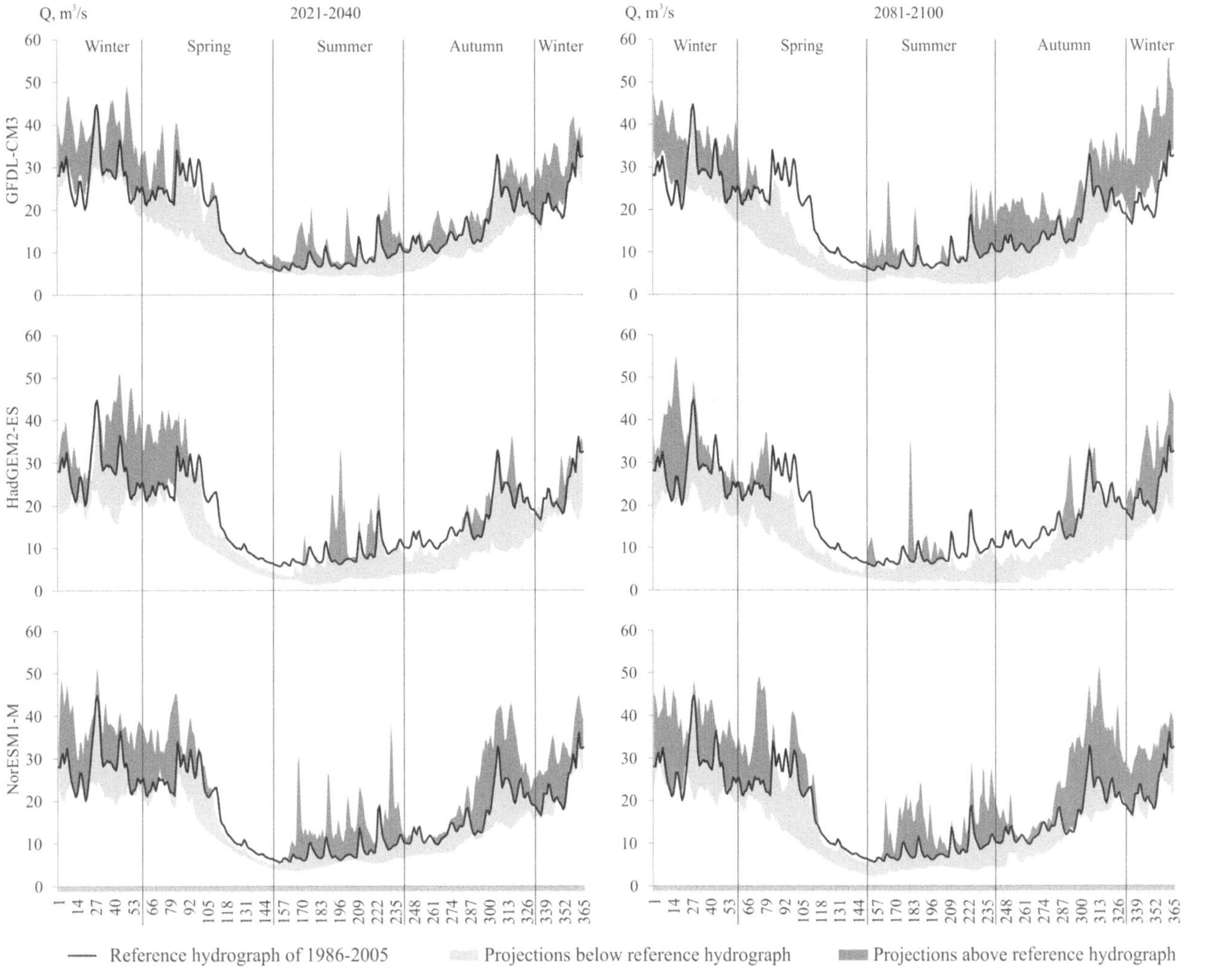

Figure 6 | Uncertainty of runoff projections of Minija River (Kartena WGS) according to GFDL-CM3, HadGEM-2ES, and NorESM1-M GCMs for the periods of 2021–2040 and 2081–2100.

seasons vary widely (Figure 7) as large seasonal differences in discharge are a prominent feature of rivers in central Lithuania. In the period 2021–2040, the annual runoff projections responsively reacted in winter and spring seasons. Therefore, their projections differed from extreme low values to values above the reference period for different scenarios. During the period 2081–2100, the projections for the mentioned seasons varied from very low discharges to values of the reference period. Only the Nor model projected higher values of the annual runoff in the seasons of winter and spring. The maximum range of possible changes in the Nevėžis River was expected in spring when runoff projections assumed very high uncertainties. This showed the vulnerability of spring floods to climate change conditions. Rising air temperature influenced the period of snow accumulation. Therefore, a lower amount of water resources was accumulated in the river basin. Consequently, the size of the spring flood decreased.

The hydrograph of the reference period (1986–2005) of the Šventoji River (southeastern hydrological region) compared with the hydrographs of other rivers (Minija and Nevėžis) has the most stable distribution (Figure 8). This form of hydrograph is typical for rivers of the LT-SE region, where groundwater feeding is dominant. Due to sandy soils, a significant part of surface runoff from snow melting and precipitation supplies groundwater which feeds the rivers during summer. Hence, water resources in this region are distributed evenly per year. The runoff

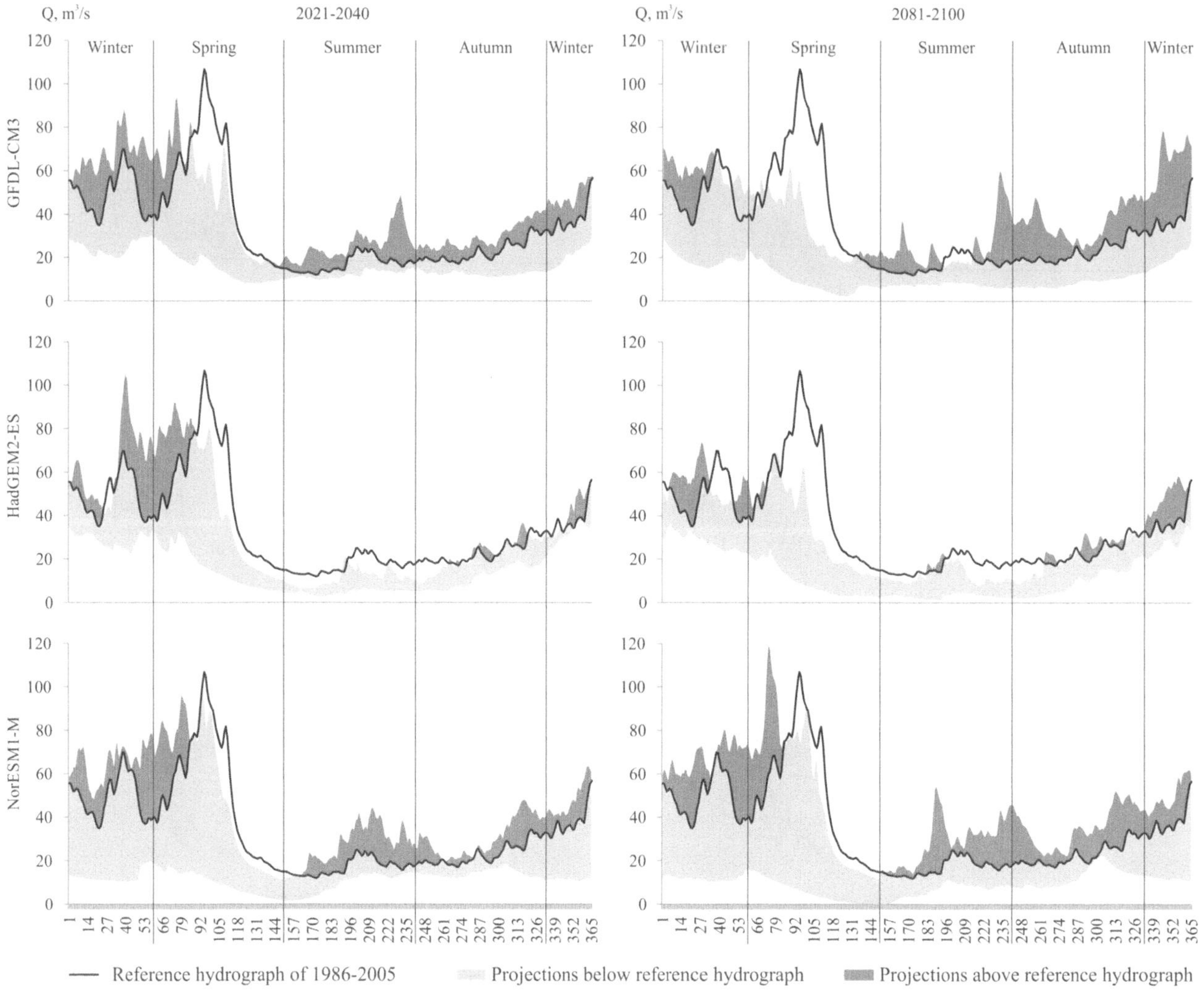

Figure 7 | Uncertainty of runoff projections of Nevėžis River (Dasiūnai WGS) according to GFDL-CM3, HadGEM-2ES, and NorESM1-M GCMs for the periods of 2021–2040 and 2081–2100.

projections of the Šventoji River (Ukmergė WGS) according to different GCMs fluctuated with a wide range (Figure 8). The most noticeable changes in the runoff were observed in the seasons of spring and winter when the decline of spring floods came together with earlier spring peaks. According to different scenarios, the projected increase of winter runoff depended on earlier snow melting. In addition to that, a relatively narrow range of projections for the summer season according to the projections of the GFDL and Nor models was identified, whereby the runoff changed from very low to very high discharges compared to the reference period. Meanwhile, all projections with the Had model provided lower values of discharges.

The variability of projections of the annual runoff was estimated according to the uncertainty sources: climate scenarios (RCPs), GCMs and SD methods. The calculations of the percentage of uncertainty sources revealed which source had the greatest impact on the wide scattering of projected runoff values in the rivers of Minija, Nevėžis and Šventoji (Table 3). In the near and far future, the largest uncertainties of runoff projections of the Minija River (Kartena WGS) were caused by the GCMs. The selected GCMs contributed 44.5 and 41% of the total spread in the ensemble of projections for the near and far future, respectively. A significant influence of SD methods was also estimated, causing the

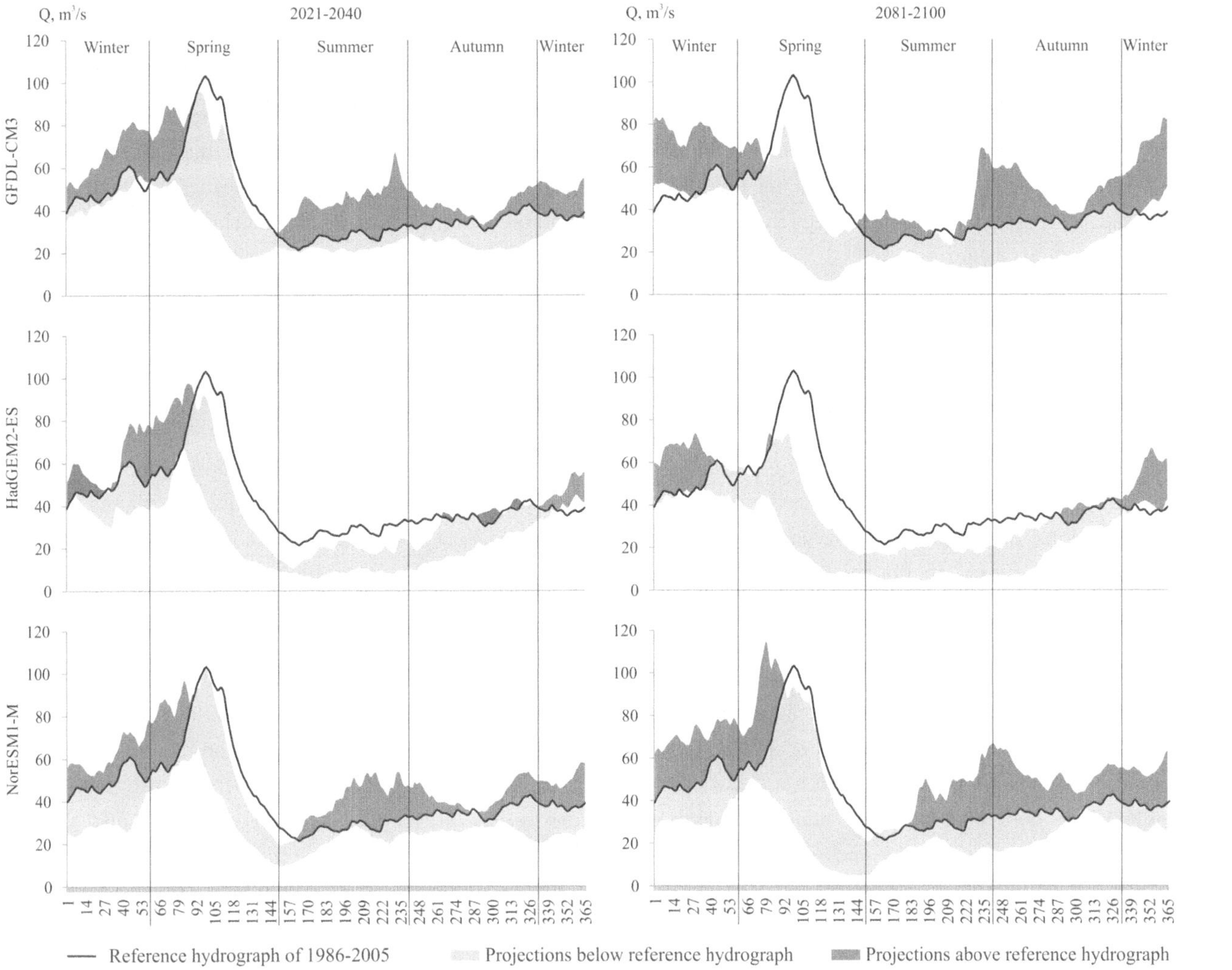

Figure 8 | Uncertainty of runoff projections of Šventoji River (Ukmergė WGS) according to GFDL-CM3, HadGEM-2ES, and NorESM1-M GCMs for the periods of 2021–2040 and 2081–2100.

Table 3 | Uncertainty (%) of annual runoff projections in rivers of Minija, Nevėžis, and Šventoji according to a different source of origin for the periods of 2021–2040 and 2081–2100

	Minija		Nevėžis		Šventoji	
	2021–2040	2081–2100	2021–2040	2081–2100	2021–2040	2081–2100
RCP	16.7	24.3	11.2	24.4	15.7	31.5
GCM	44.5	41.0	27.9	24.3	38.1	29.1
SD	38.8	34.7	60.9	51.3	46.2	39.4

uncertainties of 38.8% and 34.7% in the near and far future, respectively. The smallest dispersion of runoff projections was related to RCP climate scenarios; however, the influence of RCP increased by 7.5 percentage points in the far future compared to the near future.

In the near future, the variability of projections of annual runoff of the Nevėžis River (Dasiūnai WGS) was as high as 60.9% using SD methods, while the influence of RCP scenarios was only 11.2% (Table 3). The situation is different in the far future because uncertainties caused by SD methods decreased up to 51.3% and uncertainties of RCP increased up to 24.4%. In any case, the variability of annual runoff projections of the Nevėžis River was related to SD methods by more than 50%. Meanwhile, the accuracy of runoff projection caused by GCMs was similar in the near and far future – 27.9% and 24.3%, respectively.

The largest scatter in the annual runoff projections of the Šventoji River was determined for the SD method as well, because uncertainties related to the SD methods amounted to 46.2% in the near future. The rest of the uncertainty sources provided uncertainties of 38.1% (GCMs) and 15.7% (RCPs) (Table 3). In the far future, the influence of RCP scenarios increased; the uncertainties related to RCP scenarios reached 31.5% and were 2.4 percentage points larger than the uncertainties of GCMs. Nevertheless, the greatest scattering of annual runoff projections of the Šventoji River in the far future was caused by the SD methods, because uncertainties of SD were 39.4% compared to the other sources. The analysis of runoff of studied rivers showed the importance of the selection of GCMs and SD methods to create proper projections of river runoff, because the largest uncertainties were related to the mentioned sources of uncertainty.

The uncertainties of projections of annual runoff between the components of uncertainty sources and interrelations between them are displayed in column diagrams (Figure 9). The uncertainty of projections according to the SD methods related to RCP scenarios did not show significant differences between the used climate scenarios. Meanwhile, the SD uncertainties according to different GCMs highlighted the climate models of GFDL and Nor, which provided the largest uncertainties in the rivers of Nevėžis and Šventoji. Summary of the analysis showed that in the far future, larger uncertainties of river runoff projections of all analysed rivers were caused by RCP in comparison to the near future (2021–2040) as well as the differences between projections of RCP scenarios increasing.

The results of the uncertainty of RCP projections showed the largest uncertainties using the output of GFDL depending on the selected GCM. The smallest uncertainties of RCP projections were estimated according to the output of the Had model. In the near future, the analysis of RCP uncertainties (related to SD methods) showed a significant impact of the CF and QM methods. Meanwhile, in the far future, the influence of the CF method increased and a greater part of uncertainties was caused by the mentioned method, which affected larger uncertainties (from 0% to 13.6%) compared to the SD methods of BC and QM.

The uncertainties of GCM projections were strongly impacted by different SD methods. In the Minija River, the largest uncertainties of GCM projections were established according to the BC and QM methods. Meanwhile, in the rivers of Nevėžis and Šventoji, the QM method had the largest impact on uncertainties. The scattering of runoff projections of GCMs did not show a significant influence of different RCPs. Only the obtained uncertainties of RCP2.6 in most of the rivers and periods were higher than other scenarios.

In summary, the GCMs can be regarded as the most dominant uncertainty source (41.0–44.5%) in the Minija River, which is in the western hydrological region of Lithuania. In this region, the topography effect is strongly expressed because of the Žemaičiai Uplands. These uplands collect the greater part of moisture from air masses and have the highest annual precipitation compared to other regions. Therefore, the primary projections of precipitation are significantly related to the GCM and it is necessary to select a particular GCM for runoff projections very carefully. These results coincide with Lawrence & Haddeland (2011), where, in runoff projections of three analysed catchments in Norway the largest uncertainty sources are also related to GCM.

In the Nevėžis River (LT-C), uncertainties were linked to SD methods (51.3% and 60.9%). In this region, the lowland topography has the opposite influence to uplands and the grid cell of GCMs is sufficiently large, so SD methods, in some cases, did not properly adjust the output of GCMs to local climatic conditions of the specific area. Especially, it is important for corrections of precipitation data; therefore, the selection of SD method causes the greatest uncertainties in LT-C. Due to a large part of rivers' feeding source as snowmelt, the floods in rivers of this region are usually caused by the thick cover of snow. Since in the future an increase in air temperature is projected, the period of snow accumulation will get shorter or will be absent in some years. Accordingly, the projections of river runoff had a wide range according to various scenarios during the winter and spring seasons. In another similar study, Lawrence & Haddeland (2011) found that the estimated uncertainties in runoff projections of two river catchments which had generally been dominated by the spring

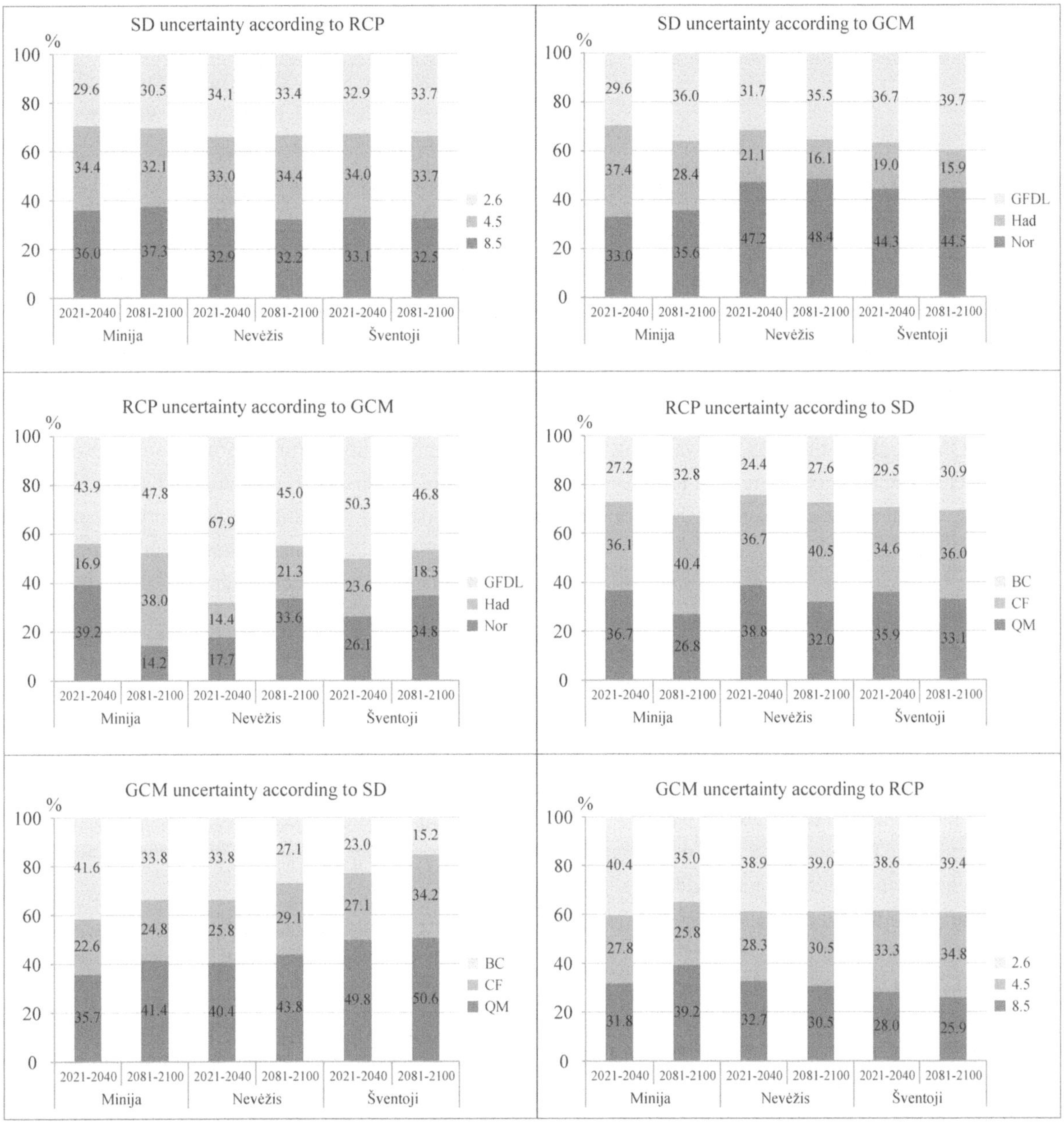

Figure 9 | Uncertainty (%) of annual runoff projections in the rivers of Minija, Nevėžis, and Šventoji between different sources of uncertainty (RCP, GCM, and SD) for the periods of 2021–2040 and 2081–2100.

snowmelt were mostly related to the SD methods (48% and 60%) as well.

In the Šventoji River (SE-LT), the influence of SD (46.2% and 39.4%) was established as well. This region is characterised by the widespread permeable sandy soils, which effectively absorb water from snow melting and later gradually release it, supplying rivers in the low-flow period. The annual discharge of rivers of southeastern Lithuanian is distributed rather equally. In the Šventoji River catchment, GCMs do not have a significant impact, therefore the importance of the SD methods increases since SDs determine the way meteorological data are adjusted for particular regional conditions. Results of Kriaučiūnienė *et al.* (2013) established that the largest uncertainties were associated with emission scenarios in the investigated rivers of Lithuania.

CONCLUSIONS

In this study, the projections of climate change impacts on hydrological processes in three Lithuanian catchments from different hydrological regions were based on scenarios from three GCMs generated by three RCP climate scenarios. The output data (T, P) of three GCMs according to RCP (2.6, 4.5 and 8.5) climate scenarios were adjusted to Lithuanian conditions by applying the SD methods of BC with variable, CF with variable and QM. Applying the HBV software, the following corrected data of T and P series were used to simulate projections of daily discharge in the near (2021–2040) and far future (2081–2100).

In the near and far future, the deviations of runoff projections from modelled runoff in the reference period varied over a wide range. In the selected rivers, the largest deviations of annual runoff projections were determined by the RCP8.5 climate scenario as well as the Had climate model. Meanwhile, the lowest deviations of river runoff projections were observed according to the output of the Nor climate model. The largest dispersion of deviations was provided by RCP2.6 and GFDL model. Such different deviations of projected runoff values require additional analysis to assess the uncertainty of each uncertainty source (RCPs, GCMs and SDs).

The GCMs were the most dominant uncertainty source (41.0–44.5%) in the runoff projections of the Minija River which belongs to the western hydrological region. In this region, the main feeding source of rivers is precipitation, which is the highest compared to other regions. Primary projections of precipitation are significantly related to the GCM, since the selected GCMs provide a wide range of the amount of precipitation in western Lithuania. Meanwhile, uncertainties of the Nevėžis and Šventoji rivers (from central and southeastern regions of Lithuania) were linked to SD methods, respectively (51.3%–60.9%) and (39.4%–46.2%). The grid cell of GCMs is quite large. Consequently, SD does not always properly adjusted GCM output data to an area with specific local conditions. Therefore, the selection of an appropriate SD method is very important, because the selected SD method must represent the climate conditions of the reference period very precisely. At the same time, accurately selected SD methods will allow the creation of better fit projections under climate change conditions.

Analysis of uncertainty sources showed the widest scattering of results related to different GCMs. The largest uncertainties of RCP projections were caused by the GFDL-CM3 climate model and the largest uncertainties of SD projections were sensitive to the NorESM1-M climate model, especially according to the QM method. Therefore, the accurate selection of GCMs and SD methods is essential for the projections with the lowest uncertainties. Understanding the uncertainty of runoff projections allows better identification of which uncertainty source has the most significant influence on the final results and consequently provides an opportunity to create more accurate runoff projections for different river catchments.

REFERENCES

Addor, N., Rössler, O., Köplin, N., Huss, M., Weingartner, R. & Seibert, J. 2014 Robust changes and sources of uncertainty in the projected hydrological regimes of Swiss catchments. *Water Resources Research* **50** (10), 7541–7562. doi:10.1002/2014WR015549.

Ahlström, A., Smith, B., Lindström, J., Rummukainen, M. & Uvo, C. B. 2013 GCM characteristics explain the majority of uncertainty in projected 21st century terrestrial ecosystem carbon balance. *Biogeosciences* **10** (3), 1517–1528. doi:10.5194/bg-10-1517-2013.

Bosshard, T., Carambia, M., Goergen, K., Kotlarski, S., Krahe, P., Zappa, M. & Schär, C. 2013 Quantifying uncertainty sources in an ensemble of hydrological climate-impact projections. *Water Resources Research* **49** (3), 1523–1536. doi:10.1029/2011WR011533.

Chen, J., Brissette, F. P., Lucas-Picher, P. & Caya, D. 2017 Impacts of weighting climate models for hydro-meteorological climate change studies. *Journal of Hydrology* **549**, 534–546. doi:10.1016/j.jhydrol.2017.04.025.

Dobler, C., Hagemann, S., Wilby, R. L. & Stötter, J. 2012 Quantifying different sources of uncertainty in hydrological projections in an alpine watershed. *Hydrology and Earth System Sciences* **16** (11), 4343–4360. doi:10.5194/hess-16-4343-2012.

Gudmundsson, L., Bremnes, J. B., Haugen, J. E. & Engen-Skaugen, T. 2012 Technical note: downscaling RCM precipitation to the station scale using statistical transformations – a comparison of methods. *Hydrology and*

Earth System Sciences **16** (9), 3383–3390. doi:10.5194/hess-16-3383-2012.

Hagemann, S., Chen, C., Haerter, J. O., Heinke, J., Gerten, D. & Piani, C. 2011 Impact of a statistical bias correction on the projected hydrological changes obtained from three GCMs and two hydrology models. *Journal of Hydrometeorology* **12**, 556–578. doi:10.1175/2011JHM1336.1.

Hawkins, E., Osborne, T. M., Hoa, C. K. & Challinor, A. J. 2013 Calibration and bias correction of climate projections for crop modelling: an idealised case study over Europe. *Agricultural and Forest Meteorology* **170**, 19–31. doi:10.1016/j.agrformet.2012.04.007.

Ho, C. K., Stephenson, D. B., Collins, M., Ferro, C. A. T. & Brown, S. J. 2012 Calibration strategies: a source of additional uncertainty in climate change projections. *Bulletin of the American Meteorological Society* **93** (1), 21–26. doi:10.1175/2011BAMS3110.1.

Hundecha, Y., Sunyer, M. A., Lawrence, D., Madsen, H., Willems, P., Bürger, G., Kriauciuniene, J., Loukas, A., Martinkova, M., Osuch, M., Vasiliades, L., Von Christierson, B., Vormoor, K. & Yücel, I. 2016 Inter-comparison of statistical downscaling methods for projection of extreme flow indices across Europe. *Journal of Hydrology* **541**, 1273–1286. doi:10.1016/j.jhydrol.2016.08.033.

Integrated Hydrological Modelling System 2005 Manual. Version 5.8. Swedish Meteorological and Hydrological Institute, Norrköping, Sweden.

Keršytė, D., Rimkus, E. & Kažys, J. 2015 Near-term and long-term climate projections for Lithuania. *Geologija. Geografija* **1** (1), 22–35 (in Lithuanian).

Kriaučiūnienė, J., Jakimavicius, D., Sarauskiene, D. & Kaliatka, T. 2013 Estimation of uncertainty sources in the projections of Lithuanian river runoff. *Stochastic Environmental Research and Risk Assessment* **27** (4), 769–784. doi:10.1007/s00477-012-0608-7.

Kundzewicz, Z. W., Krysanova, V., Benestad, R. E., Hov, Ø., Piniewski, M. & Otto, I. M. 2018 Uncertainty in climate change impacts on water resources. *Environmental Science and Policy* **79**, 1–8. doi:10.1016/j.envsci.2017.10.008.

Latif, M. 2011 Uncertainty in climate change projections. *Journal of Geochemical Exploration* **110** (1), 1–7. doi:10.1016/j.gexplo.2010.09.011.

Lawrence, D. & Haddeland, I. 2011 Uncertainty in hydrological modelling of climate change impacts in four Norwegian catchments. *Hydrology Research* **42** (6), 457–471. doi:10.2166/nh.2011.010.

Lindström, G., Johansson, B., Persson, M., Gardelin, M. & Bergström, S. 1997 Development and test of the distributed HBV-96 hydrological model. *Journal of Hydrology* **201** (1–4), 272–288. doi:10.1016/s0022-1694(97)00041-3.

Maraun, D. 2016 Bias correcting climate change simulations – a critical review. *Current Climate Change Reports* **2** (4), 211–220. doi:10.1007/s40641-016-0050-x.

Neff, B. P. & Nicholas, J. R. 2005 *Uncertainty in the Great Lakes Water Balance. U.S. Geological Survey Scientific Investigations Report 2004–5100*. USGS, Reston, VA, USA, p. 42.

Räty, O., Räisänen, J. & Ylhäisi, J. S. 2014 Evaluation of delta change and bias correction methods for future daily precipitation: intermodel cross-validation using ENSEMBLES simulations. *Climate Dynamics* **42** (9–10), 2287–2303. doi:10.1007/s00382-014-2130-8.

Šarauskienė, D., Akstinas, V., Kriaučiūnienė, J., Jakimavičius, D., Bukantis, A., Kažys, J., Povilaitis, A., Ložys, L., Kesminas, V., Virbickas, T. & Pliuraitė, V. 2018 Projection of Lithuanian river runoff, temperature and their extremes under climate change. *Hydrology Research* **49** (2), 344–362. doi:10.2166/nh.2017.007.

Shen, M., Chen, J., Zhuan, M., Chen, M., Xu, C. Y. & Xiong, L. 2018 Estimating uncertainty and its temporal variation related to global climate models in quantifying climate change impacts on hydrology. *Journal of Hydrology* **556**, 10–24. doi:10.1016/j.jhydrol.2017.11.004.

Stonevičius, E., Rimkus, E., Štaras, A., Kažys, J. & Valiuškevičius, G. 2017 Climate change impact on the Nemunas River basin hydrology in the 21st century. *Boreal Environment Research* **22**, 49–65.

Sunyer, M. A., Hundecha, Y., Lawrence, D., Madsen, H., Willems, P., Martinkova, M., Vormoor, K., Bürger, G., Kriaučiūnienė, J., Loukas, A., Osuch, M. & Yücel, I. 2015 Inter-comparison of statistical downscaling methods for projection of extreme precipitation in Europe. *Hydrology and Earth System Sciences* **19** (4), 1827–1847. doi:10.5194/hess-19-1827-2015.

Teutschbein, C. & Seibert, J. 2013 Is bias correction of regional climate model (RCM) simulations possible for non-stationary conditions? *Hydrology and Earth System Sciences* **17** (12), 5061–5077. doi:10.5194/hess-17-5061-2013.

First received 28 December 2018; accepted in revised form 1 August 2019. Available online 18 October 2019

Response of melt water and rainfall runoff to climate change and their roles in controlling streamflow changes of the two upstream basins over the Tibetan Plateau

Yueguan Zhang, Zhenchun Hao, Chong-Yu Xu IWA and Xide Lai

ABSTRACT

Located in the Tibetan Plateau, the upstream regions of the Mekong River (UM) and the Salween River (US) are very sensitive to climate change. The 'VIC-glacier' model, which links a degree-day glacier algorithm with variable infiltration capacity (VIC) model, was employed and the model parameters were calibrated on observed streamflow, glacier mass balance and MODIS snowcover data. Results indicate that: (1) glacier-melt runoff exhibits a significant increase in both areas by the Mann–Kendall test. Snowmelt runoff shows an increasing trend in the UM, while the US is characterized by a decreasing tendency. In the UM, the snowmelt runoff peak shifts from June in the baseline period 1964–1990 to May for both the 1990s and 2000s; (2) rainfall runoff was considered as the first dominant factor driving changes of river discharge, which could be responsible for over 84% in total runoff trend over the two regions. The glacial runoff illustrates the secondary influence on the total runoff tendency; (3) although the hydrological regime is rain dominated in these two basins, the glacier compensation effect in these regions is obvious, especially in dry years.

Key words | climate change, glacial runoff, snowmelt runoff, Tibetan Plateau, VIC-glacier model

Yueguan Zhang (corresponding author)
Xide Lai
Department of Water Conservancy and Hydropower Engineering,
Xihua University,
Chengdu 610039,
China
E-mail: *zhangyueguan@itpcas.ac.cn*

Yueguan Zhang
Zhenchun Hao
State Key Laboratory of Hydrology-Water Resources and Hydraulic Engineering,
Hohai University,
Nanjing 210098,
China

Chong-Yu Xu IWA
Department of Geosciences,
University of Oslo,
Oslo 0316,
Norway

INTRODUCTION

The Tibetan Plateau (TP) and the surrounding Himalayas, with an average altitude of 4,000 m a.s.l and an area of about 2.5×10^6 km^2, is the highest and most extensive highland and is also called the Third Pole (Qiu 2008). The TP exerts a profound influence on the East Asian and global climate (Lu *et al.* 2017), and is considered as a sensitive region and the amplifier for global climate change (Ma *et al.* 2006). The TP is also the source region of many major Asian rivers, such as the Brahmaputra, Mekong, Salween, Yellow and Yangtze Rivers, and is considered as the 'water towers of Asia' (Immerzeel *et al.* 2010; Zhang *et al.* 2013; Liu *et al.* 2018a). Discharge from these rivers sustains the lives of hundreds of millions of people living downstream and so preservation of these water resources is crucial for social and economic development over these regions (Immerzeel *et al.* 2009; Liu *et al.* 2018a).

The TP is characterized by an abundance of glaciers and snow cover. The glacier area over the TP is about 91,822 km^2 (Kotlyakov *et al.* 2012) and is the third largest on Earth, after the Arctic/Greenland and Antarctic regions. Snow covers the majority of the TP during winter (Immerzeel *et al.* 2009). Glacier and snow cover over the TP is very sensitive to rising temperature. Studies based on observed meteorological data have suggested a warming trend over the TP in the past several decades, particularly since the 1980s (Duan & Wu 2006; Kuang & Jiao 2016), and the increasing rate is more rapid than that of surrounding areas (Wang 2017). Under the warming climate, the glaciers on the TP have generally been retreating, with the largest shrinkage in the

doi: 10.2166/nh.2019.075

southeastern TP and the least in the interior and northwestern TP (Yao *et al.* 2004, 2012; Bolch *et al.* 2012). Simultaneously, snow coverage or snow water in the TP has also responded to climate change and experienced changes over the past period (Li *et al.* 2018). However, at present, there is no consistent conclusion about the snow cover change over the TP.

Glacier and snow are crucial water sources of rivers and greatly influence the hydrological and biological processes in the TP and the surrounding areas (Immerzeel *et al.* 2010; Kaser *et al.* 2010; Lutz *et al.* 2014; Chen *et al.* 2017; Li *et al.* 2018). The changes of glacier and snow cover due to the warming climate have exerted impacts on the water budgets over the TP (Liu *et al.* 2018a). In a warmer climate, snow will melt earlier in the year than it did before and likely affect the timing of runoff, especially in the spring when water demand for irrigation is high (Barnett *et al.* 2005). As well, changes in the amount of glacial runoff and precipitation also tend to greatly affect the volume of runoff over the TP (Yao *et al.* 2014). Yao *et al.* (2004) found that the glacial retreat in the 1990s has caused glacial meltwater runoff increasing by more than 5.5% in Northwestern China. The southeast TP, due to long-term perennial snow and glacier melting, has been shown to have among the highest total water storage depletion rates globally (Jacob *et al.* 2012; Yao *et al.* 2012; Chen *et al.* 2017).

In terms of river basin, Zhang *et al.* (2008) assessed the influence of glacier runoff and climate change on the river runoff over the Tuotuo river basin located in the source region of the Yangtze River, and indicated that a two-third increase in river runoff in the 1990s was caused by loss of ice mass as a result of warming climate. Also, Yao *et al.* (2014) have investigated the impacts of climate change and glacier on river runoff in the source region of the Yangtze River during 1986–2009, and found that the increased glacier-melt runoff due to temperature rising accounted for 17.5% of the total runoff changes and the remaining change was caused by the runoff induced from precipitation. In the Upper Brahmaputra River basin, the interannual variability of total water storage was controlled mainly by glacier mass changes driven primarily by temperature changes (Chen *et al.* 2017; Meng *et al.* 2019). In addition, the impact of climate change on water availability over the basins of Brahmaputra and Mekong Rivers was investigated by a hydrologic budget balance method, and the results indicate that increasing glacial meltwater had great effect on runoff changes during 1960–2010 (Liu *et al.* 2018b). Meanwhile, in the upper Indus and Salween basin, it is found that the regional warming affected the local hydrology due to accelerated glacial melting during the simulation period (Immerzeel *et al.* 2009; Wang & Chen 2017). Furthermore, Cao *et al.* (2006) analysed the discharge change of five large rivers over the TP during the years 1956–2000, and found that climate change had a significant effect on the seasonal runoff variation, especially that in the spring. However, most of the above studies are based on statistical analysis of the observed data to characterize the controlling roles of melt water and rainfall runoff in the TP river flow change, and comprehensive quantitative investigations into attribution of streamflow trends to the different runoff components by hydrological simulation are still relatively few. In addition, few studies are involved in quantitative analysis of the response of rainfall runoff, glacial runoff and snowmelt to recent warming climate.

In this study, a degree-day glacier algorithm (Hock 2003) linking with the variable infiltration capacity (VIC) model (Liang *et al.* 1994), here referred to as VIC-glacier model (Zhang *et al.* 2013) was employed. The investigated areas are the upstream of the Mekong (UM) and Salween (US) rivers over the TP (Figure 1). The Mekong River and Salween River are important international rivers across China and Southeast Asia. Thus, runoff variation over these upper basins not only imposes great influence on local available water resources but also further affects water allocation in the middle and lower reaches. In addition, there are many large operating and planned hydropower stations along the trunks of both rivers, which are also very sensitive to the fluctuation of the upstream water resource induced by climate change. Based on gauge observations and the VIC-glacier model simulations, this work aims to: (1) investigate the long-term and decadal changes of rainfall runoff, snowmelt and glacial runoff in the two upstream basins during 1964–2013, in order to analyse the response of different runoff components to recent climate change; and (2) quantify the roles of the three runoff components in controlling river flow trends of the two basins in the past 50 years.

STUDY AREA AND DATA

Study area

In this study, the upstream regions of the Mekong River (UM) above Changdu station and the Salween River (US)

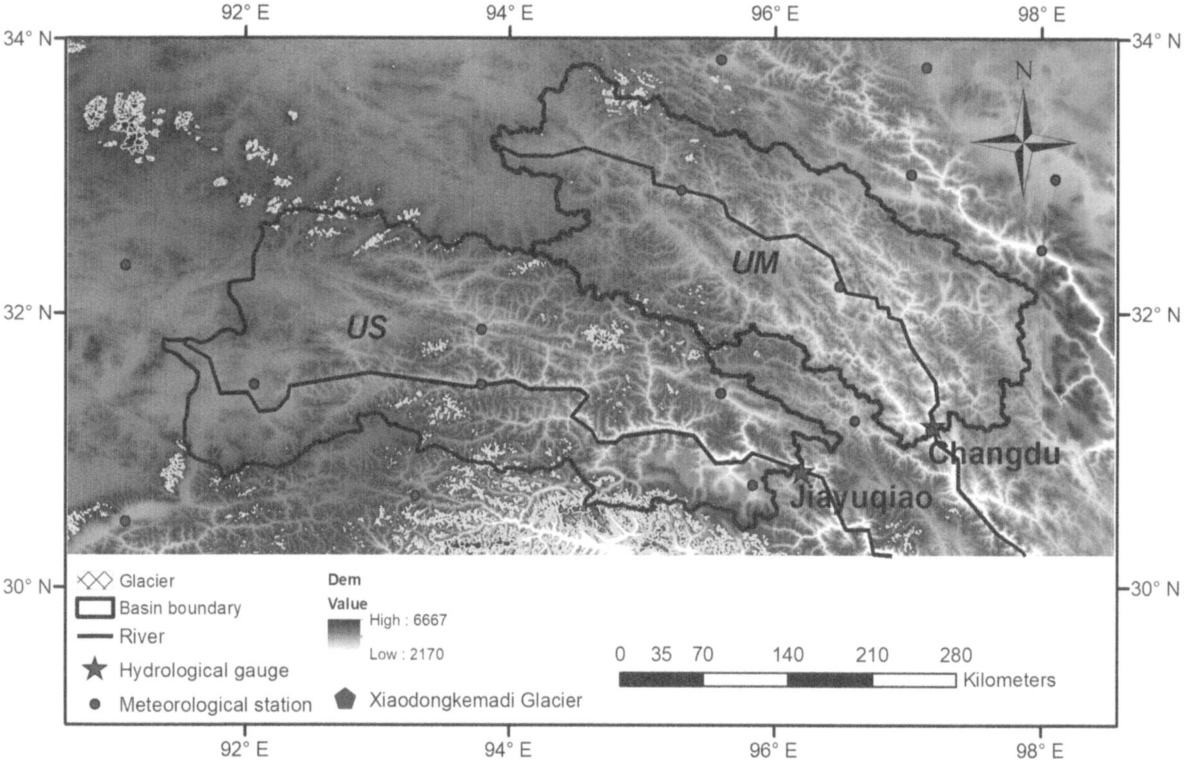

Figure 1 | Topography and boundary of the upstream of the Mekong river basin (UM) and the upstream of the Salween river basin (US) in the Tibetan Plateau.

above the Jiayuqiao hydrologic station were chosen as the study areas, respectively (Figure 1). The two upstream basins are located within 91–99°E and 30–34°N and cover a total area of about 1.21×10^5 km^2 (Table 1). The elevation of the study basins varies from about 2,170 to 6,667 m a.s.l. The hydrological gauges of Changdu and Jiayuqiao are the control outlets of the UM and US, respectively (Figure 1). The UM and US are located in the southeastern TP and belong to the Tibetan Plateau climate system, characterized by a wet and warm summer and a cold and dry winter. The UM and US are predominantly affected by the monsoon in the summer (June–August) while in winter (December–February) westerlies prevail. In addition, for the two regions, the US has the relatively large glacier coverage (about 1,152 km^2 and accounting for 1.7% of the basin area) whereas the UM has relatively less ice area (about 226 km^2 and 0.42% of the basin) (Table 1) (from the First Chinese Glacier Inventory: http://westdc.westgis.ac.cn/glacier).

Data

The atmospheric forcing data (maximum and minimum temperature, precipitation and wind speed), topography, soil and vegetation are required for running the VIC hydrological model. Other meteorological variables, such as

Table 1 | Characteristics of the two upstream basins

Basin	Gauge	Gauge location		Drainage area (km^2)	Glacier area (km^2)	Per cent of drainage area for glacier (%)
		Latitude (°)	Longitude (°)			
UM	Changdu	31.18	97.18	53,800	226.0	0.42
US	Jiayuqiao	30.85	96.20	67,740	1,151.6	1.7

vapour pressure, incoming shortwave radiation and net longwave radiation, can be calculated from daily temperature and precipitation (Kimball *et al.* 1997).

The daily meteorological data for the years 1964–2013 were collected from 17 stations of the National Climate Center of the China Meteorological Administration (CMA) in or around focus basins (Figure 1), and were fully quality controlled by CMA before releasing the data. All stations have records longer than 40 years, and 14 stations have continuous daily data spanning from January 1964 to December 2013. The dataset was created from *in-situ* observations applying solid quality control, including internal temporal and spatial consistency checks, homogeneity tests and potential outlier detection (Feng *et al.* 2004; Shen & Xiong 2016). The dataset was also regarded as the most credible meteorological dataset in China. All the station data were interpolated to the $1/12° \times 1/12°$ grids by using the inverse distance weighting (IDW) method. To account for temperature variations with elevation using the IDW method, monthly temperature lapse rates (Tlaps) or temperature gradients during 1964–2013 (Table 2) were derived by using linear regression analysis, i.e., fitting linear relationship between temperature and elevation of the weather stations. The correlation coefficient R for all months is more than 0.4, which indicates a good correlation between temperature and elevation, and meanwhile, the derived monthly temperature gradients can be applied properly in the IDW method.

Soil property data such as soil type, water-holding capacity and saturated hydrologic conductivity were obtained from the Food and Agriculture Organization (FAO) Soil Database, which provides the most detailed and globally consistent soil data (FAO 1988). The vegetation class and their parameters were derived from the University of Maryland's 1 km Global Land Cover product (http://glcf.umd.edu/data/landcover/data.shtml). Basic topography data over the study regions were obtained from GTOP30 (resolution: 1 km × 1 km) (http://eros.usgs.gov/#/Find_Data/Products_and_Data_Available/gtopo30_info).

Monthly observed streamflow data at Changdu (1964–2009) and Jiayuqiao (1980–1985) (Figure 1) were obtained from the Hydrological Bureau of the Ministry of Water Resource of China. These discharge records were used to evaluate the VIC-glacier model simulations in the two basins.

The initial glacier cover data over the UM and US were acquired from the First Chinese Glacier Inventory, which contains information about 26,000 glaciers throughout the whole country, and were digitized and archived by the 'Environmental and Ecological Science Data Center for West China' (http://westdc.westgis.ac.cn/data/ff75d30a-ee7d-4610-a5a3-53c73964a237). The glacier data were utilized to initialize the glacier percentage within each $1/12° \times 1/12°$ resolution grid. In addition, the glacier data from the Randolph Glacier Inventory 5.0 (http://www.glims.org/RGI) were used to assess the simulated glacier area changes between 1964 and 2013. Furthermore, observed glacier mass balance data for 1989–2012 at Xiaodongkemadi Glacier (33°04′N, 92°05′E) (Figure 1) have been collected (Tong *et al.* 2016). The Xiaodongkemadi glacier is near the northern part of the upstream region of the Salween River (Figure 1). It has a total area of 1.8 km^2, a length of about 2.8 km, and altitude elevation ranging between 5,380 m and 5,926 m.

In addition, the global 8-day and 0.05° Moderate Resolution Imaging Spectroradiometer (MODIS) snow products (MOD10C2) (http://nsidc.org/data/modis/index.html) during 2001–2013 were used for snow cover analysis over the two upstream basins. Previous assessments over the TP have revealed that the MODIS snow product generally has sufficient accuracy to reflect snow cover information over the TP (Pu *et al.* 2007; Li *et al.* 2018). These studies have also suggested that the MODIS data can be used to assess the snow cover dynamics in the TP.

Table 2 | Monthly temperature laps rate (Tlaps) (°C/km) and correlation coefficient (R) between temperature and elevation

Month	Jan	Feb	Mar	Apr	May	Jun	Jul	Aug	Sep	Oct	Nov	Dec
Tlaps	−3.8	−4.1	−4.4	−4.6	−4.5	−3.8	−3.9	−3.9	−3.5	−3.9	−4.2	−3.8
R	0.45	0.51	0.57	0.73	0.75	0.70	0.79	0.79	0.66	0.62	0.55	0.47

METHOD

Hydrological model

In this study, the VIC model (Liang *et al.* 1994, 1996) was employed. The VIC model is a grid-based land surface scheme which parameterizes the dominant hydrometeorological processes taking place at the land surface–atmosphere interface. The model is characterized by a mosaic representation of land surface cover and a subgrid parameterization for infiltration, which accounts for subgrid scale heterogeneities in land surface hydrologic processes (Su *et al.* 2005). The soil column comprises three soil layers, which allows the representation of the rapid dynamics of soil moisture movement during storm events and the slower deep inter-storm response in the bottom layer. Three types of evaporation are considered in the model. They are evaporation from the canopy layer of each vegetation class, transpiration from each of the vegetation classes and evaporation from bare soil. Total evaporation over the grid cell is computed as the sum of the three individual evaporation elements (Liang *et al.* 1994). For details of each part related to water balance and energy fluxes, please refer to the investigation by Liang *et al.* (1994).

However, there is no incorporation of a glacier melt module in the current official version of the VIC model to simulate melting processes for icebergs in the glaciated mountainous regions. Recently, a 'VIC-glacier' model which links a simple degree-day glacier algorithm with the original VIC model has been developed to do hydrological modelling over basins with glacier in the TP (Zhang *et al.* 2013). In this study, we utilized this VIC-glacier model. The total runoff includes the glacier meltwater from each grid which is estimated as:

$$R(i) = f \times M_i + (1 - f) \times R_{vic} \tag{1}$$

where $R(i)$ is the total runoff (mm) in grid i; f is the ratio of glacier area over grid i; R_{vic} is the estimated runoff for the ice-free part in grid i; M_i is the calculated melt runoff (mm) from the glacier part in grid i using the degree-day model.

The glacier volume is estimated using a modified equation:

$$V = 0.04\,s^{1.43} \tag{2}$$

where V is the ice volume and S is the total glacier area. The initial ice volume in the catchment was estimated from an inversion of Equation (2) using glacier surface area derived from the glacier distribution data. In our study, the VIC-glacier model was set up over the study basins at a spatial resolution of $1/12° \times 1/12°$.

Model parameters

The performances of the VIC-glacier model in the UM and US largely rely on two categories of model parameters: (1) degree-day factors (DDFs) for simulating meltwater over glacierized areas, including degree-day factors for snow (DDF_{snow}) and ice (DDF_{ice}); and (2) parameters in the VIC model for simulating runoff in non-glacierized areas. In this study, the initial values of degree-day factors for snow (DDF_{snow}) and ice (DDF_{ice}) in the UM and US were determined from the investigated results of Zhang *et al.* (2006), which indicated an average DDF_{snow} of 4.1 mm $°C^{-1}$ day^{-1} and DDF_{ice} of 7.1 mm $°C^{-1}$ day^{-1} for western China based on glacier mass balance observations. Then, the glacier area changes between the first China Glacier Inventory and Randolph Glacier Inventory 5.0, and the observed monthly flows for the two upstream basins, were utilized to calibrate the degree-day factors. The final adopted values of the DDF_{snow} and DDF_{ice} for the two basins are listed in Table 3.

The parameters of the VIC model needing calibration include the infiltration parameter (b_inf), the depth of the first and second soil layers (D_1 and D_2), and the three base flow parameters (Ws, Dsmax and Ds) (Su *et al.* 2005). The parameter b_inf, which has a common range of 0–0.4, defines the shape of the VIC curve. The first soil depth (D_1) for each grid was set to 5–10 cm, according to the investigation of Liang *et al.* (1996). The three base flow parameters, which affect the flow and storage of the water in the third layer, generally need minor adjustment during the calibration. Hence, only the infiltration-shape

Table 3 | Values of parameters adopted in the VIC model

Parameter	Description	Range	UM	US
DDF_{ice} (mm °C^{-1} day^{-1})	Degree-day factor for ice-melt	3.4–13.8	11.5	7.1
DDF_{snow} (mm °C^{-1} day^{-1})	Degree-day factor for snowmelt	3.0–7.9	6.5	4.1
D_s (fraction)	Fraction of D_{smax} where non-linear baseflow begins	0–1	0.03	0.02
D_{smax} (mm/d)	Maximum velocity of baseflow	0–50	10	10
W_s (fraction)	Fraction of maximum soil moisture where non-linear baseflow occurs	0–1	0.9	0.9
b_inf	Variable infiltration curve parameter	0–0.4	0.3	0.2
D_2 (m)	Thickness of the second soil moisture layer	0–3	1.5	1

parameter (b_inf) and the second layer depth (D_2) were targeted for intensive calibration (Table 3). The range of these parameters refers to previous studies (Kayastha *et al.* 2003; Su *et al.* 2005).

In this study, the manual calibration process, i.e., the trial and error method, was employed to calibrate DDFs and the VIC model to improve its hydrological performance in the two upstream regions over the TP. The percent bias (PBIAS) and Nash–Sutcliffe efficiency coefficient (NSE) were used as the VIC evaluation statistics during the model calibration. PBIAS and NSE are defined as follows:

$$NSE = 1 - \frac{\sum_{i=1}^{n}(Y_i^{obs} - Y_i^{sim})^2}{\sum_{i=1}^{n}(Y_i^{obs} - Y^{mean})^2} \quad (3)$$

$$PBIAS = \frac{\sum_{i=1}^{n}(Y_i^{sim} - Y_i^{obs})*100}{\sum_{i=1}^{n}(Y_i^{obs})} \quad (4)$$

where Y_i^{obs} and Y_i^{sim} are the observed data and simulated value at time i; Y^{mean} is the mean of observed data for the whole evaluating period. The final calibrated parameters used in the VIC-glacier model are shown in Table 3.

Trends and attribution analyses

In this study, the Mann-Kendall test (Mann 1945; Kendall 1975) was used to examine the monotonic trend of hydrometeorological variables for the period of 1964–2013. In this method, the standard normal statistic Z is estimated by:

$$Z = \begin{cases} (S-1)/\sqrt{Var(S)} & if \quad S > 0 \\ 0 & if \quad S = 0 \\ (S+1)/\sqrt{Var(S)} & if \quad S < 0 \end{cases} \quad (5)$$

$$S = \sum_{i=1}^{n-1}\sum_{j=i+1}^{n} \mathrm{sgn}(x_j - x_i) \quad (6)$$

$$\mathrm{sgn}(\theta) = \begin{cases} 1 & if \quad \theta > 0 \\ 0 & if \quad \theta = 0 \\ -1 & if \quad \theta < 0 \end{cases} \quad (7)$$

$$Var(S) = \left[n(n-1)(2n+5) - \sum_t t(t-1)(2t+5)\right]/18 \quad (8)$$

where t is the extent of any given tie. A positive value of Z indicates an increasing trend and vice versa. The test statistic Z is not statistically significant if $-Z_{\alpha/2} < Z < Z_{\alpha/2}$ while it is statistically significant if $Z < -Z_{\alpha/2}$ or $Z > Z_{\alpha/2}$.

In order to quantify the trend magnitude, the Theil–Sen approach (TSA) was employed (Kumar *et al.* 2009). The TSA slope β is defined as:

$$\beta = Median\left(\frac{x_j - x_i}{j - i}\right) \quad \text{where} \quad 1 < i < j < n \quad (9)$$

Positive and negative β values indicate an increasing or decreasing trend, respectively.

In this study, total runoff consists of rainfall runoff, snowmelt and glacial runoff. Thus, a change in individual runoff component will cause variation of the total runoff, and vice versa. The trend for total runoff is the sum of the trends of the three individual runoff components. Such changes can be described by the following formula:

$$\Delta R = \Delta R_r + \Delta R_s + \Delta R_g \tag{10}$$

where ΔR is the trend in total runoff for 1964–2013, ΔR_r, ΔR_s and ΔR_g are the trend for rainfall runoff, snowmelt and glacial runoff over the same period, respectively. The trend magnitude for total runoff (ΔR), rainfall runoff (ΔR_r), snowmelt runoff (ΔR_s) and glacial runoff (ΔR_g) can be calculated by Equation (9). Furthermore, the per cent contribution of trend in individual runoff variable to the tendency of the total runoff can be calculated as follows:

$$P_r = \frac{\Delta R_r}{\Delta R} * 100, \quad P_s = \frac{\Delta R_s}{\Delta R} * 100, \quad P_g = \frac{\Delta R_g}{\Delta R} * 100 \tag{11}$$

where P_r, P_s and P_g are percentage contributions of the trend in the rainfall runoff, snowmelt runoff and glacial runoff to the total runoff trend, respectively. This can be regarded as an attributing method which quantitatively describes the controlling role of the individual runoff element in the total runoff trend. The sum of per cent contributions for the three runoff constituents should equal 100%. The method utilizing per cent contribution to quantify the role of hydrological component in the change of some environmental variables has been widely employed in hydrological researches (Wu *et al.* 2014; Yao *et al.* 2014; Meng *et al.* 2019).

RESULTS

Model validation

Figures 2 and 3 provide the observed and VIC-glacier model simulated monthly time series of hydrograph for the UM (1964–2009) and US basins (1980–1985). Also, comparisons between mean monthly modelled and measured hydrograph are presented in Figure 4. Table 3 lists calibrated values for the chosen parameters in each focus region, which were determined by using trial and error techniques. In addition, the deriving evaluated statistics for the modelling efficiency over the two regions are shown in Table 4.

Generally, the simulated monthly streamflow during the focus period can capture the observed evolution and magnitude reasonably for the two upstream basins, with NSE of 0.73 and 0.86 for the UM and US, respectively (Figures 2 and 3). From Figures 2 and 3, it can be distinguished that the simulated baseflow of the VIC-glacier model is less than the observed flow. One plausible reason is due to seasonal frozen soil over the two basins which would negatively affect the VIC baseflow simulation to some extent. However, the major hydrologic characteristics, such as the timing and amplitude of the discharge peaks, the rising and recessing limb of hydrograph (Figure 4), the modelling result for the UM and US, can basically reproduce the pattern of observed flow, which further validates the good efficiency of the VIC-glacier model in terms of the streamflow simulation.

Figure 5(a) and 5(b) compare the VIC simulated and MODIS estimated mean monthly snow cover extent (fraction of basin area with snow). For both regions, the VIC model underestimates snow cover extent during all months in comparison with MODIS estimation. However, in respect of the pattern of snow dynamics related to the timing of snow accumulation and snow depletion, there is a good consistency between the VIC simulation and MODIS output, i.e., snow beginning to form accumulations around late September and starting to melt about early March for both the VIC modelling results and remote sensing estimate outcomes. Simultaneously it can be found, for any month, the difference between the VIC simulated snow cover and estimation from MODIS is no more than 15% in the UM and US. Also, with regard to modelling snow cover change, it can be understood that both the outcomes from VIC simulation and MODIS estimation contain large uncertainties (Cuo *et al.* 2013; Chen *et al.* 2017). In addition, because of the scarcity of observed data and some unpredictable processes (such as snow drift), snow melting is too complicated to fully represent, which may be the reason for underestimation of snow cover. Thus, on the whole, it can be identified that VIC modelling results can basically reflect the seasonal cycles of the satellite estimations.

Furthermore, with regard to the degree-day model in simulating glacier melt, Figure 6 illustrates the observed

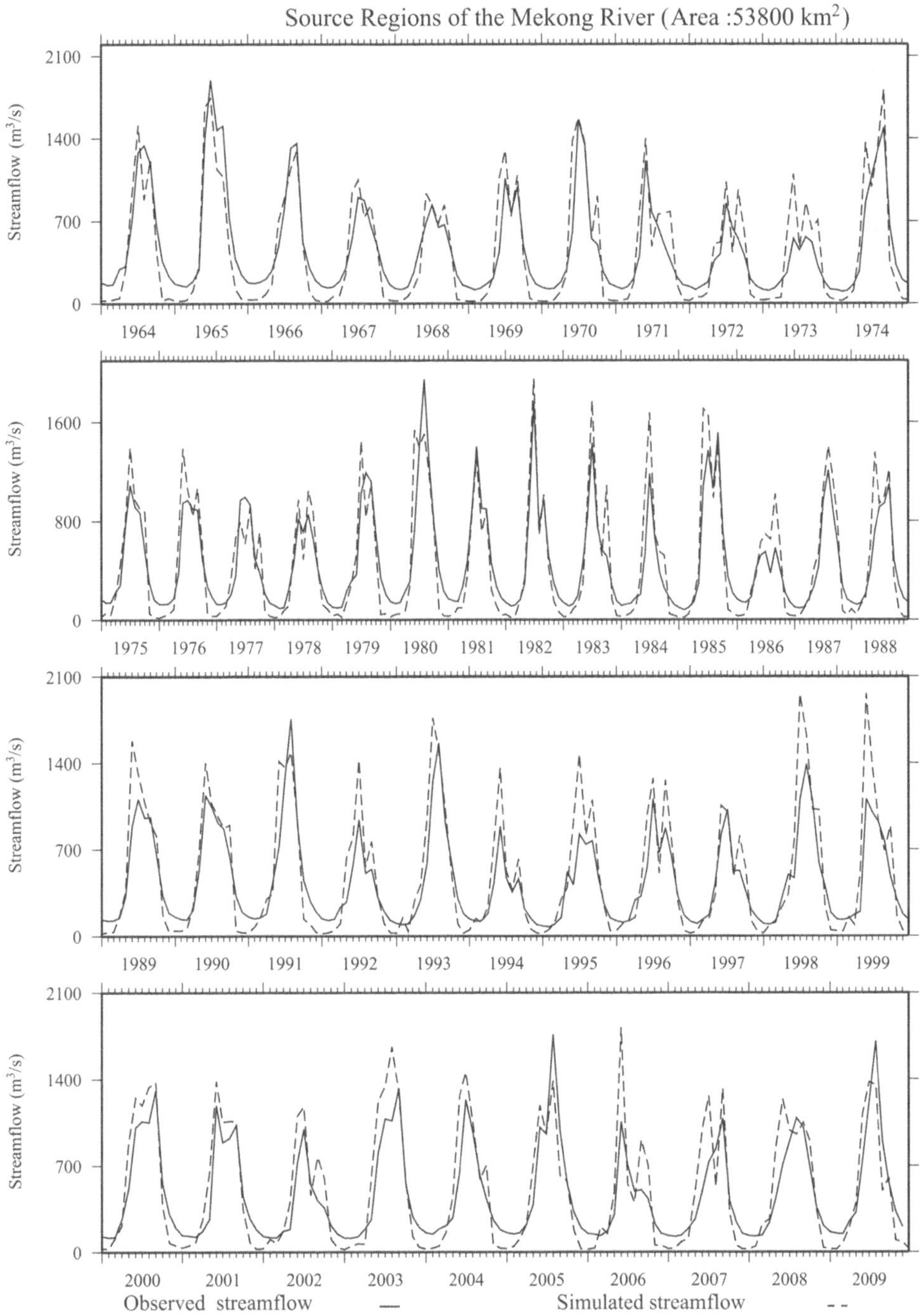

Figure 2 | Observed and modelled monthly streamflow for the upstream region of the Mekong River basin at Changdu gauge for 1964–2009.

and simulated annual mass balance of Xiaodongkemadi Glacier for 1989–2012 with a correlation coefficient of 0.81 and relative bias of −1.31%. Meanwhile, from the statistical results, the VIC-glacier model simulated a decrease in glacier areas of −26.0% in the UM and −22.0% in the US during 1968–2010, which are generally close to the observed changes of −29.0% and −20.0% in the two basins between the first Chinese Glacier Inventory and the Randolph Glacier Inventory 5.0 data, respectively. Thus, the good consistency between simulated results and measured data

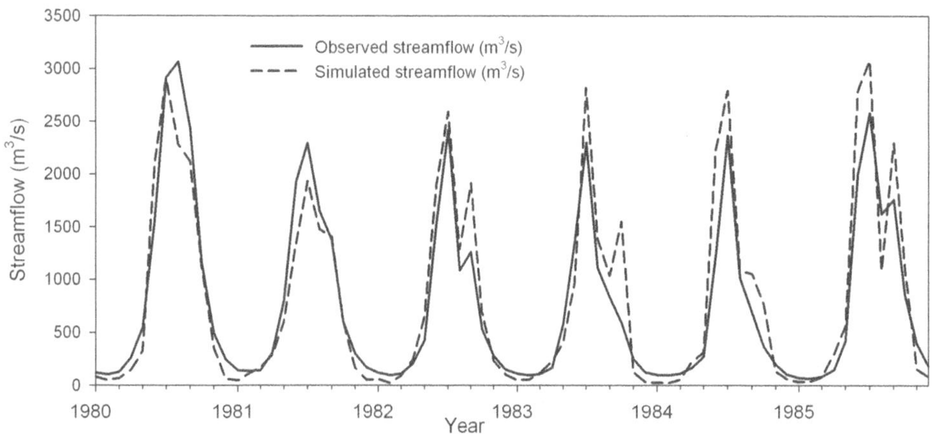

Figure 3 | Observed and modelled monthly streamflow for the upstream region of the Salween River basin at Jiayuqiao gauge for 1980–1985.

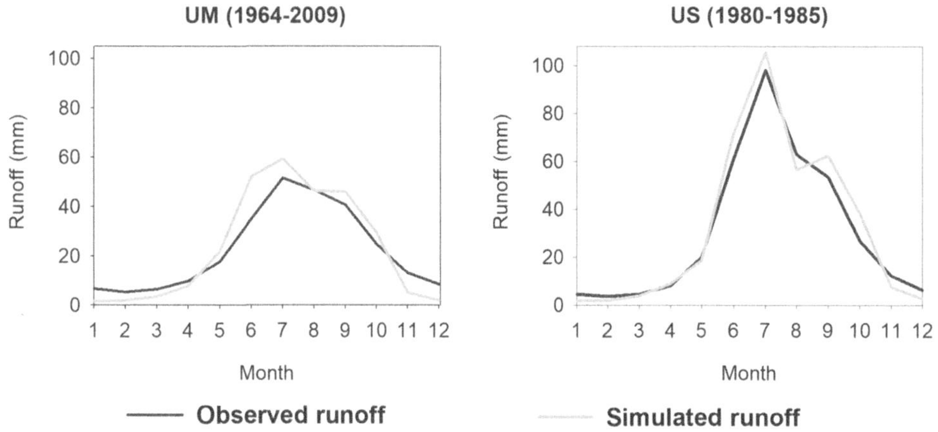

Figure 4 | Mean monthly observed and VIC simulated total runoff for the UM and US over respective period with measured flow.

Table 4 | Nash–Sutcliffe efficiency (NSE) and percent bias (PBIAS) of the simulated monthly streamflow relative to the observations for the UM (1964–2009) and US (1980–1985)

Basin	Observed annual runoff (mm)	Simulated annual runoff (mm)	NSE	PBIAS (%)
UM	278.35	279.60	0.73	0.45
US	361.95	380.26	0.86	5.1

suggests the feasibility of the VIC-glacier model in modelling glacier runoff in the study regions.

On the whole, the VIC-glacier model can reproduce the evolution and magnitude of the observed streamflow at the UM and US; the VIC performance in simulating snow cover shows acceptable results; and the modelling glacier annual mass balance and glacier area change are comparable to the observed data.

Runoff components

In this study, the total runoff consists of three components: rainfall runoff (including direct rainfall surface runoff and subsurface runoff), snowmelt runoff (surface runoff from melting snow) and glacier runoff (runoff from melting ice on the glaciated area). Figure 7 presents the seasonal distribution of the runoff components, and Table 5 gives their corresponding per cent contribution to the total annual runoff for the UM and US basins during the period of 1964–2013.

The rainfall runoff contributes to a respective 85.66% and 85.81% of total runoff for the UM and US (Table 5).

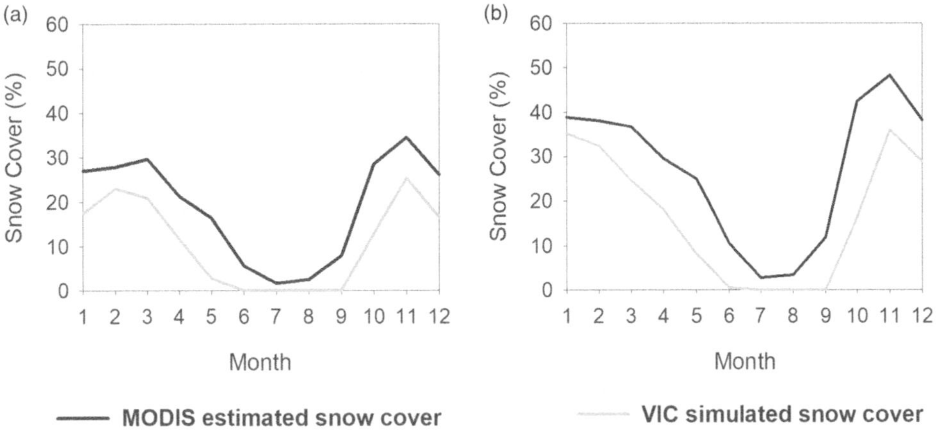

Figure 5 | Long-term mean monthly MODIS estimated snow cover versus VIC simulated snow cover during 2001–2013: (a) results for the UM; (b) results for the US.

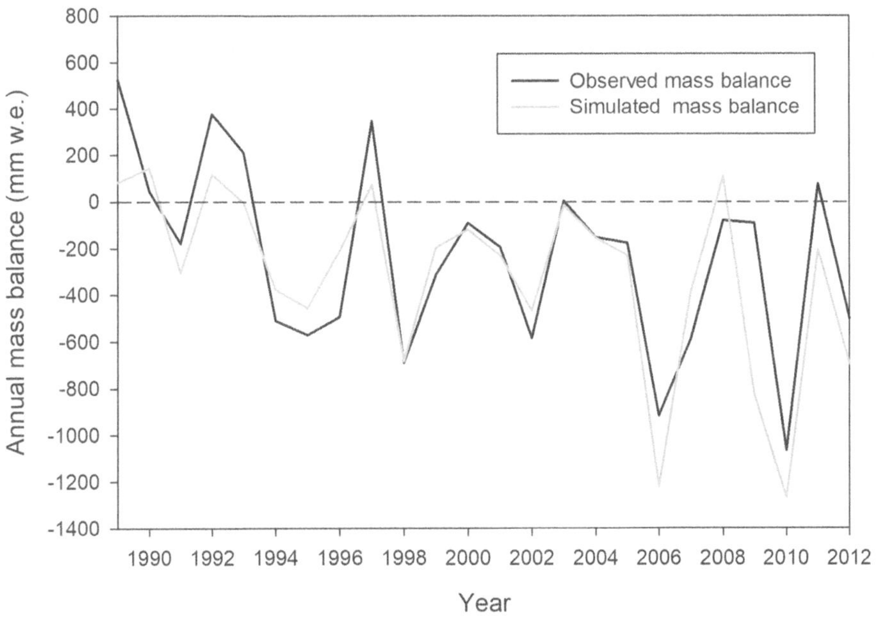

Figure 6 | Observed and simulated annual mass balance of the Xiaodongkemadi Glacier for the years 1989–2012.

This suggests that the monsoon rainfall plays an important role in the runoff generated in the two regions. The snowmelt water, which contributes about 12.37% and 6.87% of the annual total runoff in the UM and US, is an important water source especially for irrigation in the spring season. It can be identified from Figure 7 that there are two peaks in the mean monthly snowmelt runoff over the focus basins, with one in June and the other in October. The snowmelt in spring or early summer, which is from the accumulated snowpack in winter and the snow falling in spring, mostly occurs during April–June and reaches the peak in June along with the rise of temperature. With regard to the snowmelt peak in the October, it is probably ascribable to the fresh snowfall in September and October, which melts immediately after a short stay of a few days above the ground. There are similar phenomena in other regions such as the Kaidu River in northwestern China (Shen *et al.* 2018). The contribution of the glacier runoff to

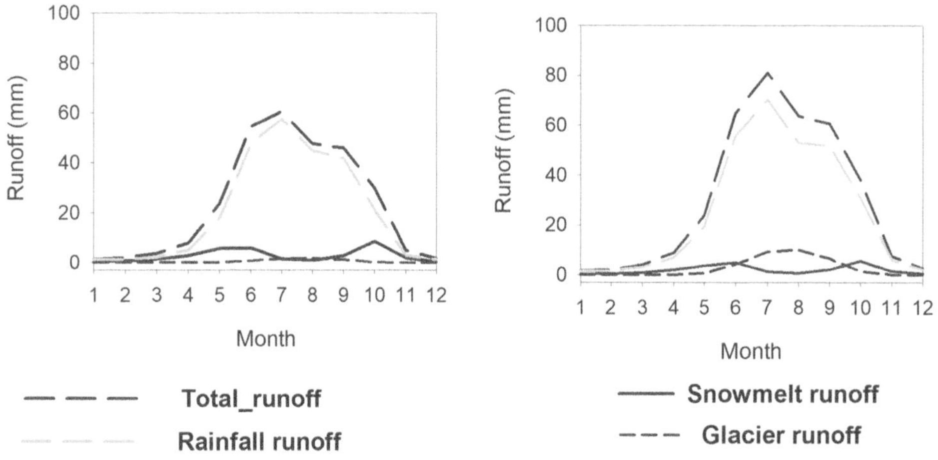

Figure 7 | Mean monthly VIC simulated total runoff and three runoff components for the UM and US during 1964–2013.

Table 5 | Contribution of simulated glacier runoff, snowmelt and rainfall runoff to total runoff during 1964–2013

Basin	Total runoff	From glacier melt (%)	From snowmelt (%)	From rainfall runoff (%)
UM	278.35 mm	1.97	12.37	85.66
US	362.00 mm	7.32	6.87	85.81

total runoff differs between the two basins, with a proportion about 1.97% over the UM and a larger ratio (7.32%) in the US, respectively (Table 5).

Changes for rainfall runoff, snowmelt, glacier-melt runoff and total runoff

Generally, annual temperature and precipitation across the two upstream basins indicate an overall increasing trend during the years 1964–2013. Both annual precipitation and temperature in the UM indicate a statistically increasing trend, at α 0.05 significance level by using Mann–Kendall test. The US is similar to the UM except for precipitation being a non-significant up trend. Figure 8 further demonstrates mean monthly precipitation and temperature for the two upstream basins over three time slices to further reveal decadal climate change occurred in the past 50 years. Under climate change, the rainfall runoff, snowmelt and glacier runoff have experienced corresponding changes.

The linear trends for annual simulated rainfall runoff, snowmelt and glacier-melt runoff for the past 50 years were derived (Table 6). Rainfall runoff generally indicates an overall increasing trend from 1964 to 2013 in both regions, of which it is significant at α 0.05 significance level in the UM by the MK test. Meanwhile, the modelling glacier-melt runoff exhibits a significant increase in both areas by the MK test. However, simulated snowmelt shows non-significant increasing trend in the UM while the US is characterized by a decreasing tendency. The total runoff in the UM displays a significant increase during 1964–2013.

To evaluate potential changes in runoff-generated regimes, we separated our results into three periods, i.e., 1964–1990 (baseline period), 1991–2000 (1990s) and 2001–2013 (2000s). Figure 9 illustrates mean monthly rainfall runoff, snowmelt, glacial runoff and total runoff for the three periods in the study basins.

Figure 9(a) and 9(b) show results of long-term average monthly rainfall runoff in three periods. In the UM, compared to the baseline period of 1964–1990, increasing rainfall runoff can be found in nearly all months for both the 1990s and 2000s, with a respective increment of 10% and 12% during the two periods. In the US, the rainfall runoff in June and August in the 1990s shows some decreases while July indicates a little increase, and the final statistical analysis suggests a slight reduction over the 1990s relative to the baseline period. However, during the 2000s, moderate increase can be observed in the warm season (April to September except for July and August) in the US and, therefore, the annual mean

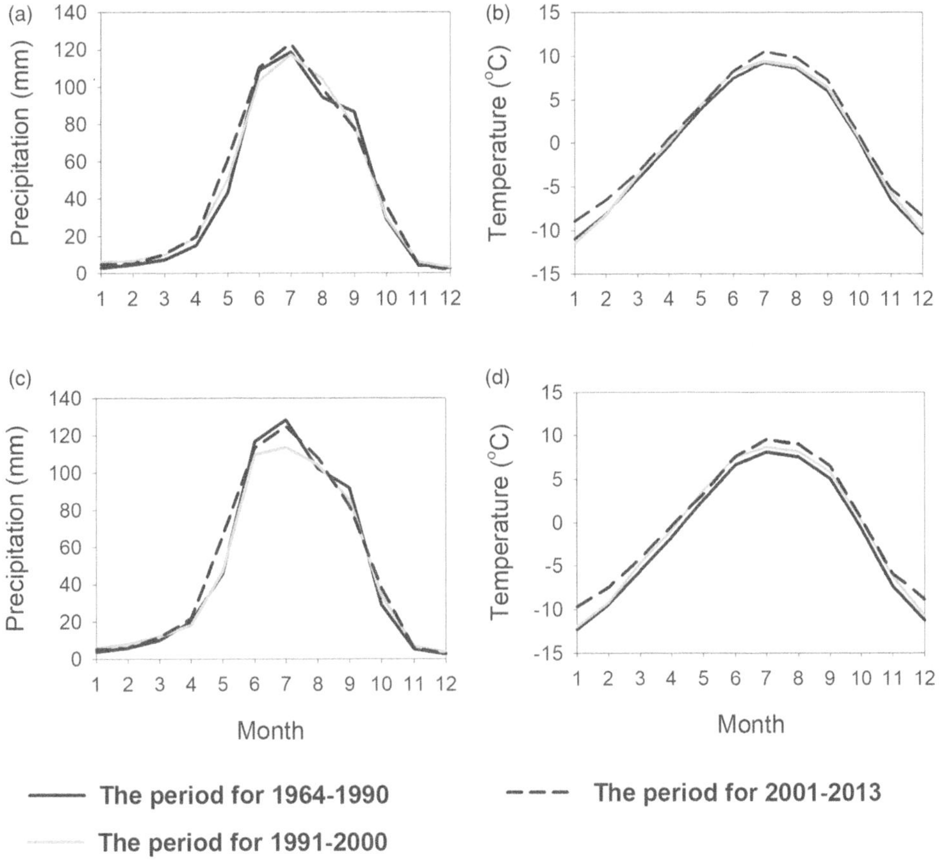

Figure 8 | Mean monthly precipitation and temperature during three periods of 1964–1990, 1991–2000 and 2001–2013: (a) and (b) are results in the UM; (c) and (d) are results in the US.

Table 6 | Trends for simulated runoff components during 1964–2013

Basin	Total runoff	Rainfall runoff	Snowmelt runoff	Glacial runoff
UM	6.845×10^7 m^3/year[a]	6.6×10^7 m^3/year[a]	0.105×10^7 m^3/year	0.14×10^7 m^3/year[a]
US	4.67×10^7 m^3/year	3.93×10^7 m^3/year	-0.4×10^7 m^3/year	1.14×10^7 m^3/year[a]

[a]Significance 0.05.

rainfall runoff exhibits an increment of 4.62% in comparison to 1964–1990.

In respect of snowmelt runoff, the long-term monthly average values in three periods over the focus areas are listed in Figure 9(c) and 9(d). In the UM, the snowmelt runoff peaks in June during 1964–1990 but, nevertheless, the peak shifts from June in the baseline period to May for both the 1990s and 2000s. In the 1990s, compared to that in the baseline period, a relatively large increment in spring is observed while the other months show a little decrease, and thus the annual snowmelt runoff in the UM suggests an increase of 8.19% relative to the baseline period. In the 2000s, there is also moderate increase during spring but considerable decrease can be noticed among other months and, therefore, the annual snowmelt runoff has been reduced by 5.63%. Meanwhile, for the US, compared to the baseline times, the months for February–April experience a small increase during the 1990s and there is also some growth during February–May in the 2000s, whereas most of the remaining months witness relatively large reductions for both the 1990s and 2000s, and therefore the annual snowmelt runoff indicates a decrease

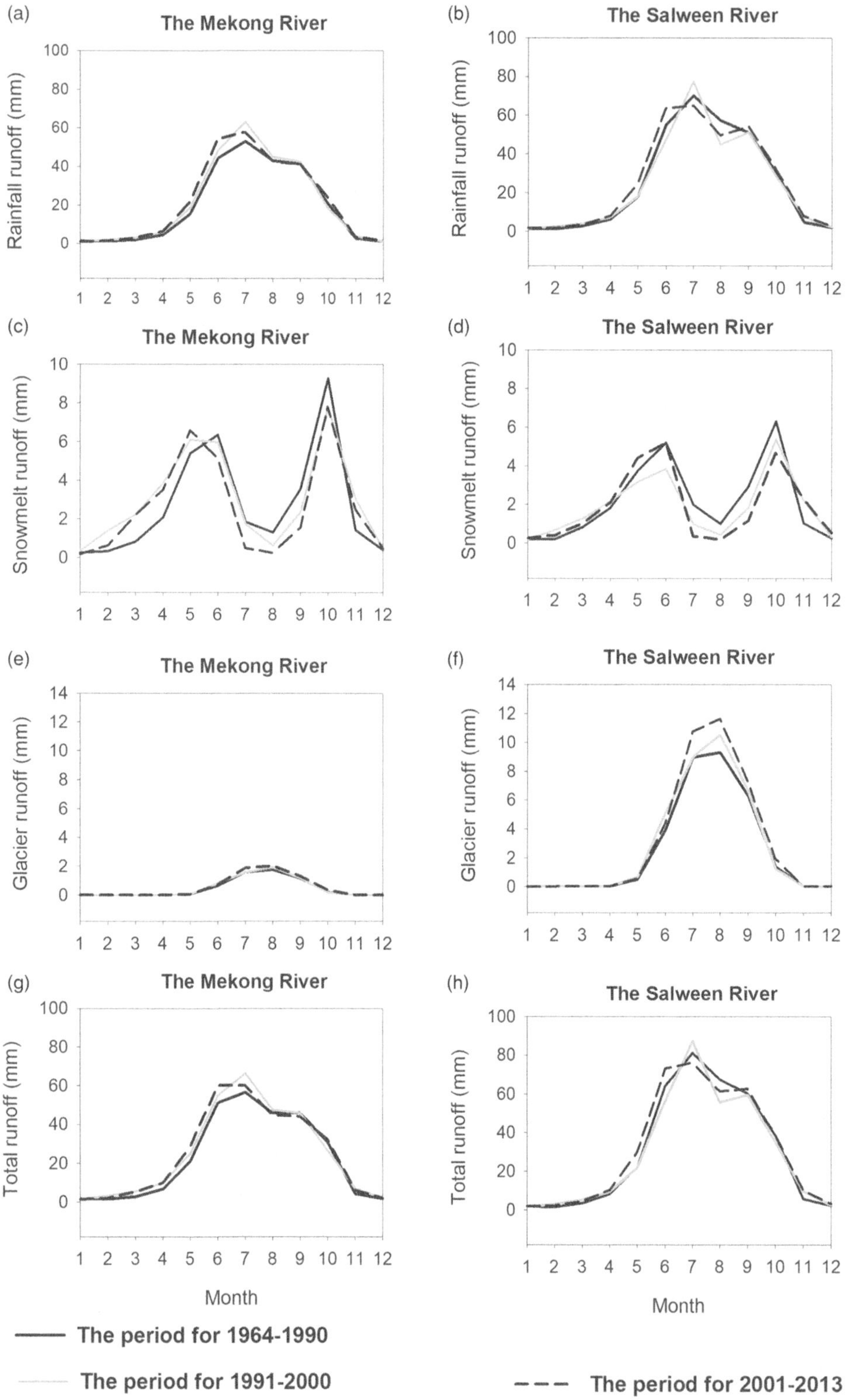

Figure 9 | Mean monthly rainfall runoff, snowmelt runoff, glacier runoff and total runoff in the study regions for three periods of 1964–1990, 1991–2000 and 2001–2013: (a), (c), (e) and (g) are the results in the UM; (b), (d), (f) and (h) are the results in the US.

Table 7 | Proportional contribution for controlling trend of total runoff to three runoff components during 1964–2013

Basin	Trend of total runoff (m^3/year)	Contribution from rainfall runoff (%)	Contribution from snowmelt runoff (%)	Contribution from glacial runoff (%)
UM	6.845×10^7	96.42	1.54	2.04
US	4.670×10^7	84.22	−8.56	24.33

of 11.55% and 11.14% for the two periods, respectively. As we can see, there is more snowmelt runoff in spring for both the 1990s and 2000s in the UM and for the 2000s in the US, compared to the baseline period. This phenomenon is probably concerned with the increasing precipitation during spring season in the 2000s over these two regions relative to 1964–1990 (Figure 8(a) and 8(c)). Due to the low temperature during spring, part of precipitation falls as snowfall and it melts quickly after a brief stay above the ground with the rise of temperature; its contribution to river runoff is suggested as 'snowfall runoff'. Therefore, the increasing precipitation may generate more snowfall in spring for the 2000s in both the UM and US, and results in a larger snowmelt runoff during this season relative to that of 1964–1990. Some other researchers also found that during the wet spring season, there is still some snowfall and the following snowmelt runoff over the investigated areas, such as the headwaters of the Yellow River (Cuo *et al.* 2013) and the upstream region of the Yarkant River (Kan *et al.* 2018).

Compared to the variation of rainfall and snowmelt runoff, changes in glacier runoff (Figure 9(e) and 9(f)) exhibit more consistency for the 1990s and 2000s across the two regions relative to 1964–1990. Glacier runoff demonstrates an increasing trend during the melting season (June to September) for both periods compared to that of baseline period. The annual average ice melt increased by about 8.4% and 9.3% for the UM and US in the 1990s in comparison with the reference period. In terms of changes for the 2000s, the glacier runoff indicates a respective increment of 19.3% and 20% in the UM and US relative to 1964–1990.

Due to changes that occurred in rainfall runoff, snowmelt and glacier runoff, the total runoff also underwent a corresponding variation during the 1990s and 2000s (Figure 9(g) and 9(h)). The variations for the 1990s and 2000s relative to the baseline period are similar to that of rainfall runoff.

The controlling roles of runoff components in the trend of river flow for the past 50 years

Based on the statistical results, both the total runoffs in the UM and US show a rising trend during 1964–2013, with an increment of $6.845\times10^7\,m^3/yr$ and $4.67\times10^7\,m^3/yr$, respectively (Table 6). The individual role of rainfall runoff, snowmelt and glacial runoff in controlling the variation of total runoff can be quantitatively derived according to Equations (10) and (11). Table 7 lists the corresponding controlling roles of each runoff constituent in the trend of runoff for the years 1964–2013.

The roles of rainfall runoff, snowmelt and glacial runoff vary from basin to basin in the tendency of total runoff. For the UM, the increasing total runoff is the compounded effect of the rainfall runoff, snowmelt runoff and glacial runoff, all of which impose a positive impact on the upward trend of total runoff. In terms of respective hydrological role, the rainfall runoff, snowmelt and glacial runoff indicate a particular contribution of 96.42% ($6.6\times10^7\,m^3/yr$ was divided by $6.845\times10^7\,m^3/yr$), 1.54% ($0.105\times10^7\,m^3/yr$ was divided by $6.845\times10^7\,m^3/yr$) and 2.04% ($0.14\times10^7\,m^3/yr$ was divided by $6.845\times10^7\,m^3/yr$) to the change trend of the total runoff for the period of 1964–2013 (Table 7). Meanwhile, for the US, the increasing rainfall runoff ($3.93\times10^7\,m^3/yr$) and glacial runoff ($1.14\times10^7\,m^3/yr$) both have a positive effect on the ascending trend of total runoff while decreasing snowmelt runoff ($-0.4\times10^7\,m^3/yr$) imposes a negative impact on the change of total runoff, and is responsible for 84.22%, 24.33% and −8.56% in order.

DISCUSSION

The hydrological role of melt runoff

In this study, the rainfall runoff contributes to more than 85% of total runoff over the UM and US and so it is the dominant runoff component, which is similar to the study of Lutz *et al.* (2014). Being located in the southeastern

Tibetan Plateau, the two basins are under the impact of the Asia monsoon and receive a large amount of precipitation during summer, and thus the hydrological regime is rain dominated. The runoff regime in other basins of the eastern TP, such as the upstream region of the Yellow River, the Yangtze River, are also mainly controlled by rainfall (Immerzeel *et al.* 2010; Su *et al.* 2016). However, although the melt runoff from glacier and snow cover is limited and no more than 15% of the total runoff over these two regions, it is still playing an important role in both water environment protection and available water resources for the local and further downstream regions. On the one hand, the water quality for the melt runoff is especially good as there is little or negligible pollution over the remote and inaccessible glacier or the snow cover, so the snow and glacial runoff could partly supply high-quality clean water resources to the downstream, and also would neutralize or mitigate the polluted water in the downstream basin. On the other hand, melt water is very important during dry years (the anti-cyclic behaviour) in summer monsoon-dominated areas such as the two focus basins in our study. When anti-cyclic behaviour occurs, the summer precipitation decreases sharply and thus the basins lack sufficient water supply from rainfall. Then, the melt water from glacier and snow cover could compensate for the potential water shortage in the local region and, to some extent, buffer the negative trend of decreasing discharge from the upper region to the downstream basin. This phenomenon is generally referred to as the glacier compensation effect (Zhao *et al.* 2015). As shown in Figure 10, there is an obvious negative correlation between glacial runoff and river total runoff during the warm season (May–October) over the two basins, which indicates that there is less glacial runoff in wet years but more glacial runoff in dry years. Taking the US as an example (Figure 10(b)), the maximum contribution for glacial runoff to river total runoff can approach 16% during an extreme dry year while the mean value for glacial runoff's contribution is less than 8%. Hence, in these two regions, the regulating capacity for glacial runoff in total runoff is important and cannot be ignored even if the glacier area is no more than 2% of the whole basin.

In the future, the changes in temperature and precipitation will be expected to seriously affect the melt characteristics and the hydrologic regimes over these high-altitude regions, including changes in runoff magnitude, shift of intra-annual patterns of streamflow and the advance of turning point for glacial runoff. Therefore, more detailed study will be dedicated to the analysis of the hydrological effect of climate change in the region.

Model uncertainty

In this study, the VIC model linking with degree-day glacier algorithm, referred to as the VIC-glacier model, was utilized to detect the hydrological regime variation under recent warming climate. Although the VIC-glacier model was calibrated by using observed data or satellite data, there is still some probable uncertainty.

As indicated in Table 7, the glacial runoff plays an important role in controlling streamflow trend over these two basins, although it is not the dominant factor for the

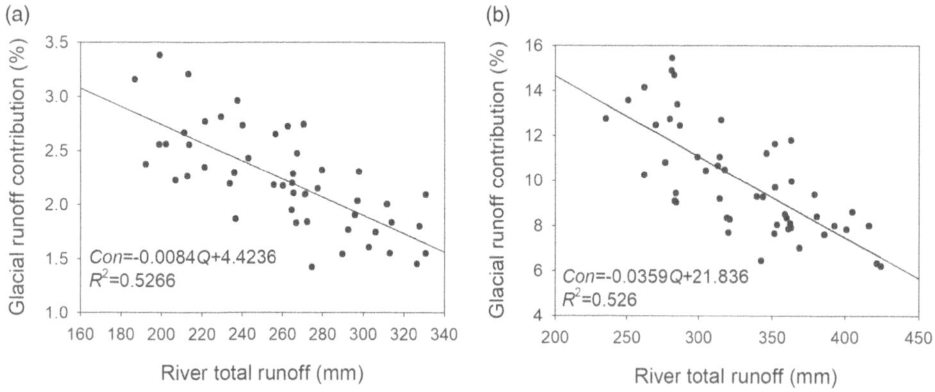

Figure 10 | Scatterplots of glacial runoff contribution (Con) versus river total runoff (Q) during the warm season (May–October) for 1964–2013 in the UM (a) and US (b).

river flow variation. Glacier melt runoff was simulated by the degree-day model and the DDFs are key parameters in this model. In this study, DDF_{ice} and DDF_{snow} were calibrated using glacier change data and observed streamflow data. Also, the mass balance observation on the Xiaodongkemadi glacier is used as auxiliary data for constraining the DDF_{ice} and DDF_{snow}. The final determined values for DDF_{ice} and DDF_{snow} are in the range of these parameters (Table 3). Also, the DDF_{ice} and DDF_{snow} in our study are comparable to other studies in the same regions. For example, the study from Su *et al.* (2016) indicates that their adopted DDF_{ice} for UM and US is 13.8 mm °C^{-1} day^{-1} and 7.1 mm °C^{-1} day^{-1}, respectively, which are comparable to our results. However, the DDFs have large spatial and temporal variation, which could greatly affect the accuracy of snow and ice melt modelling (Hock 2003; Tong *et al.* 2016). Even a single glacier could be subject to significant small-scale variations across its surface (Zhang *et al.* 2006). Meanwhile, altitude is also an important factor affecting the DDFs. Larger DDFs are found on glaciers situated at higher altitude, which may be caused by larger radiation and lower positive degree-day sum at the higher altitude (Kayastha *et al.* 2003). In our work, the two groups of constant DDFs for the UM and US basically represent an average condition for glaciers in the respective basin, and would unavoidably result in uncertainties in the simulation results. Figure 11 exhibits the sensitivity of glacial runoff to the parameters of DDFs in the UM. The mean annual glacier melt runoff would decrease/increase about 10% with the decrease/increase of one unit DDF (mm °C^{-1} day^{-1}) in this basin, which is consistent with the findings in the upstream of Yarkant basin in the Karakoram (Kan *et al.* 2018) and the Siling Co basin in the TP (Tong *et al.* 2016). The current glacier scheme (the degree-day model) has a lower complexity than other processes included in the VIC model. The extended degree-day model, in which DDFs account for the effect of aspect in mountain region and also vary with season, corresponding to the spatial and temporal variation, is currently under way to yield more realistic melt estimates and further decrease the uncertainty in the runoff modelling over the TP.

Limited to the scarcity of other data such as measured soil moisture or frozen soil depth, only the observed streamflow data are utilized to calibrate VIC soil and baseflow parameters in hydrological modelling over the ice-free region, which may induce uncertainty in runoff simulation in these areas. In the future, more observed data, including groundwater level and permafrost depth, will be collected to comprehensively calibrate hydrological parameters, and this would increase the robustness of the model and decrease uncertainty during hydrological modelling in the TP basin.

CONCLUSIONS

In this study, a coupled VIC and glacier model was applied to two catchments located in the Tibetan Plateau region to analyse the temporal evolution of the rainfall runoff, snow and glacier melt contribution to river total runoff during the

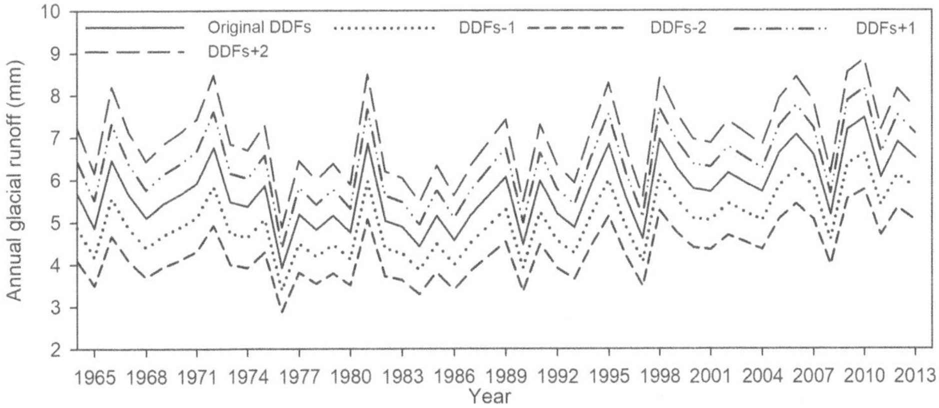

Figure 11 | Simulated annual glacial runoff with different DDFs in the UM for 1964–2013.

past 50 years, and also to evaluate their roles in controlling streamflow trends. The following conclusions were made.

During the past 50 years, the total streamflow indicates an increasing trend in the two focus areas. There is also a general increasing trend for both rainfall runoff and glacial runoff in the UM and US. In terms of snowmelt runoff, it demonstrates an increasing trend over the UM while a decreasing tendency is exhibited in the US.

The rainfall runoff was considered as the dominant factor driving changes of river discharge, which can be responsible for over 84% of changes in the total runoff over all focus catchments. Meanwhile, the glacial runoff illustrates an increasing important role in the controlling streamflow trend under continuous climate warming, in comparison to the snowmelt runoff. Findings from this study can provide beneficial reference to water resource and eco-environment management strategies for governmental policy-makers.

ACKNOWLEDGEMENTS

This work was supported by the National Natural Foundation of China (51379179) and the National key research projects of China (Grant No. 2016YFC0402704). We thank Fengge Su and Leilei Zhang for their help in the VIC data preparation over our studies areas. We also appreciate the editor and two anonymous reviewers for their constructive comments.

REFERENCES

Barnett, T. P., Adam, J. C. & Lettenmaier, D. P. 2005 Potential impacts of a warming climate on water availability in snow-dominated regions. *Nature* **438** (7066), 303–309.

Bolch, T., Kulkarni, A., Kaab, A., Huggel, C., Paul, F. & Cogley, J. G. 2012 The state and fate of Himalayan glaciers. *Science* **336** (6079), 310–314.

Cao, J., Qin, D., Kang, E. & Li, Y. 2006 River discharge changes in the Qinghai-Tibet Plateau. *Chinese Science Bulletin* **51** (5), 594–600.

Chen, X., Long, D., Hong, Y., Zeng, C. & Yan, D. 2017 Improved modeling of snow and glacier melting by a progressive two-stage calibration strategy with GRACE and multisource data: how snow and glacier meltwater contributes to the runoff of the Upper Brahmaputra River basin? *Water Resources Research* **53** (3), 2431–2466.

Cuo, L., Zhang, Y., Gao, Y., Hao, Z. & Cairang, L. 2013 The impacts of climate change and land cover/use transition on the hydrology in the upper Yellow River Basin, China. *Journal of Hydrology* **502**, 37–52.

Duan, A. & Wu, G. 2006 Change of cloud amount and the climate warming on the Tibetan Plateau. *Geophysical Research Letters* **33**, 22.

FAO – UNESCO 1988 *UNESCO Soil Map of the World, Revised Legend*. World Resources Report 60. FAO, Rome, Italy.

Feng, S., Hu, Q. & Qian, W. 2004 Quality control of daily meteorological data in China, 1951–2000: a new dataset. *International Journal of Climatology* **24** (7), 853–870.

Hock, R. 2003 Temperature index melt modelling in mountain areas. *Journal of Hydrology* **282** (1), 104–115.

Immerzeel, W. W., Droogers, P., De Jong, S. & Bierkens, M. 2009 Large-scale monitoring of snow cover and runoff simulation in Himalayan river basins using remote sensing. *Remote Sensing of Environment* **113** (1), 40–49.

Immerzeel, W. W., Van Beek, L. P. & Bierkens, M. F. 2010 Climate change will affect the Asian water towers. *Science* **328** (5984), 1382–1385.

Jacob, T., Wahr, J., Pfeffer, W. T. & Swenson, S. 2012 Recent contributions of glaciers and ice caps to sea level rise. *Nature* **482** (7386), 514.

Kan, B., Su, F., Xu, B., Xie, Y., Li, J. & Zhang, H. 2018 Generation of high mountain precipitation and temperature data for a quantitative assessment of flow regime in the Upper Yarkant Basin in the Karakoram. *Journal of Geophysical Research: Atmospheres* **123** (16), 8462–8486.

Kaser, G., Großhauser, M. & Marzeion, B. 2010 Contribution potential of glaciers to water availability in different climate regimes. *Proceedings of the National Academy of Sciences* **107** (47), 20223–20227.

Kayastha, R., Ageta, Y., Nakawo, M., Fujita, K., Sakai, A. & Matsuda, Y. 2003 Positive degree-day factors for ice ablation on four glaciers in the Nepalese Himalaya and Qinghai-Tibetan Plateau. *Bulletin of Glaciological Research* **20**, 7–14.

Kendall, M. 1975 *Rank Correlation Methods*. Griffin & Co, London, UK. ISBN 0-85264-199-0.

Kimball, J., Running, S. & Nemani, R. 1997 An improved method for estimating surface humidity from daily minimum temperature. *Agricultural and Forest Meteorology* **85** (1), 87–98.

Kotlyakov, V., Zichu, X., Khromova, T., Zverkova, N. & Chernova, L. 2012 Contemporary glacier systems of continental Eurasia. *Doklady Earth Sciences* **446**, 1095–1098.

Kuang, X. & Jiao, J. J. 2016 Review on climate change on the Tibetan Plateau during the last half century. *Journal of Geophysical Research: Atmospheres* **121**, 8.

Kumar, S., Merwade, V., Kam, J. & Thurner, K. 2009 Streamflow trends in Indiana: effects of long term persistence, precipitation and subsurface drains. *Journal of Hydrology* **374** (1), 171–183.

Li, C., Su, F., Yang, D., Tong, K., Meng, F. & Kan, B. 2018 Spatiotemporal variation of snow cover over the Tibetan Plateau based on MODIS snow product, 2001–2014. *International Journal of Climatology* **38** (2), 708–728.

Liang, X., Lettenmaier, D. P., Wood, E. F. & Burges, S. J. 1994 A simple hydrologically based model of land surface water and energy fluxes for general circulation models. *Journal of Geophysical Research* **99** (D7), 14415–14428.

Liang, X., Wood, E. F. & Lettenmaier, D. P. 1996 Surface soil moisture parameterization of the VIC-2 L model: evaluation and modification. *Global and Planetary Change* **13** (1), 195–206.

Liu, W., Sun, F., Li, Y. & Zhang, G. 2018a Investigating water budget dynamics in 18 river basins across the Tibetan Plateau through multiple datasets. *Hydrology & Earth System Sciences* **22**, 1–56.

Liu, Z., Wang, R. & Yao, Z. 2018b Climate change and its impact on water availability of large international rivers over the mainland Southeast Asia. *Hydrological Processes* **32** (26), 3966–3977.

Lu, M., Yang, S., Li, Z., He, B., He, S. & Wang, Z. 2017 Possible effect of the Tibetan Plateau on the 'upstream' climate over West Asia, North Africa, South Europe and the North Atlantic. *Climate Dynamics* **6**, 1–14.

Lutz, A., Immerzeel, W., Shrestha, A. & Bierkens, M. 2014 Consistent increase in High Asia's runoff due to increasing glacier melt and precipitation. *Nature Climate Change* **4** (7), 587–592.

Ma, Y., Zhong, L., Su, Z., Ishikawa, H., Menenti, M. & Koike, T. 2006 Determination of regional distributions and seasonal variations of land surface heat fluxes from Landsat-7 Enhanced Thematic Mapper data over the central Tibetan Plateau area. *Journal of Geophysical Research: Atmospheres* **111**, D10305.

Mann, H. B. 1945 Nonparametric tests against trend. *Econometrica: Journal of the Econometric Society* **13** (3), 245–259.

Meng, F., Su, F., Li, Y. & Tong, K. 2019 Changes in terrestrial water storage during 2003–2014 and possible causes in Tibetan Plateau. *Journal of Geophysical Research: Atmospheres* **124**, 2909–2931.

Pu, Z., Xu, L. & Salomonson, V. V. 2007 Modis/terra observed seasonal variations of snow cover over the Tibetan Plateau. *Geophysical Research Letters* **34** (6), L06706.

Qiu, J. 2008 China: the third pole. *Nature News* **454** (7203), 393–396.

Shen, Y. & Xiong, A. 2016 Validation and comparison of a new gauge-based precipitation analysis over mainland China. *International Journal of Climatology* **36** (1), 252–265.

Shen, Y. J., Shen, Y., Fink, M., Kralisch, S. & Brenning, A. 2018 Unraveling the Hydrology of the glacierized Kaidu Basin by integrating multisource data in the Tianshan Mountains, Northwestern China. *Water Resources Research* **54**, 557–580.

Su, F., Adam, J. C., Bowling, L. C. & Lettenmaier, D. P. 2005 Streamflow simulations of the terrestrial Arctic domain. *Journal of Geophysical Research: Atmospheres* **110**, D8.

Su, F., Zhang, L., Qu, T., Chen, D., Yao, T., Tong, K. & Qi, Y. 2016 Hydrological response to future climate changes for the major upstream river basins in the Tibetan Plateau. *Global Planet Change* **136**, 82–95.

Tong, K., Su, F. & Xu, B. 2016 Quantifying the contribution of glacier-melt water in the expansion of the largest lake in Tibet. *Journal of Geophysical Research: Atmospheres* **121**, 11158–11173.

Wang, S. J. 2017 Spatiotemporal variability of temperature trends on the southeast Tibetan Plateau, China. *International Journal of Climatology* **38** (5), 1953–1963.

Wang, H. & Chen, F. 2017 Increased stream flow in the Nu River (Salween) Basin of China, due to climatic warming and increased precipitation. *Geografiska Annaler: Series A, Physical Geography* **99** (4), 327–337.

Wu, Y. H., Zheng, H. X., Zhang, B., Chen, D. M. & Lei, L. P. 2014 Long-term changes of lake level and water budget in the Nam Co Lake Basin, central Tibetan Plateau. *Journal of Hydrometeorology* **15** (3), 1312–1322.

Yao, T., Wang, Y., Liu, S., Pu, J., Shen, Y. & Lu, A. 2004 Recent glacial retreat in High Asia in China and its impact on water resource in Northwest China. *Science in China Series D: Earth Sciences* **47** (12), 1065–1075.

Yao, T., Thompson, L. & Yang, W. 2012 Different glacier status with atmospheric circulations in Tibetan Plateau and surroundings. *Nature Climate Change* **2** (9), 663–667.

Yao, Z., Liu, Z., Huang, H., Liu, G. & Wu, S. 2014 Statistical estimation of the impacts of glaciers and climate change on river runoff in the headwaters of the Yangtze River. *Quaternary International* **336**, 89–97.

Zhang, Y., Liu, S. & Ding, Y. 2006 Observed degree-day factors and their spatial variation on glaciers in western China. *Annals of Glaciology* **43** (1), 301–306.

Zhang, Y., Liu, S., Xu, J. & Shangguan, D. 2008 Glacier change and glacier runoff variation in the Tuotuo River basin, the source region of Yangtze River in western China. *Environmental Geology* **56** (1), 59–68.

Zhang, L., Su, F., Yang, D., Hao, Z. & Tong, K. 2013 Discharge regime and simulation for the upstream of major rivers over Tibetan Plateau. *Journal of Geophysical Research: Atmospheres* **118** (15), 8500–8518.

Zhao, Q., Zhang, S., Ding, Y. J., Wang, J., Han, H. & Xu, J. 2015 Modeling hydrologic response to climate change and shrinking glaciers in the highly glacierized kunma like river catchment, central Tian Shan. *Journal of Hydrometeorology* **16** (6), 2383–2402.

First received 30 May 2019; accepted in revised form 8 September 2019. Available online 10 October 2019

How extreme can unit discharge become in steep Norwegian catchments?

Oddbjørn Bruland

ABSTRACT

This study presents results of observations and analysis of the flood event in Utvik on 24 July 2017. Observations during and after the event, hydraulic simulations and hydrological modelling along with meteorological observations, are used to estimate the peak discharge of the flood. Although both observations and hydraulic simulations of flood extremes are uncertain, even the most conservative assumptions lead to discharge estimates higher than 160 m^3/s at culmination of the flood from the 25 km^2-large catchment. The most extreme assumptions indicate it may have been up to 400 m^3/s, but there is also strong evidence for the discharge at culmination being between 200 and 250 m^3/s. Observations disclosed that the majority of water came from about 50% of the catchment area giving unit discharges up to 18 to 22 $m^3/s,km^2$. If the entire catchment contributed it would be from 9 to 11 $m^3/s,km^2$. This is significantly higher than previously documented unit discharges in Norway and in the range of the highest observed peak unit discharges in southern Europe. The precipitation causing this event is estimated to be three to five times higher than a 200-year precipitation taken from the intensity–duration–frequency curves for the studied region.

Key words | flash flood, hydrological extremes, Norway, unit discharge

Oddbjørn Bruland
Norwegian University of Science and Technology,
7491 Trondheim,
Norway
E-mail: *oddbjorn.bruland@ntnu.no*

INTRODUCTION

On 24 July 2017, the river Storelva in Utvik, Norway, grew from less than 1 m^3/s to extreme ranges, which led to severe detrimental consequences within a 4-hour time frame. The event was documented through onsite observations of the course of the flood by the author. The purpose of this paper, besides documenting the event, is to assess how extreme this flood was in a Norwegian and European context.

Flash floods are floods caused by heavy and excessive rainfall of duration generally less than 6 hours or sudden release of water from, for example, dam breaks or ice jams (NOAA 2017). They are of the most dangerous and common natural hazards (Barredo 2007) and, as such, the threats that have the highest impact and likelihood of occurrence (World Economic Forum 2018). In the period 1950–2005, 2,764 casualties were documented in southern and continental Europe due to flash floods, representing about 40% of all flood-related casualties in Europe (Barredo 2007). In Norway alone, the yearly costs due to floods are close to 1 billion Norwegian krones (NOK) or about 100 million € (Finans Norge 2018). This does not include the rehabilitation cost of damage in water courses and on public infrastructure. Over the last 50 years, there has been an increase in the frequency and intensity of short duration rainfall (Sorteberg *et al.* 2018) and climate change will further enhance this. It is expected that

doi: 10.2166/nh.2020.055

precipitation-dominated catchments will experience an increase of flood frequency of up to 60% and that the challenges will be particularly pronounced in small, steep rivers and streams, as well as in urban areas (Hanssen-Bauer *et al.* 2015). This will have a significant influence on the design of infrastructure and the risk level and the risk assessments all Norwegian municipalities are required to carry out. Observations of floods in small, steep rivers are very sparse and extreme local precipitation is rarely captured by a coarse network of rain gauges, thus the basis for analyses and estimates necessary to estimate the risk from floods in such areas are very limited. As combinations of high water velocities and water levels can increase the risk acceptance level from a 200- to a 1,000-year return period (Ministry of the Environment 2008), the risk related to steep rivers is particularly relevant to study. In this context, extreme events like the Utvik flood can provide useful insights into how extreme such natural hazards can also be at these latitudes.

Can the observations carried out during and after the flood in Utvik be used to estimate the discharge and can this, based on a precipitation-runoff model, be used to estimate the intensity of the precipitation and finally reveal how extreme this event was in a Norwegian and international context?

Following Borga *et al.* (2008), there are several ways to approach a post-event analysis of floods. Traces left by water, such as erosion and deposition, images and reports from eyewitnesses and other observations, not only along the flooded water courses, but within the hit region, can provide valuable information for quantifying the peak discharge and the extent of the flood, especially if these can be combined with hydrodynamic 1 or 2D modelling of flow over a dam or through a culvert. They also suggest use of radar data and rain gauge observations in the region combined with mesoscale meteorological modelling and distributed hydrological modelling. They state that successful implementation of such flash-flood response surveying methodology could transform our understanding of extreme floods and provide significant visibility for the scientific community.

In this paper, several of the methods described by Borga *et al.* (2008), adapted to the available observations, are used to document and assess the magnitude of the Utvik flood.

The aims of this paper are: to use the *in-situ* and post-observations of the flood and flow over a dam crest together with hydraulic 2D simulations to estimate the discharge at culmination; to use rain gauge and radar observations together with hydrological modelling, observed lightning and eyewitness reports to assess how this event occurred, how extreme it was and to recreate the flood hydrograph and estimate the rainfall intensities causing the event; finally, to compare the peak unit discharge estimate for this flood to unit discharges of observed and reported floods in Norway over the last decade and to floods reported in southern Europe since 1950.

Norwegian floods

Based on historical sources, Lars Roald (2013) provides an overview over major floods back to the 14th century. He mainly describes impacts and not discharges. Only at a few locations are flood discharges given. Thus, it is difficult to use these in a quantitative analysis without further knowledge and assumptions. Only since 2008, all floods observed at gauging stations in Norway and with a return period higher than ten years, have been systematically documented (Figure 1). In November 2009, south-western Norway was exposed to high precipitation and several rivers flooded (Haddeland 2009). The highest observed precipitation was 143 mm over less than 12 hours. The highest unit discharge was 2.9 $m^3/s,km^2$. Locally reported damage indicate that in some areas the intensity probably was higher (Aftenposten 2009). In October 2010, a heavy rainfall event following several days of precipitation caused flooding in numerous rivers in the south of Norway, and the highest observed unit discharge was 3.04 $m^3/s,km^2$ (Pettersson 2010). A rain event of up to 110 mm in 24 hours affected central Norway in August 2011 and caused, at some locations, 100-year floods (Pettersson 2011). The highest observed unit discharges was 1.3 $m^3/s,km^2$. In June 2011, snowmelt combined with heavy rain caused a flood with severe consequences in Kvam (NRK 2011). This event was not quantified, but is a major flash flood in a Norwegian context. The same location experienced a similar flood in 2013. Later in the summer of 2011, another event hit southern Norway with observed precipitation up to 97 mm over 2 days, and caused flooding

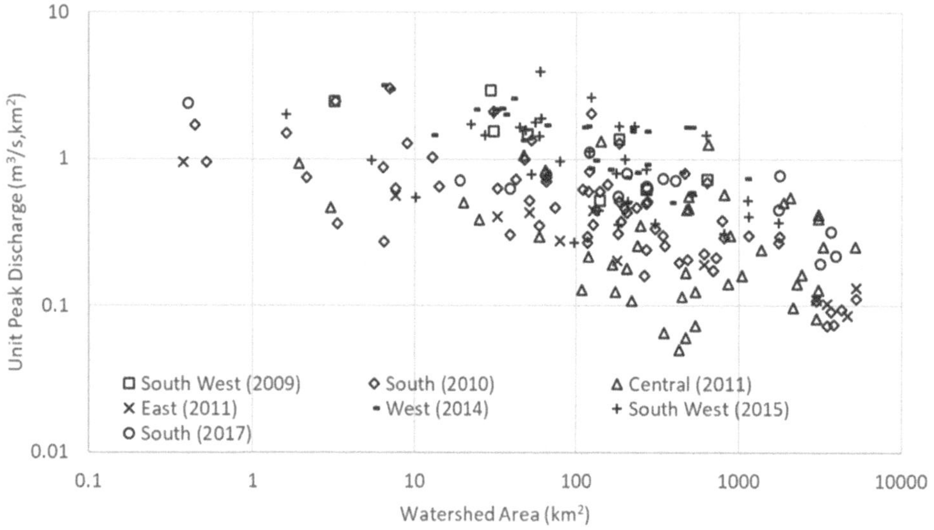

Figure 1 | Scatterplot of unit peak discharge ($m^3/s,km^2$) from registered flood events in Norway since 2008 versus watershed area (km^2) and grouped by region and year.

in regions with the highest registered unit discharge of 0.95 $m^3/s,km^2$ (Haddeland 2011). In October 2014, precipitations of 200 to 300 mm over 2 days caused severe flooding in west Norway. Damage to over 1,000 properties was reported and there was considerable damage to infrastructure (Langsholt *et al.* 2015). The highest observed unit discharge in this event was 3.2 $m^3/s,km^2$. In December 2015, south-western Norway was again exposed to floods with severe consequences. Up to 190 mm of precipitation over 24 hours was recorded and the highest observed unit discharge was 4.0 $m^3/s,km^2$ (Holmqvist 2016). A precipitation of 300 mm over 3 days and up to 173 mm over 24 hours in October 2017 caused flooding in south Norway, with a unit discharge of up to 2.4 $m^3/s,km^2$ and damage to 3,300 properties and reported repair costs of over 500 million NOK (Langsholt & Holmqvist 2017).

Stenius *et al.* (2015) compiled the maximum observed streamflow from Norwegian gauging stations with catchments smaller than 50 km^2 (Figure 2). The unit discharge shows a wide variability, ranging from 0.15 $m^3/s,km^2$ to 5.3 $m^3/s,km^2$. The highest unit discharge is observed in east Norway, but the highest average and variability is in west Norway.

All these floods have been possible to study quantitatively as they all cover gauged catchments. However, flash floods are characterized by intense rainfalls affecting small areas (Barredo 2007) and taking place very locally where there are, in most cases, no observations, as in Utvik and

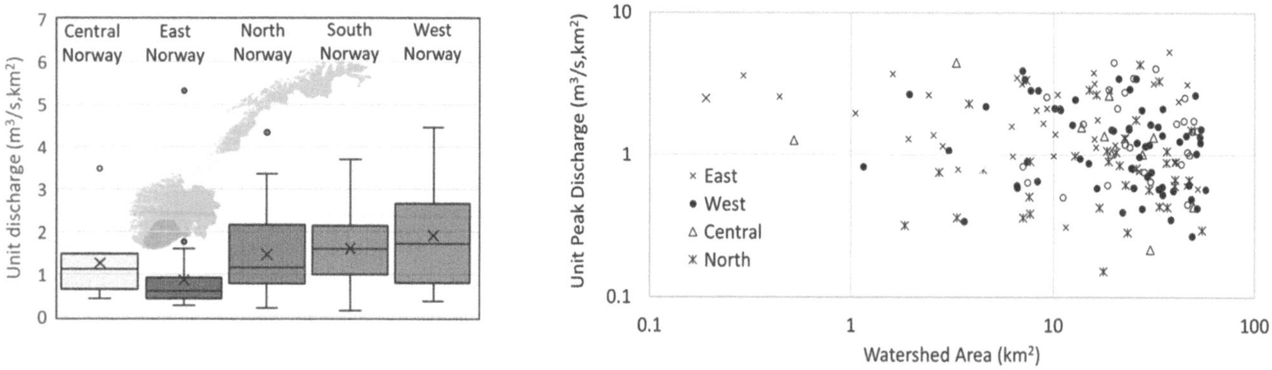

Figure 2 | Unit discharge in Norwegian catchments smaller than 50 km^2, compiled from Stenius *et al.* (2015).

Kvam described above. Since the observation network of precipitation and streamflow is usually too sparse to capture local flash floods and the flood generating precipitation, it is most likely that several of the events described above can have been far more extreme locally than the documentation indicates. In this respect, the event at Fulufjället in central Sweden is a rare exception in Scandinavia, not only in extremity, but also how well this was documented (Vedin *et al.* 1999). They investigated the meteorological situation causing the event and found that the precipitation was close to 400 mm over 24 hours and that the event had a return period of 10,000 years. Based on their observations, Lundquist (2000) estimated the peak unit discharge for this event to be 9 $m^3/s,km^2$.

European flash floods

Even in southern Europe, where flash floods annually cause close to 60 casualties, flash floods are a poorly understood and documented natural phenomenon (Gaume *et al.* 2009). Barredo (2007) studied 47 major flood events from 1950 to 2006 in the European Union, Bulgaria and Romania based on casualties and direct damage. Twenty-three of these are characterized as flash floods. He concludes that major floods have become more frequent and the damage has increased in the last decades prior to the study. Gaume *et al.* (2009) compiled data to develop a catalogue that includes the most extreme flash flood events registered between 1946 and 2007. Their collection consists of 550 extreme flash floods affecting catchments smaller than 500 km^2 in southern Europe. Based on these data, they found envelope curves for the peak unit discharges as a function of area that summarizes floods in the studied regions. Building on this study, Marchi *et al.* (2010) examined more closely the control of the watershed physiography and channel network on the flood response for a selection of the floods. In their study, they characterized initial soil moisture status, climate and the river response to identified extreme flash flood events representative of different hydro-climatic European regions, and characterized the morphological properties of the catchments, land use, soil properties and geology. The unit peak discharges they reported are in the range around 0.4 $m^3/s,km^2$ to about 20 $m^3/s,km^2$. They found an envelope curve that was in accordance with Gaume *et al.* (2009) (Equation (1)):

$$Q_u = 97.0 \cdot A^{-0.4} \tag{1}$$

where Q_u is peak discharge in $m^3/s,km^2$ and A is catchment area (km^2).

Moreover, Parajka *et al.* (2010) studied floods across the Alpine-Carpathian range (from France to Romania) and found support for a spatial and temporal clustering of floods. They suggest that extreme events in this region are often produced by one main mechanism – extreme storms during southerly circulation patterns. Bárdossy & Filiz (2005) found similar tendencies in the northern Alpine region.

The study area and the flood event

Utvik is a small village of about 400 inhabitants located at 61.8°N, 6.52°E, with the river Storelva passing through the centre of the village, flowing northwestwards and with its tail waters in the fjord (Figure 3, left). Storelva has a watershed of 25 km^2 ranging from 0 to 1,553 masl where about 25% of the landcover is forested, 62% is open alpine landscape, about 5% is marsh and 3% is glaciated. The soil types are mainly moraine deposits (Figure 3, right). The river is steep, with an average gradient of 12% and a maximum of 18%.

For the normal period 1961 to 1990, the average yearly precipitation is estimated to be 1,300 mm with 430 mm during the summer months (1 May to 30 September). The average yearly temperature is 2.3 °C, with July being the warmest month with an average of 8.7 °C. The average flow in Storelva is estimated to be about 1.6 m^3/s or 65.3 $L/s,km^2$ (Nevina 2018).

It started to rain heavily from about 4 a.m. on the morning of 24 July 2017 and continued with varying intensity until around 2 p.m. in the afternoon. The intensities were highest in the morning and in the upper part of the catchment. Following the warmest day of the summer, the discharge in Storelva was initially very low, but grew rapidly from 6 a.m. and culminated some time between 8 and 9 a.m. During this period, Storelva shifted its course at several

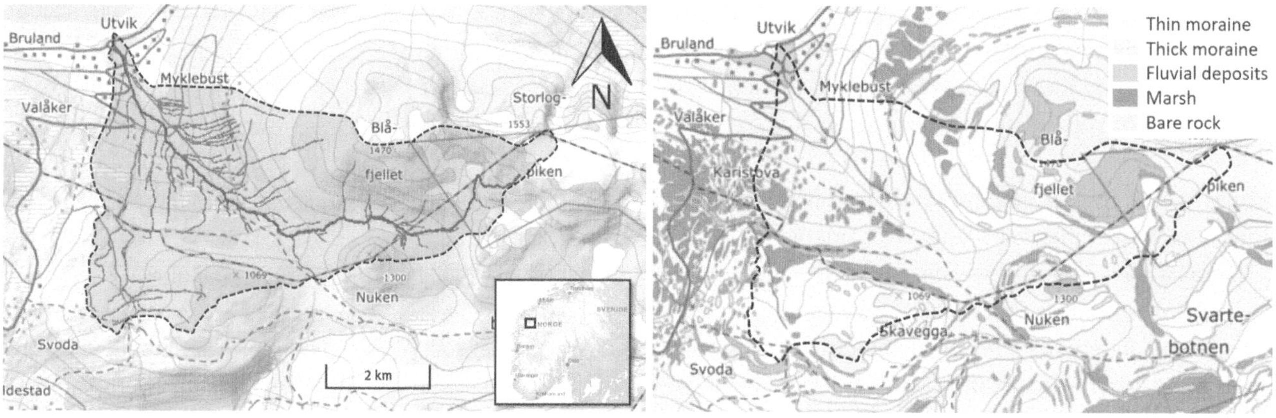

Figure 3 | Storelva watershed and watercourses are presented in the left panel. Land surface types are pictured in the right panel (NGU 2019).

locations, eroded away the county road, a bridge, and left the village isolated (Figure 4).

The total cost of damage to private property due to the flood was estimated to be at least 7 million € and the repair cost of the affected road was estimated to be at least 5 million € (Sunnmørsposten 2017). In addition, there were large costs in rehabilitating and securing the watercourse. Also, local businesses incurred the loss of millions € due to the closing down of the road (Fjordingen 2017). A 100-year-old hydropower station was also destroyed. The rebuilding of this and the loss of income is severe. Additionally, comes all the intangible costs of the devastation the flood caused to the local community, i.e., losing the historical value of several hundred-year-old buildings.

METHODS AND DATA

Hydraulic modelling and data for estimation of the peak flood discharge

Storelva is an ungauged river and, even though traces of the maximum water level during the flood are visible at several locations, it is challenging to estimate the culmination discharge since the river is steep and the topography makes it

Figure 4 | Overview picture showing the new watercourse and damage caused by the flood in Utvik (*source*: Hallgeir Vågenes, VG 2017).

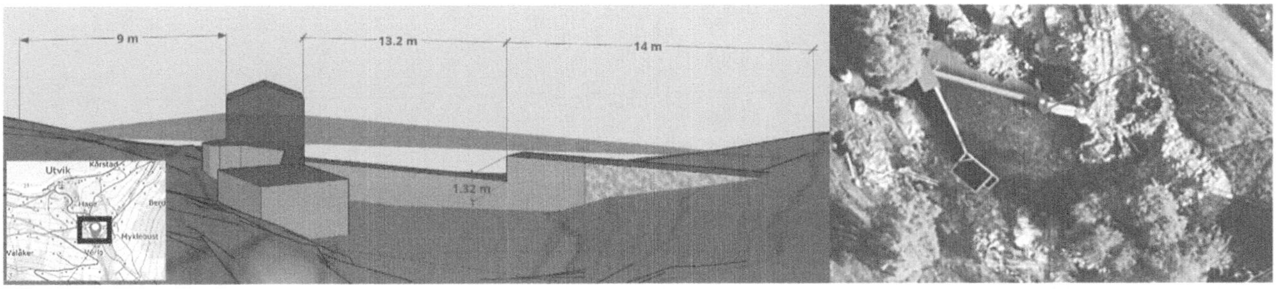

Figure 5 | Drawing (left) and picture (right) of the dam crest showing the eroded area and areal extent of the flood.

difficult to use hydrodynamic models. The best location to estimate the discharge is a dam crest (Figure 5) at about 86 masl, where it can be estimated by Equation (2):

$$Q_e = L \cdot C \cdot H_0^{\frac{3}{2}} \quad (2)$$

where Q_e is estimated discharge in m^3/s, L (m) is the length of the dam crest, C is the crest coefficient and H_0 (m) is the water level upstream of the dam crest at negligible flow velocities. For this short section of the river it is also possible to use a hydrodynamic model. During the flood, the main section of the dam was overtopped, and both sides of the dam crest were flooded as well. The water velocities over these sections are estimated by a 2D hydrodynamic model.

As displayed in Figure 5 (left), the dam crest is 13.2 m and observations showed that the river extended about 14 m to the right of the main crest and 9 m to the left. The maximum water level behind the dam crest was estimated based on the damage and traces of the flood on the wall of the aforementioned building at the left side of the crest (Figure 6) and traces in the terrain along the riverbanks upstream of the crest (Figure 5).

At maximum discharge, the water velocity upstream of the dam crest was significant. To account for this, the velocity has to be transformed into kinematic energy using the relation $v^2/2\,g$ – where v (m/s) is the velocity upstream of the crest and g (m/s^2) is gravity, and added to the potential energy, H_0, in Equation (2). The velocity upstream is estimated using Manning's equation for open channel flow, $v = M \cdot R^{2/3} \cdot I^{1/2}$, where M is Manning number ($m^{1/3}/s$) and a measure for river bed roughness (often also given as 1/n), R (m) is hydraulic radius and I (m/m) is gradient of the river. As the picture in Figure 6 shows, the flow over the crest is highly turbulent and together with a high gradient, a river bed varying from bare rock, course gravel and large stones, as well as with vegetation at the banks at high water levels, this makes water velocities difficult to estimate (Yochum *et al.* 2012).

To validate the calculations, a 2D hydrodynamic model was established for the site in Hec-Ras v5.0 (Brunner 2016b). The terrain used in the model is based on high resolution lidar measurements from after the flood (Høydedata 2019). The 2D simulation flow area (Figure 7) was selected to ensure stable inflow conditions. The gradient within the area is lower than 10%, the selected grid cell size was

Figure 6 | Illustrated water over the crest level, and picture of upstream of the crest and of the flow over the crest during the event.

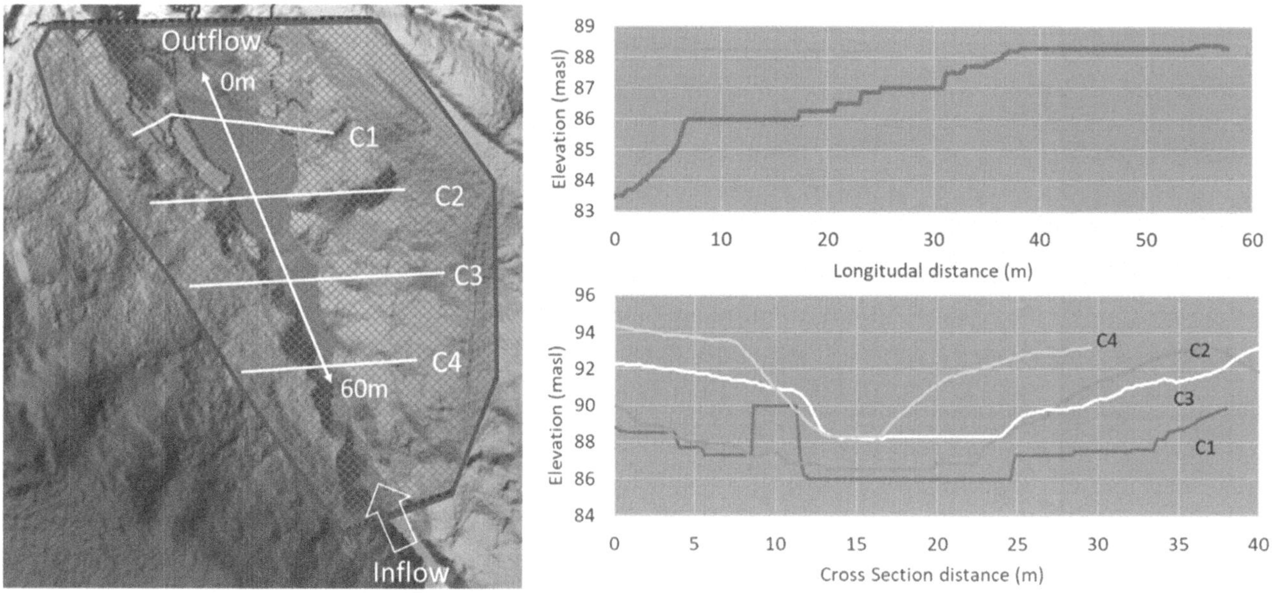

Figure 7 | 2D flow area in Hec-Ras (left). Longitude cross section and cross sections of the river stretch (right).

0.5 m and the simulation steps were 0.1 second. This is within the recommendations given by Brunner (2016a) for stable simulation conditions. The upstream Manning numbers at the peak flow are not known but can be estimated by Manning's equation at known velocities and corresponding water depths. Recently, after the flood, the mean upstream surface velocity was measured (using floating devices) to about 2.5 m/s at a discharge of about 10 m^3/s (using Equation (2)) and a water level over the dam crest of 0.3 m and between 0.3 and 0.5 m upstream of the crest. The width of the channel was from 10 to 12 m. The gradient upstream of the dam crest is about 5% at normal discharges. This gives Manning numbers of 18 to 25. During the flood, the upstream depth was from 2.3 m at the dam crest to about 4 m further upstream, and the width was 25 m to 30 m. The additional flooded area compared to the situation at 10 m^3/s will have a higher roughness due to vegetation and structures at the riverbanks. Compared to suggested roughnesses in Chow (1959), Barnes (1969) and Yochum *et al.* (2014), a Manning number for the simulated river section should be between 10 and 20. A value of 7 is included in the analysis, but this is extremely low and representative for floodplains with high and dense vegetation. Even a value of 10 is very low for this river section, but due to high bed load transport this is more likely than a value of 7. The different Manning values are used in combination with different, but stable inflows. According to Pappenberger *et al.* (2005), the roughness parameter together with geometry has the highest impact on the hydrodynamic simulation results. Using high density lidar data for the geometry leaves the roughness parameter as the most uncertain parameter. By varying the Manning number for each tested inflow, it is possible to test the sensitivity to the roughness parameter and, thus, by comparing the simulated water levels at profile C1 (Figure 7) to the highest water levels during the flood, the uncertainty of the estimated discharge.

Methods and data for estimating the precipitation

Observations from precipitation gauges and weather radar show that the precipitation was very local (Figure 8). The highest 24-hour precipitation measured in a rain gauge was 87 mm at the NIBIO station (A) at Sandane (NIBIO 2019) about 18 km west of Storelva watershed and 55 mm at Innvik (B), about 5 km to the north-east. Whereas Kroken station (C) in Stryn, 9 km further north, observed 34 mm in the same period, 17 mm of these in the morning.

The 24-hour accumulated radar observation shows that the precipitation was concentrated over the upper part of

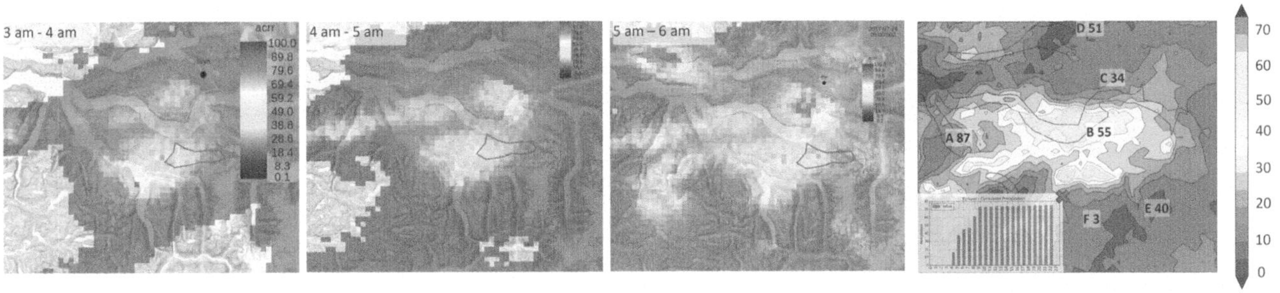

Figure 8 | Observed hourly precipitation from 3 a.m. to 6 a.m. by the radar in the three figures to the left and total observed over 24 hours by the radar and at meteorological stations in the region (right). The graph is accumulated values at the location of the cross (data acquired from Met.no).

the catchment (known as Utvikfjellet). Based on the values of the radar pixels (each of 1 km^2) covering Utvikfjellet, it is estimated that 72 mm of rain fell over 4 hours in the most intense areas. However, the rain gauge observations show that the accumulated radar values are too low. Kroken and the NIBIO station are the only ones with available hourly observations. Both showed that most of the precipitation occurred within 1 hour. The NIBIO station recorded 44 mm within 1 hour, 55 mm within 4 hours and 87 mm from 5 a.m. to 6 p.m. (Figure 9), whereas the radar pixel covering the same location gives 30 to 50 mm over the same period. At Kroken, 34 mm of rain fell compared to 10–20 mm from the radar. In Hornindal (D) and Olden (E), precipitation was 51 mm and 40 mm, respectively, while the radar shows 10 mm or less. The radar pixel covering Innvik shows 40 to 50 mm over 24 hours and this is in better accordance with an observed precipitation of 55 mm. Station F, Myklebustdalen, 5 km south of Utvikfjellet, measured only 3 mm which is coherent with the radar observations. Even though this station might be located in a rain shadow between high mountains and in a narrow valley, this observation shows how extremely local the precipitation was during this event.

Based on registered lightning (Figure 10), Bjart Eriksen, a meteorologist at Norwegian Meteorological Institute (Met.no), found that three heavy storm centres met at Utvikfjellet. The storm centres arrived at Utvikfjellet during the morning hours, first from west and north, later from south east, merged and were stable for some hours before dissolving. This indicates that the duration of the event at Utvikfjellet was longer and produced more precipitation than observations at stations in the region alone suggest. Observed precipitation and wind directions at meteorological stations in the region support this finding.

Observations of erosion along streams in the catchment and along the main river coincide with the radar observations and indicate that the precipitation was not evenly distributed

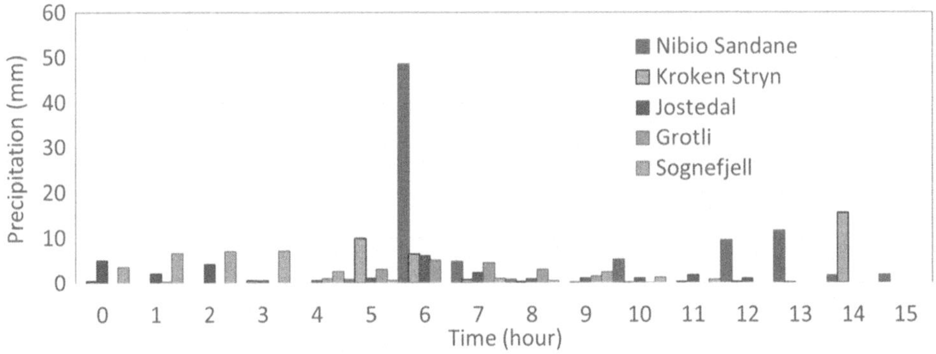

Figure 9 | Hourly observed precipitation at Nibio Sandane and at Kroken Stryn gauging stations and at stations with representative corresponding observed wind directions in Figure 10.

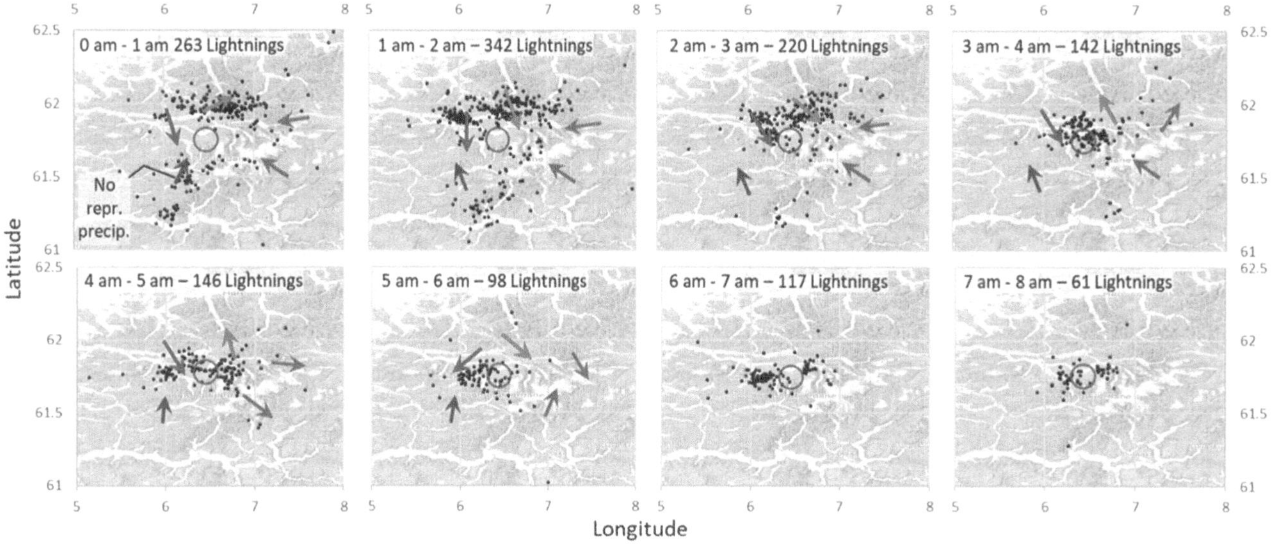

Figure 10 | Observed lightning on 24 July 2017. Arrows indicate wind directions at the location of the precipitation gauges in Figure 9. (Data from Frost.met.no and SeNorge.no).

over the 25 km^2 watershed, but rather was more concentrated in the upper part of the catchment and that this part of the catchment contributed the majority of the discharge.

Eyewitness observations in the hours between 8 and 10 a.m. revealed a 60 to 80 m-wide waterfall, not known to be observed before, in a mountain slope in the upper part of the catchment. This mountain is located where the storm centres are assumed to have coincided.

Rainfall-runoff modelling and data for estimating the peak discharge

A distributed hydrological model with a domain as shown in Figure 11 and a resolution of 0.5 km by 0.5 km was used to estimate the precipitation causing the flood. When considering the observed precipitation during the event and the topography, the observations at NIBIO Sandane and Innvik are considered the most representative for this event. These are interpolated to the domain using inverse distance weighting and an increase of precipitation with elevation of 5% per 100 masl. As the precipitation at Innvik is only a daily value, this is distributed into hourly values using the temporal precipitation distribution observed at NIBIO Sandane and Kroken. The radar images indicate that the precipitation came earlier and lasted longer at Utvikfjellet than at the surrounding precipitation stations. This is accounted for by extending the duration of the rainfall at Utvikfjellet by 2 hours compared

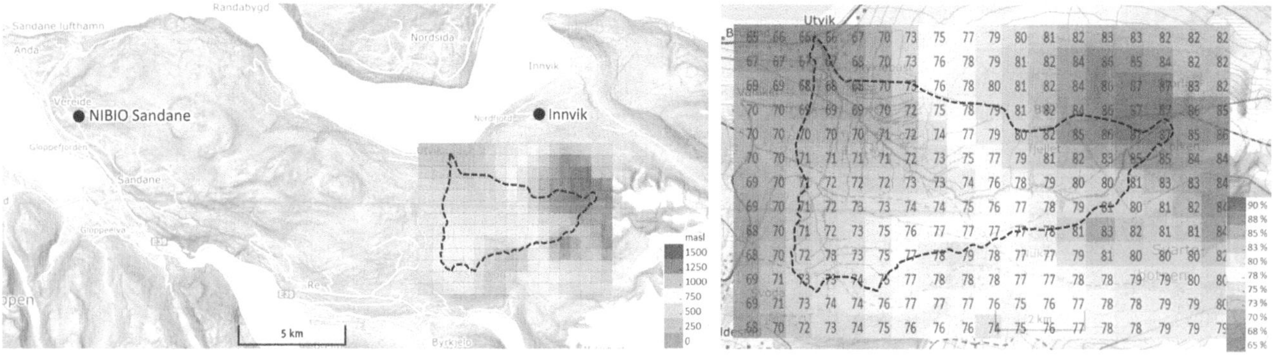

Figure 11 | Modelled region and a DTM with grid cells of 500 × 500 m overlapping the study area (left) and initial soil saturation before the event ranging from 65% to 90% (right).

to the durations at NIBIO Sandane and Stryn. The intensity of the rainfall for these hours are tuned until the timing of and discharge at culmination is in accordance with the estimated peak discharge and the observed course of the event.

The hydrological simulation for each grid cell is based on the principles summarized within the HBV model (Bergstrøm & Forsman 1973). The runoff from each grid cell is routed to and accumulated in downstream grid cells based on the slope of the grid cell. As there are no calibration data, only an estimated peak flow and a visually observed development and duration of the flood, the runoff parameters are based on previous regional model calibrations and recommendations in Stenius *et al.* (2015). The model was run on an hourly time step with a warmup period of two months.

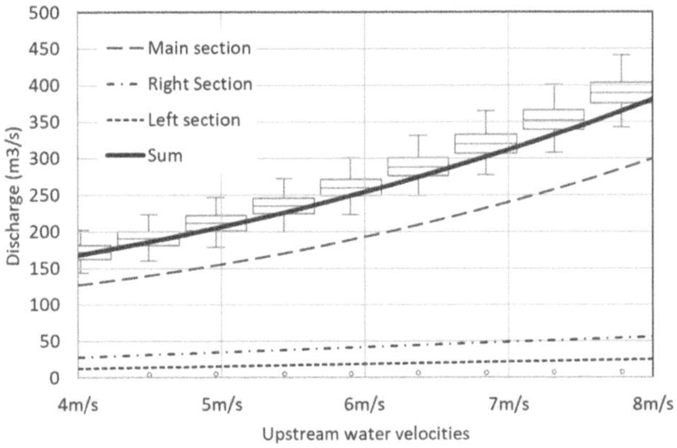

Figure 12 | Discharge (m^3/s) at different upstream water velocities (m/s) and for the different sections of the dam with main section depth of 2.36 m and 60% and 66% reduction of velocity at right and left sections, respectively. Box plot shows sensitivity to variations in depth and to velocities at the side sections.

RESULTS

Calculated discharge at the dam crest

As the dam crest was severely overtopped and there were no observations of the discharge, it was necessary to assume a velocity distribution out of the dam in order to estimate the peak discharge of the event. The assumption is based on Hec-Ras 2D simulations with discharges ranging from 130 to 270 m^3/s combined with Manning numbers ranging from 7 to 20. The results of the Hec-Ras simulations show that the velocities were reduced to 20% to 46% over the 9 m-long section at the left bank side and to 25% to 70% over the 14 m-long section at the right bank, compared to the velocities at the main section. The velocities at the main section simulated by Hec-Ras ranged from 4 to 8.9 m/s on average over the cross section C1.

Manning equation with Manning numbers, M, of 10 to 20 gives velocities of 5.3 to 8 m/s for gradients of 5 to 10%. By using Equation (2) with velocities v from 4 m/s to 9 m/s, depth d at the main section of 2.36 m and average velocity reductions from the Hec-Ras simulations of 60% and 66% velocity reduction at the right and left side sections, respectively, the discharges, Q_e, will be from 165 to 380 m^3/s (Figure 12). The sensitivity to uncertainties in depth and velocity distributions are tested by varying the depth at the main section with ±10% and the velocities at three sections in accordance with the variations found by the Hec-Ras simulations. This gave mean discharge values ranging from 172 m^3/s for the lowest estimated upstream velocities to 390 m^3/s for the highest and standard deviations of 9% to 6%, respectively (Figure 12).

Figure 12 shows that the major volume passes the main section when compared to the left and right sections and thus, that the sum discharge is not strongly influenced by uncertainty in the discharges at the side sections.

Hec-Ras gave simulated water levels at the left side of the dam crest between 1.93 m and 2.86 m for discharges between 130 and 230 m^3/s and Manning numbers between 7 and 20 (Figure 13). The right graph of Figure 13 shows that inflows larger than 160 m^3/s give water levels higher than 220 cm and that Manning number of 20 will give water levels of maximum 243 cm also for inflows up to 260 m^3/s.

The simulated water surface profile along the crest shows a significant variation between the different combinations of inflow and roughness factor (Manning number). The lowest roughnesses give the highest water velocities (Figure 14) and most variable water surface levels, up to 1.5 m, along the crest. Except from the highest discharge combined with the lowest roughness, the water levels over the crest were stable over time in the simulations.

The simulated flooded area in Hec-Ras for inflows from 150 to 230 m^3/s with roughness of $M = 10$ is shown in Figure 15. With some exceptions, the simulated water-covered area was in coherence with the area where traces of the flood and erosion due to high water velocities were

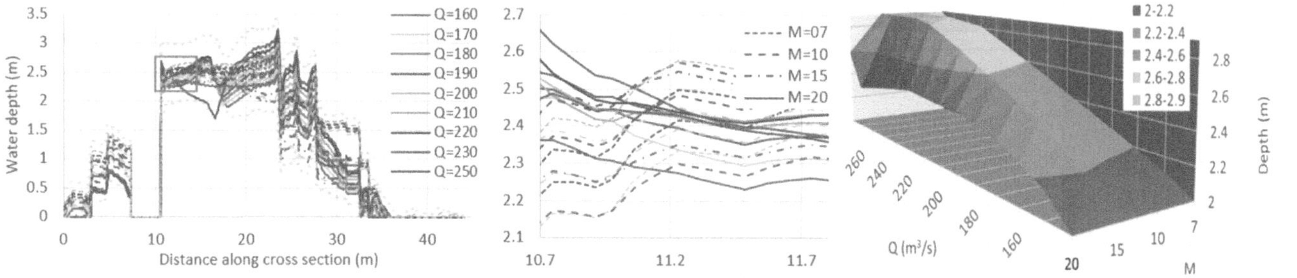

Figure 13 | Resulting water depths at dam crest from Hec-Ras simulation with discharges from 130 to 250 m^3/s and Manning number from 7 to 20. Left graph shows results that are within ±15 cm of observed water levels and maximum and minimum water levels for all simulations as the dotted grey line. The centre graph shows an enlargement of the marked segment. The right graph shows the relation between simulated depth at the left side of the crest, inflow in m^3/s and Manning number.

evident. The upper part of the simulated section is a narrow, tree-covered gorge (particularly at the left bank), that is not fully captured in the digital terrain model. This can explain the discrepancies at the left bank of the upper part of the river section.

At the right bank, there is an elevation where the erosion was significant. It is not likely that the water level in the dam was this high, but that an increasing upstream water level diverted a part of the flow in this direction which caused this erosion. None of the simulations were able to recreate this as erosion was not included, but the water level and velocities must have been high around this elevation. Simulations with $Q = 200\ m^3/s$ to $230\ m^3/s$ gives about 50 to 80 cm higher water level around this elevation than simulations with $Q = 180\ m^3/s$ and lower.

Simulated precipitation and discharge

To reconstruct the hydrograph of floods in the range found plausible by the calculations and the Hec-Ras simulations, the areal precipitation needed to be from 80 to 140 mm in total over the three most intense hours (Figure 16, right). In the simulations, this was achieved by increasing the observed precipitation the hour before and the hour after the observed maximum. The base case with no increase of the precipitation gave a peak discharge of 143 m^3/s. A total increase of the precipitation from 14 to 50 mm over the 3 hours gave discharges from 164 up to 243 m^3/s. The increased duration of the rainfall is coherent with the indicated intensities and duration observed by the radar (Figure 8). The accumulated estimated 3-hour precipitation at Utvikfjellet is up to two times higher than precipitation accumulated from the radar data, but the difference is less than the difference between precipitation observed by rain gauges and by the radar. The gauges showed 1.5 to 5 times higher 24-hour precipitation than estimates from the radar pixels ($1 \times 1\ km^2$) covering the gauge locations.

The runoff generation and accumulation from grid cell to grid cell and finally to the water courses within the region is also coherent with field observations and where

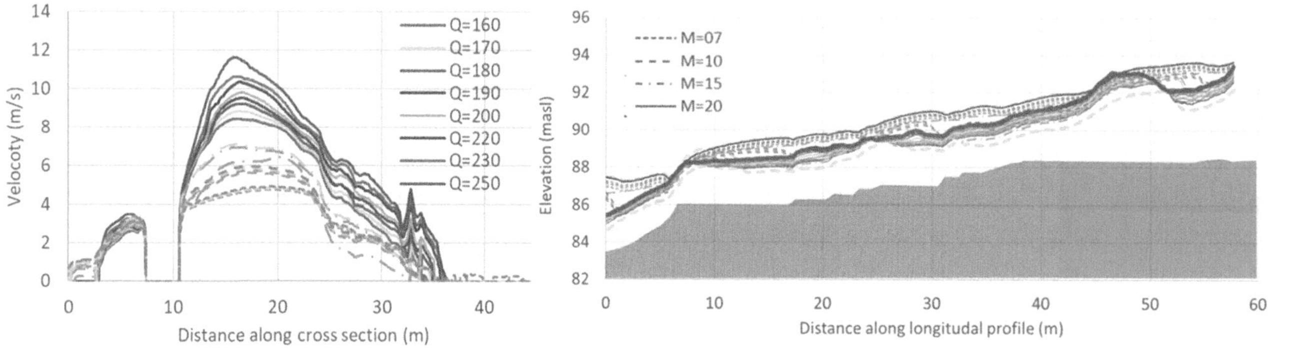

Figure 14 | Water velocities (left) and longitudinal profile of water surface profiles from Hec-Ras simulations. Type of line indicates Manning number used in simulations.

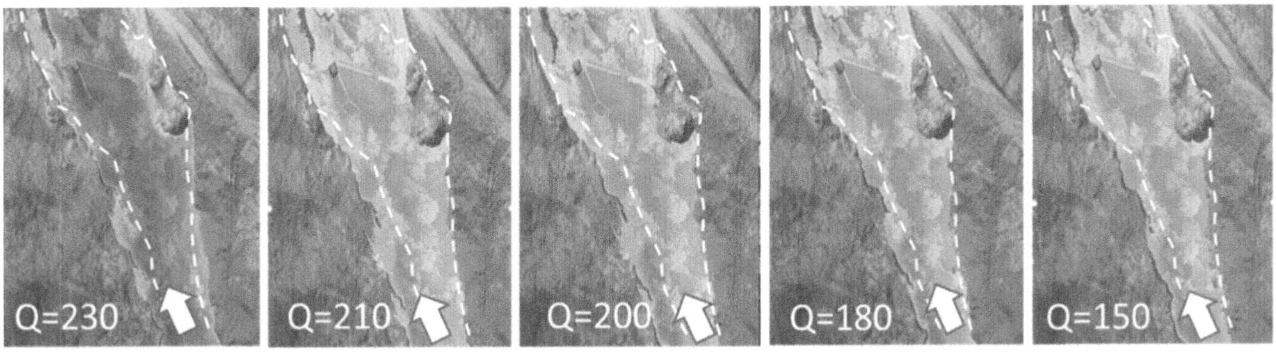

Figure 15 | Simulated water-covered area in Hec-Ras compared to areas with significant erosion observed.

damage due to flooding occurred. In addition to the damage along the main course of Storelva, there was significant damage along the two rivers, denoted a and b, east of Storelva catchment and along the tributary river (c) to Storelva (Figure 16, centre).

The damage was of the same order in the upper part of these rivers as further downstream. This indicates that discharge was relatively higher in the upper part of their catchments than the simulation shows. Another observation supporting the indication of locally very high rainfall intensities and runoff generation, is the observed presence of the waterfall in the mountain slope in the upper part of the catchment. The watershed generating the runoff for this waterfall is 0.15 km^2, with mostly bare rock and areas with thin soil layers. Visibility of a waterfall depends on how it is cascading and the discharge in L/s per meter waterfall (Simensen *et al.* 2011). The database of waterfalls (World Waterfall Database 2019) provides pictures, average discharge and width for different waterfalls and gives thereby an indication of discharge per meter waterfall versus visibility. The average discharge of some selected waterfalls comparable to the one observed at Utvikfjellet is about 200 L/s per metre width. To create discharges like this in this waterfall, it is necessary to have a precipitation of over 100 mm prior to the event over the concentration time of the drainage area. According to the definition of concentration time given by Stenius *et al.* (2015), this is less than an hour for this watershed. This indicates that the rainfall intensities in this area must have been extremely high. This can explain the severe local damage in the upper parts of these rivers.

Observations along the Storelva river and in its catchment indicate that the upper part of the catchment contributed to most of the discharge also for Storelva (Figure 17, left). Based on these observations and the accumulated radar observations, the major contributing area is estimated to be about 50% of the catchment. A discharge of 200 m^3/s and a contributing area in the range of 100% to 50% of the catchment, gives a unit discharge of this event from 9 $m^3/s,km^2$ to over 17 $m^3/s,km^2$. In the most intense areas, the unit peak discharge can have been as high as 20 $m^3/s,km^2$.

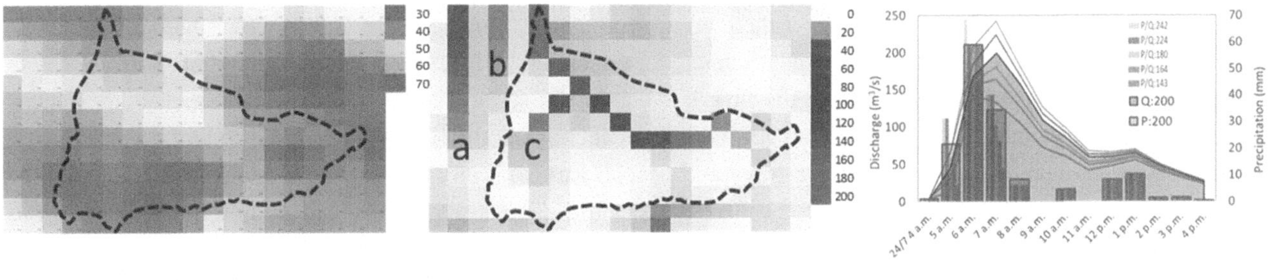

Figure 16 | Precipitation distribution (left) at maximum intensity (5–6 a.m.), maximum runoff (7–8 a.m.) (centre), and precipitation intensities for different simulated peak discharges with corresponding hydrographs (right).

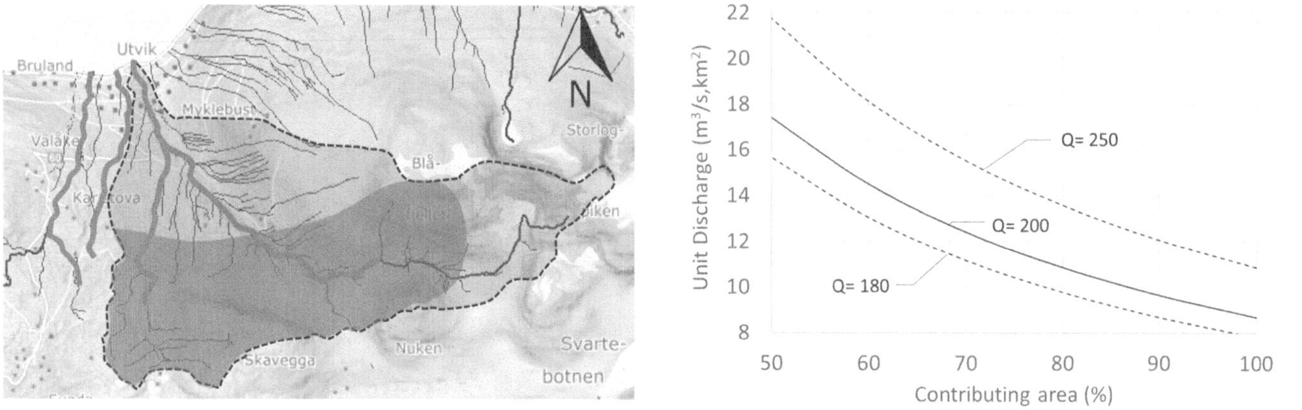

Figure 17 | River stretches with high erosion and the estimated contributing area are presented on the left-hand side, and peak unit discharge (m^3/s, km^2) depending on the contributing area (%) for culmination discharges of 180, 200 and 250 m^3/s on the right-hand side.

DISCUSSION

There are several uncertainties related to the calculation of the unit peak discharge for the flood in Utvik in July 2017. The dam crest is not ideal for the estimation as it is not perpendicular to the flow and, due to the high water level, the flow passes through sections without a defined crest. As the picture in Figure 6 shows, the flow was very turbulent, which makes it difficult to define maximum water levels and water velocities and as Figure 12 shows, the discharge estimates are dependent on the upstream water velocity. In addition, unknown Manning numbers and a significant sediment load and bed load at peak discharge influences also the estimates of water velocities (USDA 2007). However, even though the river stretches far outside its original course and the bed load is significant, with a Manning of 10 which is low for a river with a clear defined channel, some pools and a rough bed (Chow 1959; Barnes 1969; Yochum *et al.* 2014), an upstream slope between 5 and 10% and a depth of more than 2 m, it is not likely that the water velocity was any lower than 5 m/s. Even if there are uncertainties regarding the flow through the side sections, the calculations show that more than 70% of the flow is where the crest is well defined and, and even with very conservative assumptions on the upstream water level and water velocities, the discharge would need to be higher than 200 m^3/s.

The calculations using the dam crest formula (Equation (2)) give discharges that are coherent with the Hec-Ras simulations. According to Pappenberger *et al.* (2005), the uncertainties of hydrodynamic simulations are mainly related to the representation of the topography and the roughness, and according to Brunner (2016a), the 2D simulation in Hec-Ras is reliable at slopes lower than 10%. The slope is between 5 and 10% for the actual river section, the topography is detail mapped by lidar and the sensitivity to roughness is tested with Manning numbers from 7 to 20. For inflows of 160 to 250 m^3/s Hec-Ras gave water levels of ±15 cm compared to the observed depth at the same section of the river. As Figure 6 shows, there is significant damage higher up at the wall than the indicated maximum water level during the flood. Furthermore, the erosion extends higher up at the riverbanks than the simulated water levels. It is therefore likely that the water level at culmination used to determine the discharge is conservative. Compared to pictures of reference rivers for roughnesses (Barnes 1969; Yochum *et al.* 2014), the roughness of the simulated river section is most likely in the range 10 to 20, which gives upstream velocities higher than 5 m/s. Hec-Ras gives velocities from 5.39 to 6.25 with these roughnesses. From Figure 12 this gives discharges from 223 to 267 m^3/s. The sensitivity to water level and velocity reductions at the side sections indicate that these estimates can vary with up to ±9%. As the water level estimate used in the dam crest formula and comparison with Hec-Ras simulations is conservative, the upper range of the confidence interval, indicated in Figure 12, is more likely than the lower, at least for the lowest probable discharges.

Also, the rainfall-runoff simulation supports a peak discharge higher than 200 m^3/s. The radar images show that the duration and intensity of the rainfall was higher at Utvikfjellet than anywhere else in the region. The registered damage in the region due to flooding also indicates this. Based solely on the observations at rain gauges and a normal increase in precipitation due to elevation, the peak discharge reached 143 m^3/s. A rainfall duration and intensity more coherent with site observations and in accordance with the radar observations, give peak discharges higher than 200 m^3/s. To give a peak discharge over 200 m^3/s and a duration in accordance with the observed course of the event, the areal precipitation had to be at least 60 mm in the most intense hour and 114 mm over 3 hours. In the upper part of the catchment, the intensity was up to at least 144 mm over 3 hours. For the simulations giving peak discharges of 230 m^3/s, the areal precipitation was over 130 mm and the highest intensities up to 170 mm over 3 hours. Field observations and observed damage along rivers show that the rainfall and runoff generation was more concentrated than the simulations suggest. This indicates that the rainfall intensity in the area giving the major contribution to the flood, can have been even higher than the highest simulated intensities.

According to the IDF curves for the region (Norsk Klimaservicesenter 2019), intensities between 114 and 170 mm over 3 hours is about three to five times higher than a 200-year return period rain event (P_{200}) for similar duration. How this extreme precipitation could occur is a question open for discussion, but it is reasonable to believe, based on wind directions in the region and observations of lightning in the hours prior to the flood, that three storm centres, each producing rainfalls of 40 mm/hour or more, merged at Utvikfjellet. This can explain the longer duration and higher intensities here than in the surrounding areas. Furthermore, post-processed rerun from MetCoOp Ensemble Prediction System shows a southerly located low pressure meeting a high pressure in the north and northeast (Figure 18), somewhat similar to the pattern reported by Parajka *et al.* (2010) and Bárdossy & Filiz (2005) that caused extreme floods across the Alpine-Carpathian range. This was also the case for the event at Fulufjället in 1997 (Vedin *et al.* 1999). The weather as they described it, with very warm humid air prior to the event, was also comparable to the weather prior to the event in Utvik.

The radar observations support that the most intense area of precipitation covered a limited part of the catchment and that the total rainfall during the event was at least 70 mm in this region. Comparing the radar images with

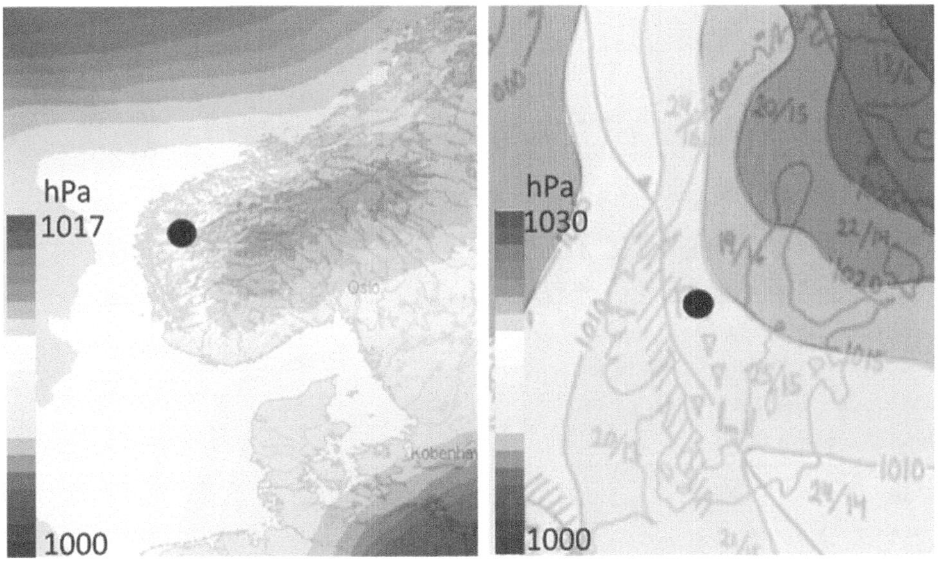

Figure 18 | Sea surface air pressure at 3 a.m. in the morning on 24 July 2017 prior to the Utvik flood to the left (Met.no 2019) and prior to the Fulufjället flood on 30 August 1997 from Vedin *et al.* (1999) to the right. The black dots indicate the location of Utvik and Fulufjället.

local observations confirms that the radar is underestimating the precipitation. Although the radar underestimates the intensity of the precipitation, it is able to locate the event quite well. The radar observations and the suggested extent of the contributing area based on observed extent of erosion along the river coincide quite well and reinforces the assumption of a contributing area down to 50% of the entire catchment. With a peak discharge between 200 and 250 m^3/s, this gives an estimated unit discharge from 9 to 13 $m^3/s,km^2$ if the entire area contributed up to 17 to 23 $m^3/s,km^2$ if the major contribution came from 50% of the catchment. In the most intense area, it will have probably been higher. This is significantly higher than any documented peak unit discharges in Norway so far and even higher than the peak unit discharge reported for the 22 km^2-large catchment at Fulufjället in 1997 (Lundquist 2000) that was estimated to be a 10,000-year event. Both these events are in the same range as peak unit discharges for extreme floods documented for southern Europe (Figure 19).

Even when taking the identified uncertainties into consideration, the findings from the flood in Utvik in 2017 suggest that previous observed peak unit discharges in Norway are low compared to what can, in fact, be expected. Although extraordinary claims require extraordinary evidence, all observations during and after the event indicate that this event is, beyond doubt, very extraordinary. Traces of the flood give a strong indication of the maximum flood level. Even with conservative estimates of this and the Manning coefficients, the peak discharge estimates become higher than 200 m^3/s and the corresponding peak unit discharge is in the range from 9 to 17 $m^3/s,km^2$. Hydraulic simulations support this conclusion. The most conservative estimate possible of the water level and velocity distributions would give discharges down to 160 m^3/s. Even this would give significantly higher unit discharges than previously observed in Norway. Radar images of the precipitation compared to rain gauge observations support the conclusion that the accumulated precipitation must have been at least 1.5 times higher than the 70 mm the radar gives at the most intense area of the catchment.

In a post-event flood frequency analysis, Q_{200} for Storelva in Utvik is suggested to be 140 m^3/s (5.68 $m^3/s,km^2$) and increased to 193 m^3/s (7.95 $m^3/s,km^2$) when including climate correction (Leine 2017b). However, an estimate of the areal precipitation causing a flood of this size is, according to the IDF curves for the region, at least two to three times higher than a 200-year return period rain event

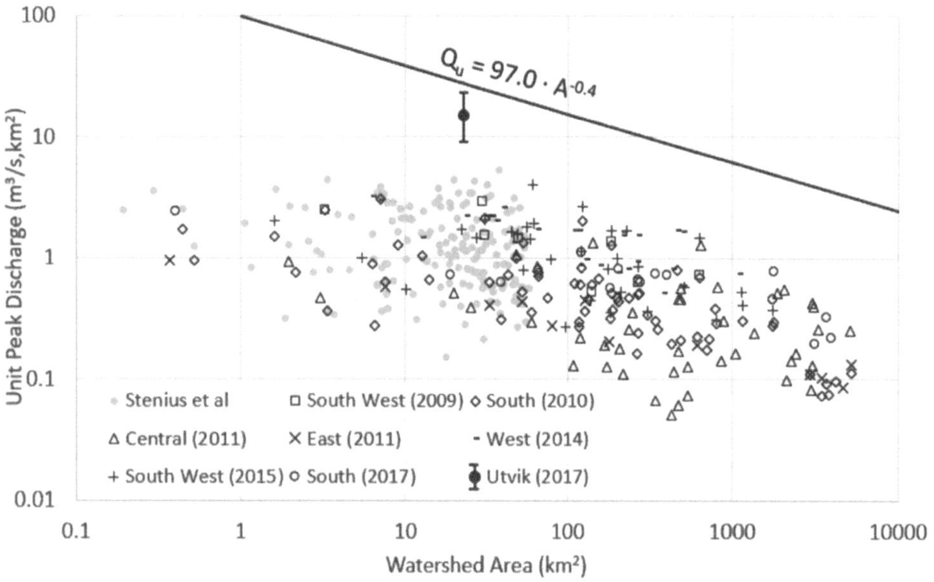

Figure 19 | Unit peak discharges observed in Norway compared to the estimated most likely range of unit peak discharges for the Utvik flood, including the envelope curve suggested by Marchi *et al.* (2010).

(P_{200}) for similar duration. This indicates that the Utvik flood event was significantly higher than a 200-year event. Prior to this event, Q_{200} for a neighbouring, hydrologically similar, catchment was estimated to be 2.82 $m^3/s,km^2$, and 3.95 $m^3/s,km^2$ including climate correction (Leine 2017a). Besides exposing the uncertainty of the Q_{200} estimates, this indicates also that the suggested Q_{200} for Storelva in Utvik is too high. Flood frequency analysis has significant impact on dimensioning and thus the cost of future infrastructure and also where people can live. Uncertainties like those identified in this case, illustrate the need of more information to base these analyses on to make them more reliable.

CONCLUSIONS

The Utvik flood is one of very few flash floods in Norway that are documented and quantified based on onsite observations during and after the flood combined with hydraulic and hydrological modelling. Analysis based on these observations and presented in this paper, shows that the discharge at the culmination of the Utvik flood most likely was in the range between 200 and 250 m^3/s corresponding to a unit discharge of 9 $m^3/s,km^2$ to 11 $m^3/s,km^2$. Assuming that the main contribution of the flood came from 50 to 100% of the catchment area, the peak unit discharge was from 9 $m^3/s,km^2$ to 22 $m^3/s,km^2$. Hydrological analysis based on gauge and radar observations tuned to the estimated peak discharge and observed flood propagation, shows that the areal precipitation causing the event probably was higher than 114 mm over 3 hours and between 140 and 170 mm in the most intense areas of the catchment. It is also found that the peak unit discharges for the Utvik flood are significantly higher than for previous floods observed at gauging stations in Norway and comparable to the most intense flash floods observed in southern Europe.

Floods like this have a high societal impact and this paper documents how extreme they can also become at these latitudes. Their impact is not only through the damage they cause, but also indirectly as they influence design criteria for infrastructure. In respect to how important but uncertain estimates of design floods (Q_{200}) are in rivers like Storelva, as also documented herein, this paper tries to point out that there is clearly a need for more information about floods in small, steep and fast responding catchments in order to have a better basis for future decision-making in regard to infrastructure and societal and economical optimized mitigation measures.

This paper may also indicate that we are experiencing a new hydrological regime that makes previous observations less relevant and thus new ones are more urgently needed.

ACKNOWLEDGEMENTS

Discussion with Bjart Eriksen and Jørn Kristiansen at the Norwegian Meteorological Service (Met.no), as well as their input on precipitation, have been valuable for this analysis. Acknowledgement also to Adina Moraru who gave constructive feedback on the analysis and the presentation.

REFERENCES

Aftenposten 2009 *Flomkaos i Egersund* (Floodcaos in Egersund) Aftenposten, November 20, 2009. https://www.aftenposten.no/norge/i/jB5Jw/flomkaos-i-egersund

Bárdossy, A. & Filiz, F. 2005 Identification of flood producing atmospheric circulation patterns. *Journal of Hydrology* **313**, 48–57. https://doi.org/10.1016/j.jhydrol.2005.02.006.

Barnes, H. H. 1969 Roughness characteristics of natural channels. *Journal of Hydrology* **7** (3), 354. https://doi.org/10.1016/0022-1694(69)90113-9.

Barredo, J. I. 2007 Major flood disasters in Europe: 1950–2005. *Natural Hazards* **42** (1), 125–148. https://doi.org/10.1007/s11069-006-9065-2.

Bergstrøm, S. & Forsman, A. 1973 Development of a conceptual deterministic rainfall-runoff model. *Hydrology Research* **4** (3), 147–170. https://doi.org/10.2166/nh.1973.0012.

Borga, M., Gaume, E., Creutin, J. D. & Marchi, L. 2008 Surveying flash floods: gauging the ungauged extremes re-analysis of radar-based precipitation estimates of the flash-flood producing storm. *Hydrological Processes* **22** (18), 3883–3885. https://doi.org/10.1002/hyp.7111.

Brunner, G. W. 2016a *HEC-RAS River Analysis System 2D Modeling User's Manual.* US Army Corps of Engineers, Davis, CA. www.hec.usace.army.mil

Brunner, G. W. 2016b *HEC-RAS River Analysis System Hydraulic Reference Manual.* US Army Corps of Engineers, Davis, CA. www.hec.usace.army.mil

Chow, V. T. 1959 *Open-Channel Hydraulics*. McGraw-Hill, New York, USA. http://web.ipb.ac.id/~erizal/hidrolika/Chow

Finans Norge 2018 Norsk Naturskadestatistikk (Nask) (Natural Damage Statistics). https://www.finansnorge.no/statistikk/skadeforsikring/Naturskadestatistikk-NASK/. Last accessed 30 December 2019.

Fjordingen 2017 *Bestilte og betalar asfaltering i Utvik sjølv* (Ordered and paid for asphalt in Utvik at own initiative) Fjordingen, August 10, 2017. https://www.fjordingen.no/nyhende/2017/08/10/Bestilte-og-betalar-asfaltering-i-Utvik-sjølv-15136446.ece.

Gaume, E., Bain, V., Bernardara, P., Newinger, O., Barbuc, M., Bateman, A., Blaškovičová, L., Blöschl, G., Borga, M., Dumitrescu, A., Daliakopoulos, I., Garcia, J., Irimescu, A., Kohnova, S., Koutroulis, A., Marchi, L., Matreata, S., Medina, V., Preciso, E., Sempere-Torres, D., Stancalie, G., Szolgay, J., Tsanis, I., Velasco, D. & Viglione, A. 2009 A compilation of data on European flash floods. *Journal of Hydrology* **367** (1–2), 70–78. https://doi.org/10.1016/j.jhydrol.2008.12.028.

Haddeland, I. 2009 *Flommen på Sør-og Vestlandet november 2009* (The flood in South and West Norway November 2009) NVE Report 3–2009. http://publikasjoner.nve.no/dokument/2010/dokument2010_03.pdf

Haddeland, I. 2011 *Flommen Østafjells juli* 2011 (The flood in East Norway July 2011) NVE Report 9–2011 http://publikasjoner.nve.no/dokument/2011/dokument2011_09.pdf.

Hanssen-Bauer, I., Førland, E. J., Haddeland, I., Hisdal, H., Mayer, S., Nesje, A., Nilsen, J. E. Ø., Sandven, S., Sandø, A. B., Sorteberg, A., Ådlandsvik, B., Andreassen, L. M., Beldring, S., Bjune, A., Breili, K., Dahl, C. A., Dyrrdal, A. V., Isaksen, K., Haakenstad, H., Haugen, J. E., Hygen, H. O., Langehaug, H. R., Lauritzen, S.-E., Lawrence, D., Melvold, K., Mezghani, A., Ravndal, O. R., Risebrobakken, B., Roald, L., Sande, H., Simpson, M. J. R., Skagseth, Ø., Skaugen, T., Skogen, M., Støren, E. N., Tveito, O. E. & Wong, W. K. 2015 *Klima i Norge 2100. Kunnskapsgrunnlag for Klimatilpasning Oppdatert i 2015 (Climate in Norway 2015. Knowledge Base for Climate Adaptation Updated in 2015)*. Norwegian Environment Agency, Trondheim, Norway.

Holmqvist, E. 2016 *Flommen i Rogaland og Agder desember 2015* (The Flood in Rogaland og Agder December 2015) NVE Report 5–2016. www.nve.no.

Høydedata 2019 Høydedata. https://hoydedata.no/LaserInnsyn/ (accessed 18 December 2019).

Langsholt, E. & Holmqvist, E. 2017 *Flommen På Sørlandet 30.9–3.10.2017* (The flood in South Norway 30.9–3.10.2017) NVE Report 80–2017. www.nve.no.

Langsholt, E., Roald, L. A., Holmqvist, E. & Fleig, A. 2015 *Flommen På Vestlandet Oktober* 2014 (The flood at West Norway October 2014) NVE Report 11–2015 www.nve.no.

Leine, A.-L. Ø., 2017a *Flomberegning for Storelva i Utvik (087.4Z)* (Flood estimates for Storelva Utvik (087.42)) NVE Report 94–2017. www.nve.no.

Leine, A.-L. Ø. 2017b *Flomberegning for Storelva* (Flood estimates for Storelva) NVE Report 51–2017. www.nve.no.

Lundquist, D. 2000 What is an extreme flood and for whom? In: *The Extremes of the Extremes: Extraordinary Floods* (A. Snorrason, H. Finnsdottir & M. E. Moss, ed.). IAHS Press, Reykjavik.

Marchi, L., Borga, M., Preciso, E. & Gaume, E. 2010 Characterisation of selected extreme flash floods in Europe and implications for flood risk management. *Journal of Hydrology* **394** (1–2), 118–133. https://doi.org/10.1016/j.jhydrol.2010.07.017.

Met.no 2019 TdsStaticCatalog. http://thredds.met.no/thredds/catalog.html (accessed 26 December 2019).

Ministry of the Environment 2008 Planning and Building Act (2008), June. https://www.regjeringen.no/en/dokumenter/planning-and-building-act/id173817/

NEVINA 2018 NEVINA Nedbørfelt-Vannføring-INdeks-Analyse (NEVINA Catchment-Discharge-Index-Analysis). http://nevina.nve.no/. Last accessed 17 April 2019.

NGU 2019 NGU Løsmasser – Nasjonal løsmassedatabase (NGU Soil cover – National Soil Cover database). http://geo.ngu.no/kart/losmasse_mobil/. Last accessed 18 April 2019.

NIBIO 2019 NIBIO – AgroMetBase. https://lmt.nibio.no/agrometbase/getweatherdata.php (accessed 30 December 2019).

NOAA, NWS 2017 Flash Flooding Definition. NOAA's National Weather Service. 2017. https://www.weather.gov/phi/FlashFloodingDefinition

Norsk Klimaservicesenter 2019 Norsk Klimaservicesenter. https://klimaservicesenter.no/faces/desktop/index.xhtml (accessed 1 April 2019).

NRK 2011 - Skader for 200 millioner kroner – NRK Norge – Oversikt over nyheter fra ulike deler av landet [WWW Document]. https://www.nrk.no/norge/–skader-for-200-millioner-kroner-1.7671294 (accessed 20 August 2019).

Pappenberger, F., Beven, K., Horritt, M. & Blazkova, S. 2005 Uncertainty in the calibration of effective roughness parameters in HEC-RAS using inundation and downstream level observations. *Journal of Hydrology* **302** (1–4), 46–69. https://doi.org/10.1016/j.jhydrol.2004.06.036.

Parajka, J., Kohnová, S., Bálint, G., Barbuc, M., Borga, M., Claps, P., Cheval, S., Dumitrescu, A., Gaume, E., Hlavkova, K., Merz, R., Pflaundler, M., Stancalie, G., Szolgay, J. & Blosch, G. 2010 Seasonal characteristics of flood regimes across the Alpine-Carpathian range. *Journal of Hydrology* **394** (1–2), 78–89. https://doi.org/10.1016/j.jhydrol.2010.05.015.

Pettersson, L.-E. 2010 *Flommen i Sør-Norge oktober 2010* (The flood in South Norway October 2010) NVE Report 10–2010. www.nve.no

Pettersson, L.-E. 2011 *Flommen i Trøndelag august 2011* (The flood in Trøndelag August 2011) NVE Report 12–2011. http://publikasjoner.nve.no/dokument/2011/dokument2011_12.pdf.

Roald, L. A. 2013 *Flom i Norge. (Flood in Norway)*. Tom & Tom, Norway.

Simensen, T., Hiller, P. H. & Vaskinn, K. 2011 *Vassdrag, Vannføring og Landskap. (Water Course, Discharge and Landscape)*. NVE-Report 1-2011, Trondheim, Norway. http://publikasjoner.nve.no/rapport_miljoebasert_vannfoering/2011/miljoebasert2011_01.pdf

Sorteberg, A., Lawrence, D., Dyrrdal, A. V., Mayer, S. & Engeland, K. 2018 Climatic Changes in Short Duration Extreme Precipitation and Rapid Onset Flooding – Implications for Design Values. https://cms.met.no/site/2/klimaservicesenteret/rapporter-og-publikasjoner/_attachment/13537?_ts=163df95ff7b

Stenius, S., Glad, P. A., Wang, T. K. & Væringstad, T. 2015 Veileder for Flomberegninger i Små Uregulerte Felt. (Guide for flood calculations in small unregulated fields) NVE. www.nve.no

Sunnmørsposten 2017 Prislapp for Flommen i Sogn Og Fjordane: 120 Millioner Kroner. (Price tag for the flood in Sogn og Fjordane: NOK 120 million) Sunnmørsposten, Norway.

USDA 2007 Natural Resources Conservation Service-Stream Restoration Design (National Engineering Handbook 654). NRCS. https://www.nrcs.usda.gov/wps/portal/nrcs/detail/national/water/manage/restoration/?cid=stelprdb1044707

Vedin, H., Eklund, A. & Alexandersson, H. 1999 The rainstorm and flash flood at Mount Fulufjallet in August 1997: the meteorological and hydrological situation. *Geografiska Annaler, Series A: Physical Geography* **81**(3), 361–368. https://doi.org/10.1111/1468-0459.00065.

VG 2017 Flere bygninger flyttet av flommen i Utvik: – Dette er forferdelig (Several buildings moved by the flood in Utvik: – This is terrible) VG, 24 July 2017. https://www.vg.no/nyheter/innenriks/i/LVG6V/flere-bygninger-flyttet-av-flommen-i-utvik-dette-er-forferdelig

World Economic Forum 2018 *The Global Risks Report 2018 – Reports – World Economic Forum*. Geneva, Switzerland.

World Waterfall Database 2019 *World Waterfall Database*. https://www.worldwaterfalldatabase.com/ (accessed 27 December 2019).

Yochum, S. E., Bledsoe, B. P., David, G. C. L. & Wohl, E. 2012 Velocity prediction in high-gradient channels. *Journal of Hydrology* **424–425**, 84–98. https://doi.org/10.1016/J.JHYDROL.2011.12.031.

Yochum, S. E., Comiti, F., Wohl, E., David, G. C. L. & Mao, L. 2014 Photographic guidance for selecting flow resistance coefficients in high-gradient channels. *Gen. Tech. Rep.* RMRS-GTR-323. U.S. Department of Agriculture, Forest Service, Rocky Mountain Research Station, Fort Collins, CO, 91 p.

First received 19 April 2019; accepted in revised form 4 March 2020. Available online 13 April 2020

Features and causes of catastrophic floods in the Nemunas River basin

Vytautas Akstinas, Diana Meilutytė-Lukauskienė, Jūratė Kriaučiūnienė and Diana Šarauskienė

ABSTRACT

The Nemunas River basin falls within the territories of five different countries – Belarus, Lithuania, Russia, Poland and Latvia. In general, the beginning of spring floods highly depends on rapid rise of air temperature, heavy precipitation and sudden snow melting in the analysed basin. In this paper, the conditions of formation and consequences of two catastrophic floods in 1958 and 1979 in the Nemunas River basin were studied regarding the hydrometeorological parameters (maximum snow water equivalent before the beginning of flood and precipitation amount during the flood) as well as runoff coefficients for each selected catastrophic flood. Differences between the main drivers and evolution of these floods were analysed. Spatial distribution of maximum snow water equivalent and precipitation, as well as runoff coefficient in different parts of the river basin, were identified as having the most significant impact on the formation of the studied catastrophic floods.

Key words | catastrophic flood, flood volume, Nemunas River, runoff coefficient, snow water equivalent

Vytautas Akstinas
Diana Meilutytė-Lukauskienė (corresponding author)
Jūratė Kriaučiūnienė
Diana Šarauskienė
Laboratory of Hydrology,
Lithuanian Energy Institute,
Breslaujos st. 3, LT-44403 Kaunas,
Lithuania
E-mail: *diana.meilutyte-lukauskiene@lei.lt*

INTRODUCTION

According to numerous scientific studies, including Intergovernmental Panel on Climate Change (IPCC 2013), more frequent and destructive floods all over the world may happen as a consequence of climate change. Beyond the fact that a number of catastrophic weather phenomena continue to grow and often result in awesome catastrophic floodings, they generate huge economic losses and kill people. For obvious reasons, huge floods are attracting significant attention of the mass media, as well as scientific society. Tweed (2011) describes the term *catastrophic flooding* as exceptional or rare floods with high magnitude. In general, catastrophic floods can be characterised by abruptness of water level rise and increase of intensity of flood phenomenon that are followed by enormous monetary losses and fatalities. Economic flood exposure is simulated to increase by about 200% between 2010 and 2050 (Jongman *et al.* 2012), whereas the number of flood-affected people may increase five-fold by the end of the 21st century (Hirabayashi *et al.* 2013).

The historical records of catastrophic floods reveal that their increased occurrence is mostly caused by extreme precipitation events. However, natural causes of fluvial floods are not limited to increased precipitation due to higher temperatures; snowmelt processes and soil conditions prior to flooding are also of high significance. Berghuijs *et al.* (2016) exposed the primary drivers of flooding across the contiguous United States and found that for most catchments soil moisture-dependent precipitation excess, snowmelt and rain-on-snow events are much better predictors of the flooding responses. Many other studies also confirm that floods are the result of a complex interaction between pre-event meteorological characteristics and hydrological catchment conditions (Nied *et al.* 2014; Beniston & Stoffel 2016;

doi: 10.2166/nh.2019.147

Woldemeskel & Sharma 2016). Hence, flood magnitude is determined by a certain flood-prone combination of hydrometeorological patterns before the event. An attempt to quantify multi-continental changes in the frequency and magnitude of extreme floods revealed that the key drivers of extreme floods strongly vary between catchments (Berghuijs *et al.* 2017).

Existing studies in Lithuania have mostly concentrated on spring flood phenomena in the Nemunas River basin (Stankunavicius *et al.* 2007; Rimkus *et al.* 2013; Meilutytė-Lukauskienė *et al.* 2017). However, only a few very old written sources about the catastrophic floods in the Nemunas River basin are available (Kolupaila 1932). Some information on water resources dynamics in the Nemunas River basin, which influenced extreme events, can be found in Korneev *et al.* (2015). Therefore, the aim of this research is to analyse the main drivers and conditions of the formation of two (1958 and 1979) catastrophic floods in the Nemunas River basin and to find the most unfavourable combination of hydrological and meteorological factors which may cause catastrophic floods in river catchments of Eastern Europe.

STUDY AREA AND DATA

The Nemunas River is the 14th longest river in Europe and the 4th longest in the Baltic Sea drainage basin. The river basin covers an area of around 98,200 km^2 and it mainly includes the territories of Belarus and Lithuania, whereas small parts of this basin fall within Russia (Kaliningrad district), Poland and Latvia (Figure 1). The length of the Nemunas River is 937 km, i.e., 436 km flows in Belarus (from the springs), 116 km of this river coincides with state borders between Lithuania and Belarus as well as Lithuania and Russia (Kaliningrad district) and the other 359 km – in Lithuania. Three types of climate (marine, transitional and continental) on the regional scale are detected in the territory of the Nemunas River basin. These types of

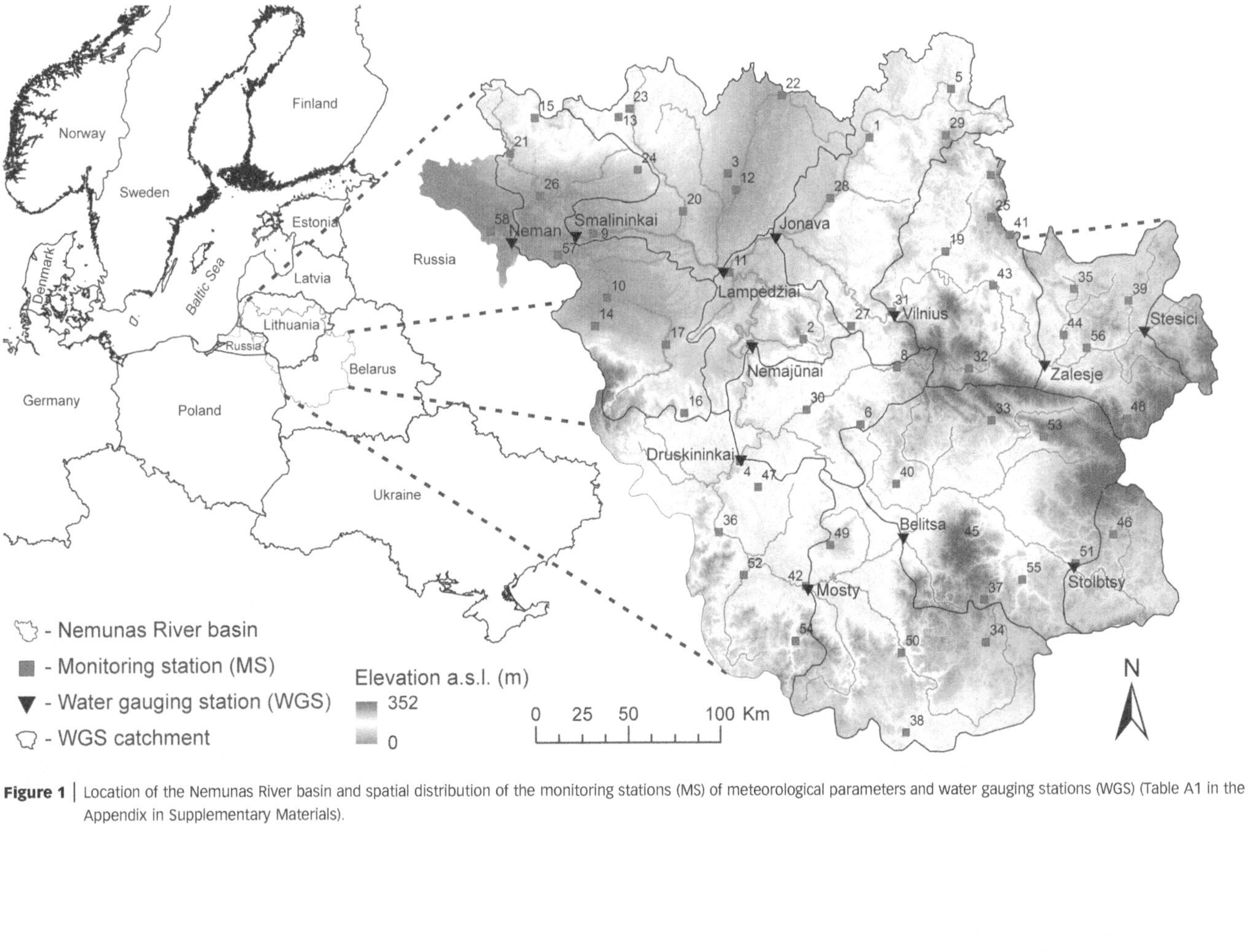

Figure 1 | Location of the Nemunas River basin and spatial distribution of the monitoring stations (MS) of meteorological parameters and water gauging stations (WGS) (Table A1 in the Appendix in Supplementary Materials).

climate highly depend on the distance of the Baltic Sea and local topography. The area with the highest precipitation ratio (700–900 mm) is located in the downstream part of the selected river basin. The amount of precipitation (450–650 mm) slightly decreases moving from downstream to the upstream as well as increase in amplitude of the air temperature due to the Baltic Uplands, which together with the distance from the Baltic Sea have significant impact on the distribution patterns of different meteorological parameters. Consequently, these local conditions cause more intensive accumulation of snow cover in the upper part of the selected river basin. The main reasons for the floods in the Nemunas River basin are a sudden snow melting combined with intense rainfall; also there is a probability of floods caused by dam failure and landslides. However, the local floods in the Nemunas River basin can happen suddenly because of ice jams. Such a distribution of factors was confirmed by investigation of the Environmental Protection Agency of Republic of Lithuania (EPA 2012), where the main causes of catastrophic floods are identified: the snow melting and ice jam events (75%), heavy rainfall (15%) and others factors (10%).

In this basin, the hydrological observation network consists of 73 water gauging stations (WGS). Eighteen of them are situated in the territory of Belarus, 52 stations in the territory of Lithuania, and 3 of them in the Kaliningrad district (Russia). Smalininkai WGS (water gauging station) has one of the longest series of water level (since 1812) and discharge continuous observations in Europe. The network of monitoring stations (MS) of meteorological observations comprises 26 meteorological stations: 13 stations in Belarus and 13 in Lithuania. The MSs include those stations which observed casual meteorological parameters (*T*, *P*, *SWE*, etc.) as well as stations where only *SWE* was measured.

The evaluation of impact of meteorological and hydrological parameters on formation of the catastrophic floods was carried out using long-term series of daily discharge data from 12 WGS (Table 1), as well as data of daily precipitation (*P*, mm), monthly air temperature (*T*, °C) and decadal (i.e., measured every tenth day) snow water equivalent (*SWE*, mm) from 58 MS.

METHODS

The beginning of spring flood in the Nemunas River basin mainly depends on climatic conditions (air temperature, precipitation, snow melting), whereas the end of flood may be influenced by many different elements (such as size, form and slopes of the basin, snow reserve in the basin, density of river network, etc.). In Figure 2, a scheme of flood formation in the Nemunas River is displayed, where the beginning of spring flood with abrupt increase of discharges, its course and culmination is presented. Sudden increase of

Table 1 | Characteristics of WGS of the Nemunas River basin

No.	Country	River	WGS	Distance from the mouth, km	Basin area, km^2	Watercourse slope, ‰	Forests, %
1	Belarus	Nemunas	Stolbtsy	854	3,070	0.93	35
2	Belarus	Nemunas	Belica	673	16,700	0.54	27
3	Belarus	Nemunas	Mosty	598	25,600	0.47	28
4	Lithuania	Nemunas	Druskininkai	450	37,100	0.40	27
5	Lithuania	Nemunas	Nemajūnai	339	42,800	0.37	27
6	Belarus	Neris	Stesici	455	1,230	1.30	48
7	Belarus	Neris	Zalesje	353	6,840	0.68	41
8	Lithuania	Neris	Vilnius	165	15,200	0.47	36
9	Lithuania	Neris	Jonava	39	24,600	0.45	27
10	Lithuania	Nemunas	Lampėdžiai	206	71,400	0.35	27
11	Lithuania	Nemunas	Smalininkai	112	81,200	0.54	18
12	Russia	Nemunas	Neman	72	91,600	0.54	20

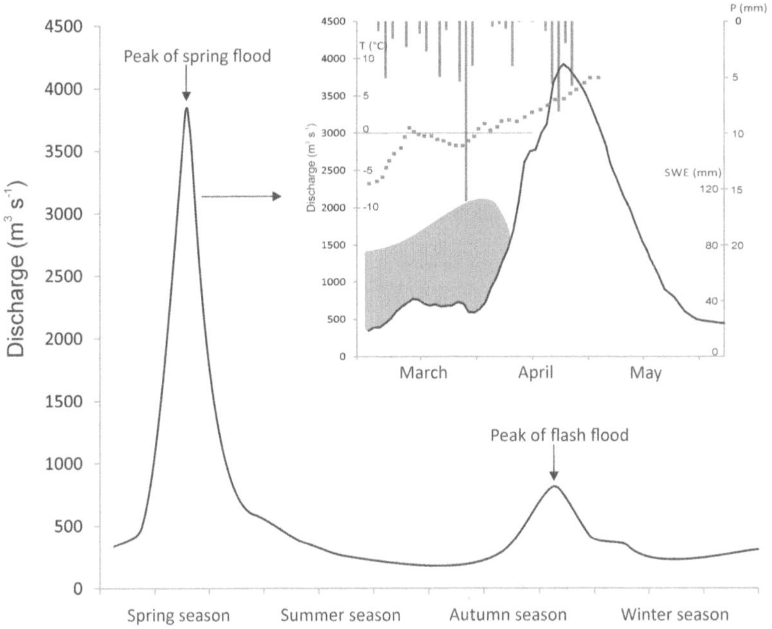

Figure 2 | Scheme of flood formation in the Nemunas River basin.

air temperature (above zero) together with a high rate of precipitation influences decline in thickness of the snow cover. Then, intensive snow melting causes abrupt increase of the water level in the river and these conditions give rise to the spring flood in the river basin.

A general scheme of the research methodology is presented in Figure 3. The first step in this research was probability distribution analysis in order to find out the floods of rare probability in the selected river basin. After selection of catastrophic floods, assessment of the main hydrological (water level during the flood (h_{flood}, cm), daily discharge during the flood (Q_{flood}, $m^3 s^{-1}$)) and meteorological (maximum snow water equivalent before the flood (SWE_{max}, mm) precipitation amount during the flood (P_{flood}, mm) and air temperature (T, °C)) characteristics was carried out. In the following step, the number of investigated variables was reduced by keeping the most important ones: Q_{flood}, SWE_{max} and P_{flood}. SWAT BF (Soil and Water Assessment Tool) tool was used for determination of the part of the surface runoff ($Q_{surface}$, $m^3 s^{-1}$) by eliminating the runoff part of groundwater feeding. Volume of catastrophic flood (V_{flood}, km^3) in the Nemunas River was calculated from the data of $Q_{surface}$. The flood volume in the WGS catchment (an area between two WGS, i.e., from upstream WGS to outlet WGS – $V_{WGSflood}$, km^3) was assessed as well. In

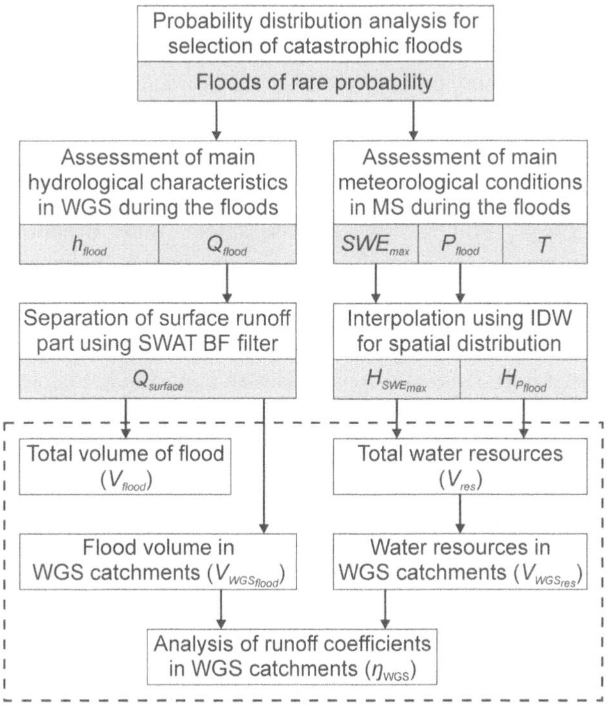

Figure 3 | Research concept for perception of main drivers of two selected catastrophic floods.

parallel, IDW (inverse distance weighted) method was used for analysis of the spatial distribution of height of maximum snow water equivalent before the beginning of the flood

(H_{SWEmax}, mm) and height of precipitation amount during the flood (H_{Pflood}, mm). After that, the total amount of water resources (V_{res}, km^3) from SWE_{max} and P_{flood} in the Nemunas River basin was estimated. Additionally, water resources in the WGS catchment (V_{WGSres}, km^3) were calculated within the scale of separate WGS catchments. Finally, analysis of runoff coefficients in WGS catchments η_{WGS} was carried out for the evaluation of surface runoff conditions during both catastrophic floods.

IDW (inverse distance weighted) interpolation (using ArcGIS Spatial Analyst extension) was applied in this study for the creation of isoline maps of certain indices on the catchment scale. The IDW method defines MS values, when, according to linearly weighting, the neighbouring MS values are established. The weight of monitoring stations is based on the function of inverse distance. The surface point values of SWE_{max} and P were interpolated according to the dependent variable of the closest MS. IDW method also was used to estimate values of the ungauged areas, which were calculated according to the surrounding MS data. The values of the closest monitoring stations had more influence on the unmeasured areas than stations further away. Accordingly, the total amount of water resources from SWE_{max} and P_{flood} in the Nemunas River basin was estimated.

SWAT (Soil and Water Assessment Tool) Baseflow Filter (BF) program (https://swat.tamu.edu/software/) was used to separate a part of the groundwater feeding and surface runoff from the data of historical observations (i.e., hydrograph). This software provides an opportunity to investigate the influence of the surface processes on volume of the spring flood. The methodology of hydrograph separation is described by Arnold & Allen (1999) in more detail. In this research, SWAT BF tool was used for determination of the part of the volume of catastrophic flood (V_{flood}) in the Nemunas River, which was caused by the surface runoff in the basin. Eliminated groundwater feeding from the daily discharge during the flood (Q_{flood}) allows assessment of the discharge part from surface runoff during the flood ($Q_{surface}$). Surface runoff during the flood gives an opportunity to estimate the interaction between surface processes (precipitation and snow melting) and catastrophic flood runoff. V_{flood} in the Nemunas River basin was compared with water resources in the WGS catchment (V_{WGSres}).

Calculations of hydrometeorological variables. In the Nemunas River basin, V_{res} consists of water from SWE_{max} together with P_{flood}, which determine V_{flood}. Volume of the selected flood was calculated by using the equation:

$$V_{flood} = \sum_{n}^{i=1} Q_{surface_i} \cdot t \tag{1}$$

where $Q_{surface_i}$ is daily discharge during the flood without groundwater feeding (estimated from output of SWAT BF) (m^3 s^{-1}), t is daytime (s), i is from 1 to n, n is flood duration expressed by days (the time period from the beginning until the end of spring flood, i.e., from the first day of sudden increase of hydrograph until the last day of sharp decrease in hydrograph after maximum discharge of the spring flood).

The volume of water resources V_{res} in the Nemunas River basin was calculated as:

$$V_{res} = (\bar{H}_{SWE_{max}} \cdot S_{basin}) + (\bar{H}_{P_{flood}} \cdot S_{basin}) \tag{2}$$

where $\bar{H}_{SWE_{max}}$ is average height of maximum snow water equivalent (calculated from area of the whole basin) before the beginning of the spring flood (mm), which was selected from decadal (i.e., measured every tenth day) data of snow water equivalent, $\bar{H}_{P_{flood}}$ is average height of precipitation amount (calculated from area of the whole basin) during the spring flood (mm), which was calculated by assessing the time period from abrupt rise of the river discharge until the maximum peak of catastrophic flood, S_{basin} is area of selected basin (km^2). Average heights of SWE_{max} and P_{flood} were estimated from the isoline maps.

In analysis of surface runoff processes, the runoff coefficient, an important input parameter in hydrologic modelling, characterised as the ratio of runoff volume and rainfall volume, is widely used. In the present study, the runoff coefficient η is defined as a portion of accumulated water resources that directly becomes a part of the volume of catastrophic flood. A runoff coefficient was calculated for each WGS catchment (η_{WGS}) and showed the ratio between $V_{WGSflood}$ and accumulated water resources from SWE_{max} and P_{flood} (V_{WGSres}):

$$\eta_{WGS} = \frac{V_{WGS_{flood}}}{V_{WGS_{res}}} \tag{3}$$

Thus estimation of η_{WGS} shows which part of the river basin has the greatest weight on the volume of catastrophic flood and spatial differences in the surface runoff conditions between WGS catchments.

ArcGIS (ArcMap, version 10.5, http://desktop.arcgis.com/en/) software was used for mapping of the research results.

RESULTS AND DISCUSSION

Hydrological characteristics of investigated floods

This research concentrates on two catastrophic floods. One of them (1958) occurred before the construction of Kaunas Hydro Power Plant (Kaunas HPP), and the second one after that (1979). These two floods were among the largest in this basin and both of them have available series of daily data and other related variables (meteorological data) which are necessary for the investigation. The flood of 1958 (one of the biggest floods in this basin of the last 200 years) affected large territories, covered three countries and reached a historical peak discharge (6,580 m^3/s at Smalininkai WGS in Lithuania). The flood peak in 1979 (the last biggest in this basin after the flood in 1958) reached 5,300 m^3/s (at Neman WGS in Russia) and caused a great deal of damage as well. In 1959, Kaunas HPP (227 km from the mouth of the Nemunas River) was set into operation and, since then, floods of the magnitude of that in 1958 have not been recorded. HPP significantly changed the hydrological regime of the river and conditions of flood formation below the HPP dam. Analysis of the probability distribution of the 200-year data set of the Nemunas River (at Smalininkai WGS) (Figure 4) showed flood peaks of rare probability (1% and 10%) in 1958 and 1979, respectively. These probabilities were calculated from the data series of annual maximum discharge of the period of 1812–2017.

Both floods took place in the period March to April. The peak discharge of flood in 1958 was 2.5 times greater than the average of maximum discharges at Smalininkai WGS, while in 1979 it was 1.5 times greater (Figure 5).

In 1958, spring started almost a month later than usual, i.e., 10–16 April (mean annual date of floods in the Nemunas River is 18–22 March). A few days later, the catastrophic flood hit the Nemunas River basin (Figure 6). The peak discharge of this flood exceeded 7,000–8,000 m^3/s. The 74,000 hectares of Kaliningrad District and 57,000 hectares of Lithuanian territory were flooded (Ginko 1982). Many structures were demolished, many dams were washed out, protected areas suffered from the overflowing waters as well (Figure 7). During this flood, the Lithuanian cities of Kaunas, Alytus, Balbieriškis, Prienai, Druskininkai and Birštonas were inundated; many houses and streets were damaged. Over the observation period, the highest water level h_{flood} was recorded at Druskininkai WGS (10.93 m on 24 April). In Kaunas city, the factories and

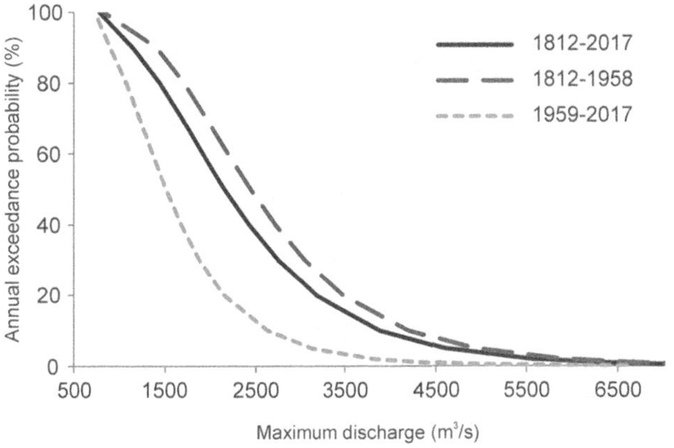

Figure 4 | The annual exceedance probability of maximum discharge in the Nemunas River at Smalininkai WGS for three different periods: the whole observation period (1812–2017), before construction of Kaunas HPP (1812–1958) and after (1959–2017).

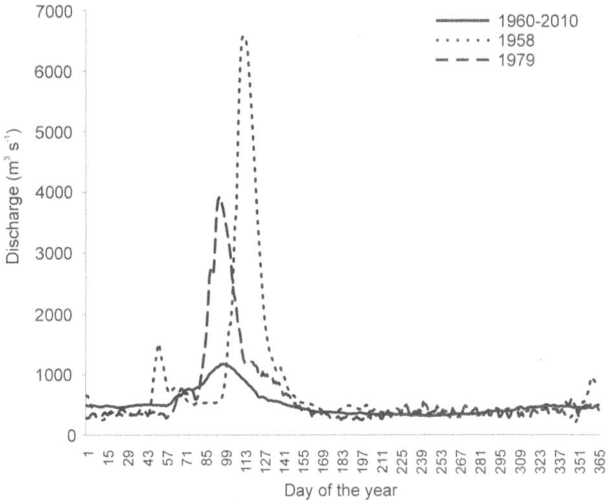

Figure 5 | Hydrograph of the Nemunas River at Smalininkai WGS in 1958 and 1979.

Figure 6 | The confluence of the Nemunas and Neris during the flood of 1958 in Kaunas city (photo by Stanislovas Lukošius, Kaunas city Museum Collection).

Figure 7 | Comparison of Kaunas (Lithuania) street view during the flood of 1958 (left) (photo by Stanislovas Lukošius, Kaunas city Museum Collection) with the same street view in 2018 when the river water level was close to average annual value (right) (photo by Vytautas Akstinas).

churches were flooded, and construction work of Vilijampolė Bridge (at that time it was under construction) was disturbed. In Belarus, h_{flood} of 7.10 m was observed at Michaliski WGS on 21 April. Grodno city experienced the biggest losses during this disaster (Briliovski 2012). In this city, h_{flood} rose to 8.63 m and many houses, cellars, a beer factory, port and city water pumping station were destroyed.

In March 1979, another large flood hit the large areas in the Nemunas River basin. This flood happened on 20–25 March, i.e., at the usual time. Although the spring season weather was not uncommon, in February it was extremely cold and a great deal of snow fell. During this flood, the maximum discharge at Neman WGS was 5,300 m^3/s (4 April). This time, the municipalities were ready for the flood and a significant amount of water (from the Neris River – the major tributary of the Nemunas) was detained in the water reservoir of Vileika; and large parts of the population along the rivers were successfully evacuated. However, an area of 30,000 hectares was flooded and many communication lines were damaged (Ginko 1982). The oldest WGS station of the Nemunas (Smalininkai WGS) was almost destroyed and temporarily no measurements were taken there (Jablonskis & Lasinskas 2011). During this flood, the Lithuanian cities of Kaunas, Smalininkai and Druskininkai were inundated. The highest h_{flood} was observed in the WGS of Druskininkai and Smalininkai (7.28 m on 7 April and 7.40 m on 5 April, respectively). In Druskininkai, the sanatoriums were flooded (and had to be closed); the Nemunas levee was breached (Ginko 1982). During this flood, the Belorussian city of Grodno suffered the most (Sajapin 2012). Many streets, houses, a beer factory, main bus station and church were inundated and damaged.

Characteristics of the meteorological conditions before catastrophic floods in 1958 and 1979

In general, the weather in the winter season of 1957/1958 was not unusual, i.e., the amount of precipitation as well as temperatures were very similar to average values. However, the spring season was extremely cold, especially March (air temperature was very low). For example, in March at Raseiniai MS, the mean temperature was equal to −5.2 °C (whereas, the annual mean of 1950–2015 is −0.7 °C). Thus, the beginning of spring could be regarded as exceptionally cold and having favourable conditions for formation of snow cover before the flood. Meanwhile, the winter season of 1978/1979 was completely different compared with 1957/1958. The winter season of 1978/1979 was very cold and air temperature in the Nemunas River basin territory was abnormally low. The mean values of air temperatures in December of 1978 varied from −7.6 °C (Ivacevici MS) to −11.6 °C (Lyntupy MS), whereas in January of 1979 it was from −7.7 °C (several MS) to −9.3 °C (Lyntupy MS) and in February from −6.9 °C (Tauragė MS) to −9.3 °C (Naroc MS). Therefore, the winter of 1978/1979 was cold and the major part of water resources was accumulated in snow cover, which resulted in such extensive flooding.

Interaction between precipitation and air temperature determines the form of precipitation – liquid (rainfall), freezing (drizzle) or frozen (snowfall). A large amount of snow was accumulated in the basin area after intense snowfall over the period from December 1957 to March 1958. For example, in February and March, the precipitation amount was significantly bigger than the annual mean (1950–2015) in three MS (Figure 8). In the winter season of 1978/1979, the amount of precipitation was distributed unequally: in January/February, it was close to the annual mean, whereas in March it was particularly high – 73.6 mm (at Lyntupy MS).

Estimated values of maximum snow water equivalent before the floods of 1958 and 1979 differed from each other. In 1958, SWE_{max} was very high and in some parts of the Nemunas River basin it exceeded 200 mm. The highest value of SWE_{max} was estimated in March and April (207 mm at Novogrudok MS in April, 189 mm at Varėna MS in March). Meanwhile, in the winter season of 1978/1979, accumulated SWE_{max} was high as well – in some MS greater than 140 mm. The largest SWE_{max} was identified in February at the monitoring stations of Tauragė and Novogrudok (144 mm and 159 mm, respectively). Abundant precipitation (snowfall and rainfall) and accumulation of thick snow cover could lead to such significant floodings in the Nemunas River basin. Therefore, analysis of spatial distribution of SWE_{max} and P_{flood} was performed.

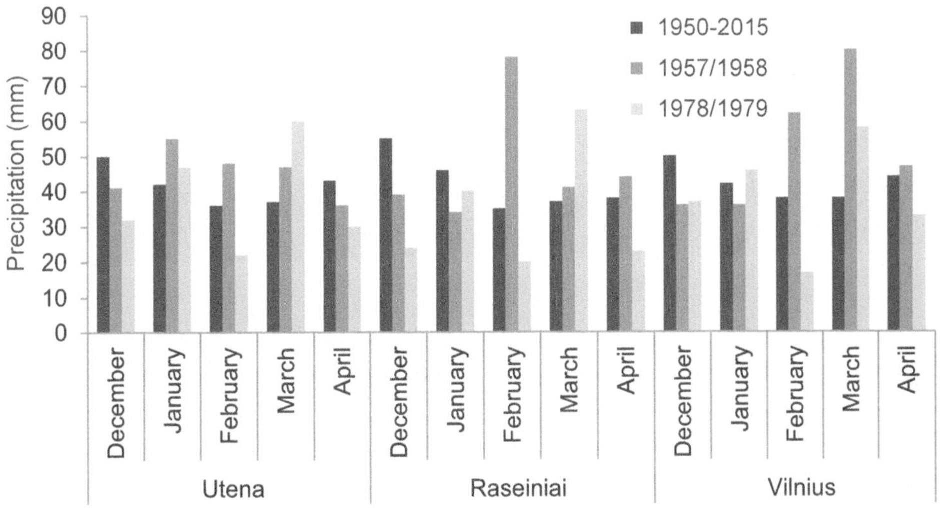

Figure 8 | Monthly distribution of the annual mean (1950–2015) and the monthly amount of precipitation from December (1957 and 1978) to April (1958 and 1979) in three MS of the Nemunas River basin.

Spatial distribution of meteorological conditions in the Nemunas River basin

The analysis of spatial distribution of SWE_{max} and P_{flood} was accomplished for the selected floods of 1958 and 1979 in the Nemunas River basin. The largest amount of water resources was accumulated in the snow cover before the flood of 1958 (Figure 9) when the isoline of SWE_{max} of 100 mm divided the basin in two different parts. These parts are characterised by distinct water resources and distribution of SWE_{max} (Figure 9(a)). The largest resources of SWE_{max} were concentrated in the southeastern part of the basin and they slightly decreased in the northwest (closer to the Baltic Sea). Consequently, formation of the flood of 1958 highly depended on the snow melting processes in the southeastern part of the Nemunas River basin; whereas the resources of SWE_{max} were smaller before the flood of 1979. They were concentrated in the northeastern and central parts of the basin (Figure 9(b)). This area was separated by the isoline of 100 mm and this division coincided with uplands in the analysed basin. These differences in distribution of the SWE_{max} are among the many factors which have led to major differences between the catastrophic floods of 1958 and 1979.

The spatial distributions of P_{flood} of 1958 and 1979 are displayed in Figure 10. During the flood of 1958, the amount of rainfall was also greater (same as maximum SWE_{max}) than in the flood of 1979. In the major part of the Nemunas River basin, the amount of precipitation exceeded 50 mm, whereas in the central part it was even greater than 60 mm. Such large amount and even distribution of precipitation indicated a significant impact of P_{flood} on formation of the catastrophic flood in the analysed basin compared with the flood of 1979. Meanwhile in 1979, P_{flood} reached up to 40 mm in the largest part of the basin. Only in the northern part was the increase of precipitation observed. During formation of the catastrophic floods in the Nemunas River basin, after snow melting, rainfall immediately becomes the major part of surface runoff. These conditions are formed due to the soils, which are already waterlogged before the cold period; and all moisture surplus is draining together with surface runoff due to the frozen soils. Therefore, the significant impact of precipitation amount on the magnitude of catastrophic flood is obvious.

Distribution of V_{WGSres} and $V_{WGSflood}$ in different WGS catchments (the whole analysed basin was divided into separate catchments and each catchment was described by the outlet WGS) in 1958 and 1979 is shown in Figure 11. The largest differences of $V_{WGSflood}$ between the two analysed floods were identified in the southeastern part of the basin (in the catchment of Belitsa WGS). These differences originated due to the high volume of water resources from

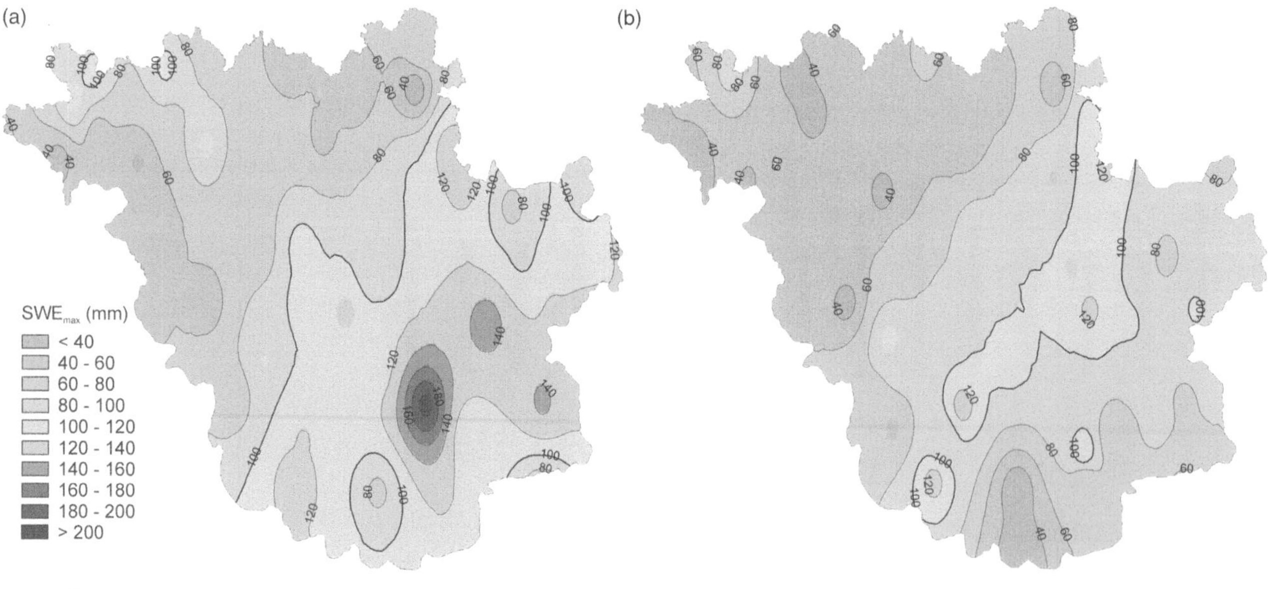

Figure 9 | The isolines of SWE_{max} (mm) before the floods in the Nemunas River basin in 1958 (a) and 1979 (b).

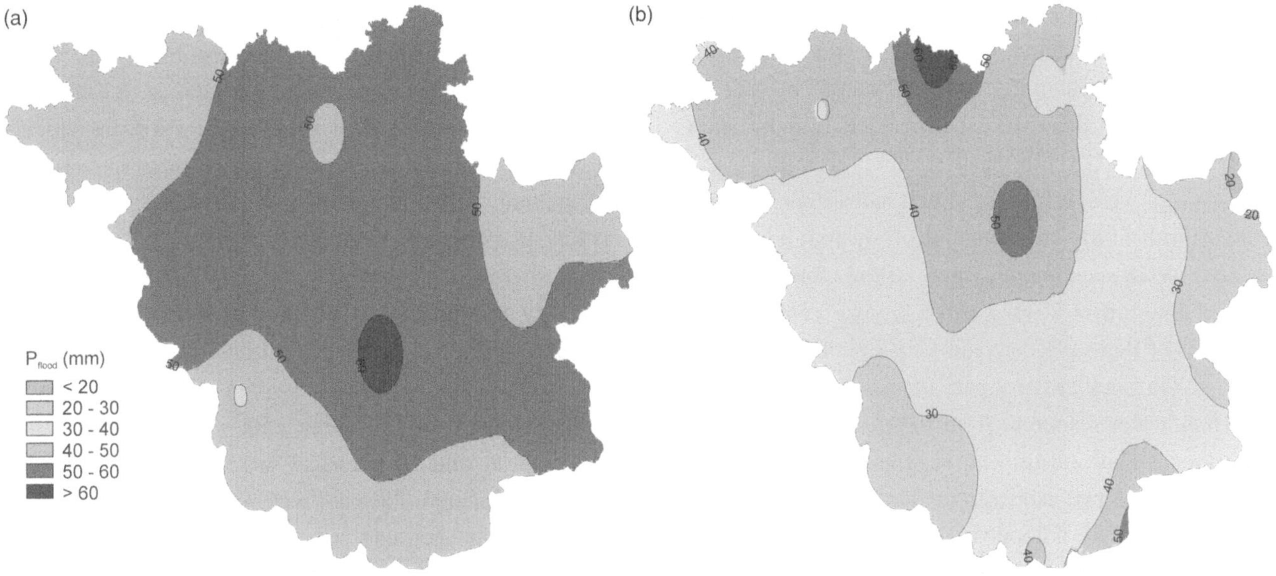

Figure 10 | The isolines of P_{flood} (mm) in the Nemunas River basin in 1958 (a) and 1979 (b).

SWE_{max} (Figure 9(a)) in 1958 and smaller input of water from P_{flood} (Figure 10(b)) in the southeastern part of the basin during the flood of 1979. Also, significant distinctions between $V_{WGSflood}$ of 1958 and 1979 were established in the catchments of Zalesje WGS and Vilnius WGS, where differences of V_{WGSres} were not as large as differences in flood volume. The influence of Vileika Reservoir and the water system of Vileika-Minsk (built in 1976) are reflected in the

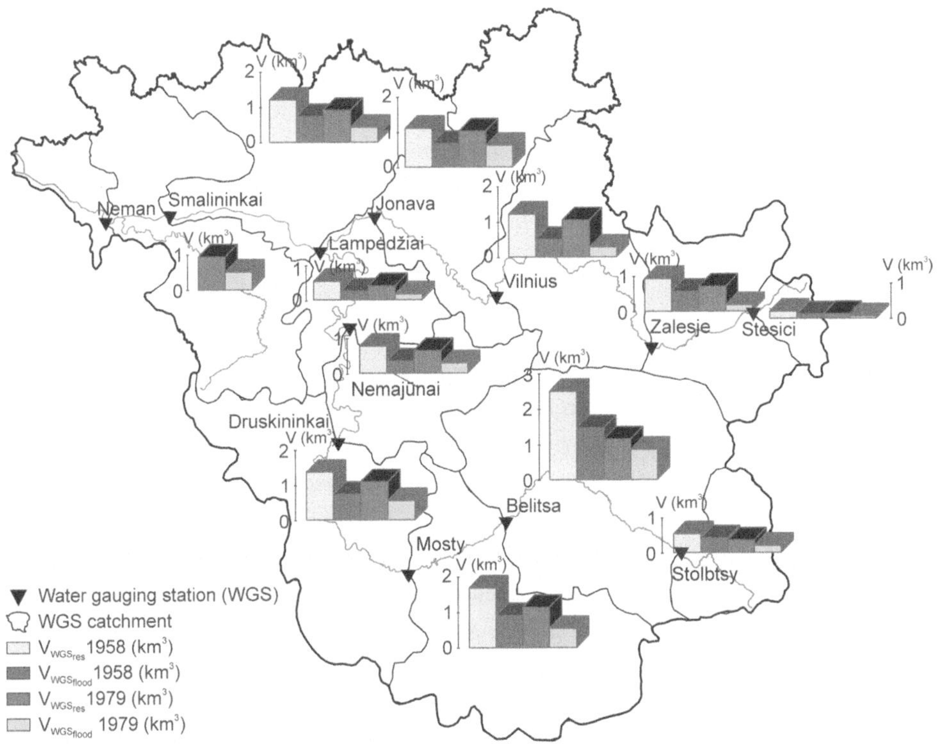

Figure 11 | Volumes of floods $V_{WGSflood}$ and water resources V_{WGSres} in the Nemunas River basin at different WGS catchments (in 1958 and 1979).

differences obtained in Belarus, because these constructions collected part of the flood water in the reservoir; water losses during the pumping and infiltration in Vileika-Minsk water system also have to be considered. In the catchment of Jonava WGS, the smallest differences in V_{WGSres} and $V_{WGSflood}$ were estimated. Such consistent patterns may be a result of the large amount of precipitation in the mentioned WGS catchment (Figure 10(b)). Summarising, the largest amounts of V_{WGSres} and $V_{WGSflood}$ were detected in the WGS of Belitsa, Mosty and Druskininkai, which are located in the southeastern part of the Nemunas River basin. Thus, water resources from these WGS catchments had the most significant impact on flood formation during the analysis of the two catastrophic floods.

The development of flood volume and increase of water resources from Stolbtcy WGS to Neman WGS in the investigated basin are shown in Figure 12. The changes of V_{flood} and V_{res} according to different WGS indicated the most significant increase of volumes in different sections of the basin. At Smalininkai WGS, V_{res} of 1958 and 1979 floods were estimated as 12.27 km^3 and 9.40 km^3, respectively. The significant inflow part of V_{res} was from the Neris River, which is the right tributary of the Nemunas River between Nemajūnai and Lampėdžiai WGS. Also, large

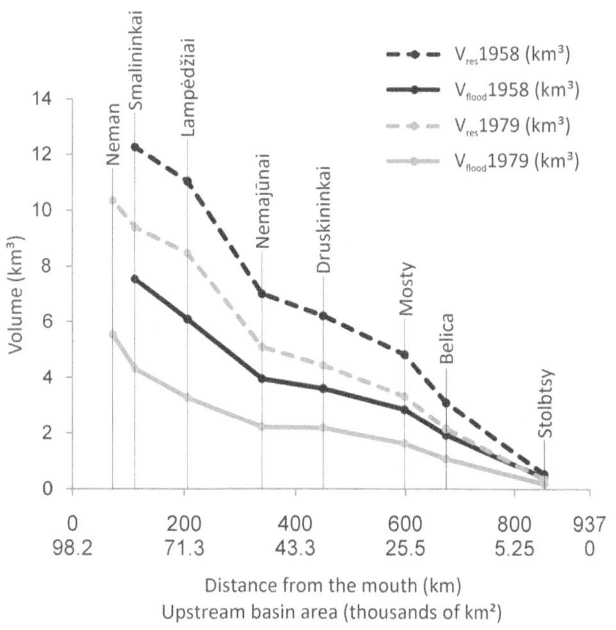

Figure 12 | The increase of V_{flood} and V_{res} at different WGS in the Nemunas River basin in 1958 and 1979.

increases of V_{flood} were estimated in the upper reach of the analysed basin (catchments of Belitsa, Mosty and Druskininkai WGS). The impact of V_{res} on floods in 1958 and 1979 is clearly expressed at Lampėdžiai WGS by sudden increase of this variable. At outlet WGS (Smalininkai), the total volume of catastrophic floods V_{flood} consisted of 7.54 km^3 in 1958 and 4.33 km^3 in 1979.

Variability and spatial distribution of runoff coefficients in WGS catchments of the Nemunas River basin

The variability of runoff coefficients η highly depends on meteorological and hydrological factors at the catchment scale. These factors influence processes of the surface runoff and differences between runoff conditions, which are expressed by the runoff coefficient. Maximum discharge Q_{max}, height of maximum snow water equivalent SWE_{max}, precipitation amount during the flood P_{flood} and runoff coefficient η_{WGS} were calculated in catchments of different WGS for both analysed catastrophic floods (Table 2). Estimated η_{WGS} indicated the part of SWE_{max} and P_{flood} that turned into the flood volume of different WGS catchments. The highest η_{WGS} (0.78) of the catastrophic flood of 1958 was estimated at the catchment of Stolbtsy WGS, meanwhile the average values of η_{WGS} fluctuated in the range of 0.51–0.62. During the flood of 1979, the η_{WGS} were lower and varied from 0.28 to 0.57. The highest η_{WGS} was obtained in Stolbtsy WGS catchment for both floods, but the lowest η_{WGS} were determined at WGS catchments of the Neris River – in the WGS of Vilnius and Zalesje (0.28 and 0.29, respectively). These differences revealed the parts of the Nemunas River basin that had the largest weight on flood volume.

The isoline map of annual η in the Nemunas River basin was created according to the studies of Jablonskis & Janukėniene (1978) and Makarevic (2017) (Figure 13). This map was used for comparison of obtained runoff coefficients η_{WGS} and the annual values of η. The obtained η_{WGS} were higher than the annual values of η in almost all analysed WGS catchments. This tendency confirmed the significance of surface runoff on magnitude of flood volume V_{flood}, because a larger part of water resources V_{res} directly transformed into V_{flood}. Several WGS catchments of the southeastern part of the basin had much higher η_{WGS} than the annual runoff coefficients, especially in the catchment of Belitsa WGS. Here,

Table 2 | Q_{max}, height of SWE_{max}, P_{flood} and η at catchments of different WGS during the floods of 1958 and 1979

			Maximum discharge Q_{max} (mm)		Average height of SWE_{max} (mm)		Average height of P_{flood} (mm)		Runoff coefficient η	
River	**WGS**	**WGS catchment area (km^2)**	**1958**	**1979**	**1958**	**1979**	**1958**	**1979**	**1958**	**1979**
Neris	Stesici	1,228	12.2	6.4	113.4	84.1	49.3	22.8	0.59	0.44
Neris	Zalesje	6,162	12.4	3.3	103.8	88.5	48.3	28.3	0.61	0.29
Neris	Vilnius	7,893	8.4	3.6	106.9	97.8	53.2	40.9	0.51	0.28
Neris	Jonava	9,244	7.3	4.7	68.2	67.1	53.1	45.4	0.54	0.39
Nemunas	Stolbtsy	3,182	18.3	9.6	119.2	78.1	51.4	34.0	0.78	0.57
Nemunas	Belitsa	13,935	12.2	6.8	130.0	94.0	53.8	35.7	0.62	0.50
Nemunas	Mosty	10,866	9.7	5.3	109.8	70.8	48.8	36.3	0.59	0.49
Nemunas	Druskininkai	10,126	7.4	4.5	86.4	77.5	47.3	30.4	0.58	0.49
Nemunas	Nemajūnai	5,113	7.0	4.2	98.4	87.6	55.1	40.3	0.57	0.44
Nemunas	Lampėdžiai	3,940	6.4	3.8	79.7	62.5	55.4	39.8	0.55	0.39
Nemunas	Smalininkai	9,841	7.0	4.2	72.5	48.8	52.3	45.7	0.61	0.46
Nemunas	Neman	10,332	–	5.0	–	53.2	–	38.7	–	0.54

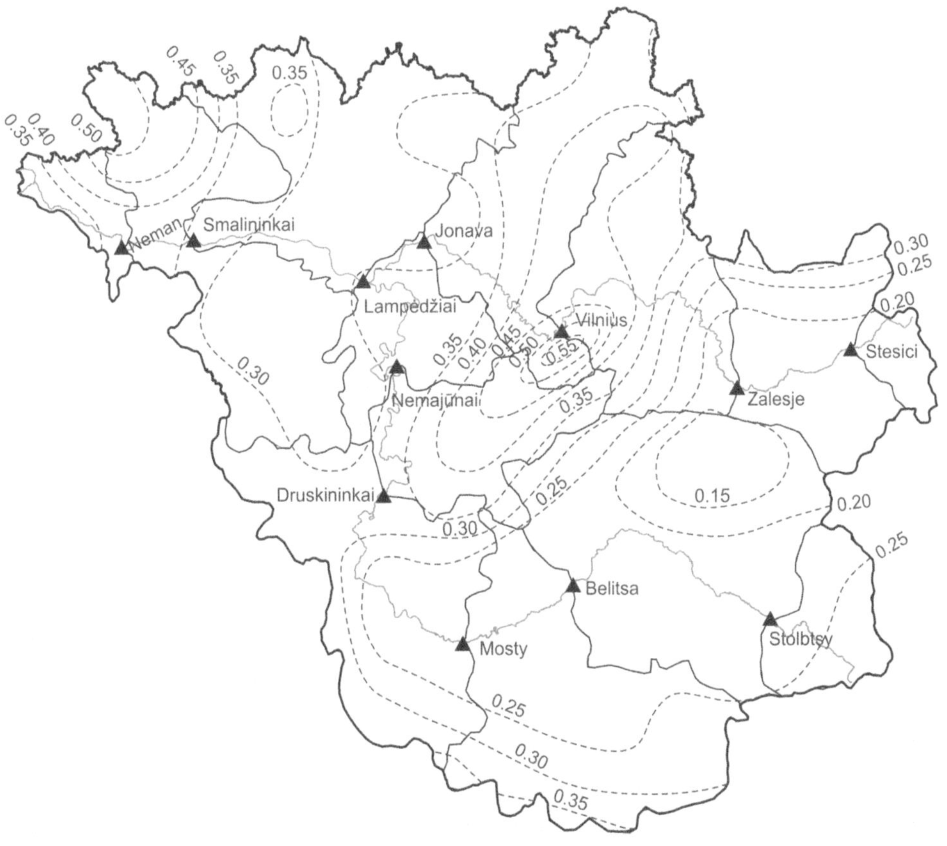

Figure 13 | Spatial distribution of annual coefficient of surface runoff in the Nemunas River basin according to Jablonskis & Janukėniene (1978) and Makarevic (2017).

the annual mean of η fluctuated in the range of 0.15–0.25, while the obtained η_{WGS} was 0.62 and 0.50 during the catastrophic floods of 1958 and 1979, respectively. These variations of η could be explained by a high difference in elevation as well as steeper slopes of the catchments in the southeastern part of the analysed basin.

The calculated runoff coefficients η_{WGS} during the catastrophic floods in the Nemunas River basin related to accumulated water resources from SWE_{max} and P_{flood} correspond with the studies from other research. Usually, the runoff coefficients are small in the catchments of Austrian rivers, but these coefficients increase significantly with event rainfall including snowmelt (Merz *et al.* 2006). Meanwhile, in the research of Alpine areas of Austria, runoff coefficients ranged from 0.2 to 0.6 during the flood of 2013 (Blöschl *et al.* 2013). Runoff coefficient computed for the catchments of the eastern Italian Alps confirmed an increase of η with event snowmelt floods and it is relatively low for rain floods (Norbiato *et al.* 2009).

CONCLUSIONS

In this study, the main drivers of the formation of catastrophic flood were analysed in the Nemunas River basin (situated in Eastern Europe). During the catastrophic floods of 1958 and 1979, the strong dependence of the flood severity on the distinctive combination of meteorological factors and hydrological characteristics was revealed.

The large amount of accumulated maximum snow water equivalent (up to 207 mm in the southeastern part of the Nemunas River basin and up to 120 mm in the central part, respectively, in 1958 and 1979) before the floods was the main factor which caused both catastrophic floods in the Nemunas River basin. Such conditions could occur due to a long period of negative temperatures during the cold season. The impact of excessive precipitation (average height of P_{flood} was 51.6 mm in 1958, while in 1979 it was 36.3 mm) during both analysed floods was significant as well, because this precipitation interacted with snow melting and consequently it was transformed into flood volume. Hence, the magnitude and interaction between these meteorological parameters resulted in the following volumes of catastrophic floods in the lower reaches of the basin at Smalininkai WGS: 7.54 km^3 (in 1958) and 4.33 km^3 (in 1979).

In order to find which part of the basin has a greater input on the flood volume formation, this basin was divided into separate WGS catchments. The runoff coefficient (η_{WGS}) was expressed as a ratio between the volume of catastrophic flood in WGS catchment and the volume of water resources in WGS catchment. η_{WGS} provides an opportunity to evaluate surface runoff processes in each WGS catchment, i.e., it enables the detection of the areas having the greatest weight on the volume of catastrophic flood and spatial differences of flood runoff formation. The highest values of η_{WGS} were estimated in the catchments of Stolbtcy (0.78 and 0.57) and Belitsa (0.62 and 0.50) WGS during the floods of 1958 and 1979, respectively, and these WGS were located in the upper reaches of the Nemunas River basin. The reason for the mentioned consistent patterns could be related to the different physical-geographical factors (steeper slopes, different soils, forest area, etc.).

The findings of the study extended our knowledge and have direct practical relevance for regional flood management, because the obtained results showed the significance of hydrometeorological processes on formation of catastrophic floods in the separate parts (WGS catchments) of the Nemunas River basin. Such subdivision of the analysed river basin highlighted the relevant WGS catchments with the highest weight on volume of selected catastrophic floods. These particular parts of WGS catchments should be investigated further in the future, to see if the similar meteorological conditions would be repeated with the obtained distribution patterns in the analysed river basin. Moreover, the warning systems and preventative actions as a consequence of catastrophic floods should be adapted and improved in each WGS catchment. The lack of available data of other floods with catastrophic status in the Nemunas River basin produces some limitations on the final results. Accordingly, more detailed investigation is required in smaller scale of the selected basin for better understanding of the flood formation process and identification of other possible drivers of catastrophic flooding.

SUPPLEMENTARY MATERIAL

The Supplementary Material for this paper is available online at https://dx.doi.org/10.2166/nh.2019.147.

REFERENCES

Arnold, J. G. & Allen, P. M. 1999 Automated methods for estimating baseflow and groundwater recharge from streamflow records. *Journal of the American Water Resources Association* **35** (2), 411–424. DOI: 10.1111/j.1752-1688.1999.tb03599.x.

Beniston, M. & Stoffel, M. 2016 Rain-on-snow events, floods and climate change in the Alps: events may increase with warming up to 4 °C and decrease thereafter. *Science of the Total Environment* **571**, 228–236. DOI: 10.1016/j.scitotenv.2016.07.146.

Berghuijs, W. R., Woods, R. A., Hutton, C. J. & Sivapalan, M. 2016 Dominant flood generating mechanisms across the United States. *Geophysical Research Letters* **43**, 4382–4390. DOI: 10.1002/2016GL068070.

Berghuijs, W. R., Aalbers, E. E., Larsen, J. R., Trancoso, R. & Wood, R. A. 2017 Recent changes in extreme floods across multiple continents. *Environmental Research Letters* **12**, 114035. DOI: 10.1088/1748-9326/aa8847.

Blöschl, G., Nester, T., Komma, J., Parajka, J. & Perdigao, R. A. P. 2013 The June 2013 flood in the Upper Danube Basin, and comparisons with the 2002, 1954 and 1899 floods. *Hydrological Earth System Sciences* **17**, 5197–5212. DOI: 10.5194/hess-17-5197-2013.

Briliovski, M. 2012 The Nemunas river. *Rodnaja Priroda (Native Nature)* **5**, 25–43.

Environmental Protection Agency of Republic of Lithuania 2012 *Preliminaraus Potvynių Rizikos Vertinimo Ataskaita (Preliminary Report on Floods Risk)*. https://e-seimas.lrs.lt/portal/legalAct/lt/TAD/TAIS.417035?jfwid=q86m1vvfu (accessed 15 November 2018).

Ginko, S. 1982 *Katastrofos Upių Pakrantėse (Catastrophes in River Shores)*. Mokslas, Vilnius, Lithuania, pp. 154–158, 165–167.

Hirabayashi, Y., Mahendran, R., Koirala, S., Konoshima, L., Yamazaki, D., Watanabe, S., Kim, H. & Kanae, S. 2013 Global flood risk under climate change. *Nature Climate Change* **3** (9), 816–821.

IPCC 2013 Detection and attribution of climate change: from global to regional. In: *Climate Change 2013: The Physical Science Basis. Contribution of Working Group I to the Fifth Assessment Report of the Intergovernmental Panel on Climate Change* (T. F. Stocker, D. Qin, G.-K. Plattner, M. Tignor, S. K. Allen, J. Boschung, A. Nauels, Y. Xia, V. Bex & P. M. Midgley, eds). Cambridge University Press, Cambridge.

Jablonskis, J. & Janukėniene, R. 1978 *Lietuvos Upių Nuotėkio Kaita (Change of Runoff in Lithuania Rivers)*. Mokslas, Vilnius, Lithuania, pp. 36–38.

Jablonskis, J. & Lasinskas, M. 2011 Nemuno stacionariems hidrometriniams tyrimams 200 metų (The 200th year anniversary of the stationary hydrometric measurements in the Nemunas River). *Energetika* **57** (1), 34–43.

Jongman, B., Ward, P. J. & Aerts, J. C. J. H. 2012 Global exposure to river and coastal flooding: long term trends and changes. *Global Environmental Change* **22** (4), 823–835.

Kolupaila, S. 1932 Nemuno nuotakis per 121 metus (1812–1932) (The Nemunas River discharges over 121 years). *Kosmos* **1**, 7–12.

Korneev, V. N., Volchak, A. A., Hertman, L. N., Usava, I. P., Anufriev, V. N., Pakhomau, A. V., Rusaya, I. E., Bulak, I. A., Bahadziazh, E. P., Dubenok, S. A., Zavyalov, S. V., Rachevsky, A. N., Rimkus, E., Stonevičius, E., Šepikas, A., Buijs, P., Crema, G., Denisov, N. B. & Koeppel, S. 2015 *Strategic Framework for Adaptation to Climate Change in the Neman River Basin*. https://www.unece.org/fileadmin/DAM/env/documents/2016/wat/04Apr_6-7_Workshop/Strategy_of_Adaptation_to_Climate_Change_ENG_for_print.pdf (accessed 15 November 2018).

Makarevic, A. A. 2017 *Гидравлика и инженерная гидрологи (Hydraulics and Engineering Hydrology)*. http://elib.bsu.by/bitstream/123456789/181537/1/Makarevich1.pdf (accessed 12 November 2018).

Meilutytė-Lukauskienė, D., Akstinas, V., Kriaučiūnienė, J., Šarauskienė, D. & Jurgelėnaitė, A. 2017 Insight into variability of spring and flood events in Lithuania. *Acta Geophysica* **65**, 89–102. DOI: 10.1007/s11600-017-0009-x.

Merz, R., Bloschl, G. & Parajka, J. 2006 Spatio-temporal variability of event runoff coefficients. *Journal of Hydrology* **331**, 591–604. DOI: 10.1016/j.jhydrol.2006.06.008.

Nied, M., Pardowitz, T., Nissen, K., Ulbrich, U., Hundecha, Y. & Merz, B. 2014 On the relationship between hydro-meteorological patterns and flood types. *Journal of Hydrology* **519**, 3249–3262. DOI: 10.1016/j.jhydrol.2014.09.089.

Norbiato, D., Borga, M., Merz, R., Bloschl, G. & Carton, A. 2009 Controls on event runoff coefficients in the eastern Italian Alps. *Journal of Hydrology* **375**, 312–325. DOI: 10.1016/j.jhydrol.2009.06.044.

Rimkus, E., Stonevičius, E., Korneev, V., Kažys, J., Valiuškevičius, G. & Pakhomau, A. 2013 Dynamics of meteorological and hydrological droughts in the Neman river basin. *Environmental Research Letters* **8**, 045014. DOI:10.1088/1748-9326/8/4/045014.

Sajapin, V. 2012 *Крупные наводнения в истории Гродно (Catastrophic Flooding in Grodno City)*. http://grodnonews.by/category/istoriya/news10115.html (accessed 9 May 2018).

Stankunavicius, G., Valiuskevicius, G., Rimkus, E., Bukantis, A. & Gulbinas, Z. 2007 Meteorological features behind spring runoff formation in the Nemunas River. *Boreal Environment Research* **12** (6), 643–651.

Tweed, F. 2011 Catastrophic flooding. In: *Encyclopedia of Snow, Ice and Glaciers* (V. P. Singh & U. K. Haritashya, ed.). Springer, Netherlands, pp. 112–113.

Woldemeskel, F. & Sharma, A. 2016 Should flood regimes change in a warming climate? The role of antecedent moisture conditions. *Geophysical Research Letters* **43**, 7556–7563. DOI: 10.1002/2016GL069448.

First received 16 November 2018; accepted in revised form 16 June 2019. Available online 18 July 2019

Risk assessment for areas prone to flooding and subsidence: a case study from Bergen, Western Norway

Guri Venvik, Ane Bang-Kittilsen and Floris C. Boogaard

ABSTRACT

Bergen city centre is prone to both subsidence and flooding. With a predicted increase in precipitation due to climate change, a higher proportion of rainfall becomes surface runoff, which results in increased peak flood discharges. In addition, it has been predicted that sea-level rise and increasing storm surges will result in coastal flooding. In this study, the dual hazards of flooding and subsidence are analysed to exemplify possible risk assessment maps for areas most prone to the combination of both. Risk assessment maps are a support tool to identify areas where mitigation of subsidence and adaptation for surface water management will be most efficient and measures can be implemented. The results show that dual hazard assessment, like that described in this paper, can be a useful tool for decision-makers when prioritizing areas to implement measures such as Sustainable Urban Drainage Systems.

Key words | flooding, groundwater, InSAR, risk assessment, subsidence, surface water

Guri Venvik (corresponding author)
Ane Bang-Kittilsen
Geological Survey of Norway (NGU),
P.B. 6315 Torgarden, 7491 Trondheim,
Norway
E-mail: *guri.venvik@ngu.no*

Ane Bang-Kittilsen
Faculty of Engineering, Department of Civil and Environmental Engineering,
Norwegian University of Science and Technology,
Trondheim,
Norway

Floris C. Boogaard
NoorderRuimte, Centre of Applied Research and Innovation on Area Development, Hanze University of Applied Sciences,
Zernikeplein 7, P.B. 3037, 9701 DA Groningen,
The Netherlands

INTRODUCTION

It is expected that 60% of the world's population will be living in urban areas by 2030, and most of this area has yet to be built (UN 2016). The pace of urban growth may be overwhelming and exert tremendous pressure on the catchment hydro(geo)logy in general and urban drainage in particular (Marsalek *et al.* 2006). The built urban infrastructure, with asphalt and concrete-covered ground surfaces, alters hydrologic abstractions and water flow found in natural catchments (Bolund & Hunhammar 1999). It has been predicted that climate change will increase precipitation (Hanssen-Bauer *et al.* 2017), and a higher proportion of rainfall will become surface runoff, which, in turn, will result in increased peak flood discharges and degraded water quality (Haughton & Hunter 1994). In addition, the sea level is predicted to rise by up to 1 m by 2090 (Hanssen-Bauer *et al.* 2017). Changes in the urban environment due to growth in addition to climate change put the urban water cycle out of balance, thereby affecting other surface and subsurface processes, such as flooding and subsidence.

Urban areas are, to a large extent, built environments, and from that view constitute a unique environmental challenge. As Pregnolato *et al.* (2017) point out, cities are particularly vulnerable to flooding and rapid and intense rainfall due to the impermeable surfaces that dominate areas with high concentrations of people, buildings and infrastructure. As a result of the increasing flood damage in Europe, there has been a shift in attention from flood protection to flood risk management (Albano *et al.* 2018), where risk assessment with tools, such as maps, are central. This

doi: 10.2166/nh.2019.030

shift is also valid for risks other than flooding, such as subsidence.

Both pluvial and coastal flooding can be related to subsidence (Dixon *et al.* 2006; Miller *et al.* 2008; Yin Yu & Wilby 2016). For the Bryggen Wharf, in central Bergen, western Norway, there is a strong link between water and subsidence, due to reduction in water in the subsurface cultural layers, as well as lowering of the groundwater levels leading to the decay of organic layers as well as historical wooden foundations and thereby subsidence (de Beer *et al.* 2012; de Beer & Seither 2015; Matthiesen *et al.* 2015; Rytter & Schonhowd 2015b). Other geological processes commonly linked to subsidence include tectonic structures, land and rock slides, gravitation (Berardino *et al.* 2003; Lauknes *et al.* 2010; Eriksen *et al.* 2017) and subsidence due to groundwater depletion (Chaussard *et al.* 2014; Castellazzi *et al.* 2016; Motagh *et al.* 2017).

In order to provide communities with urban infrastructures that are durable and well-functioning, climate change impact and adaptability assessments are vital (Pregnolato *et al.* 2017). Flood modelling is a useful tool for planning floodways, identifying areas for mitigation measures and for bringing awareness of water issues into decision-making processes in urban areas (Fletcher *et al.* 2013; Albano *et al.* 2017; Boogaard *et al.* 2017a, 2017b; Lyu *et al.* 2018). Hence, risk assessment mapping can be further used for identifying areas for the implementation of Sustainable Urban Drainage Systems (SuDS), such as swales, to infiltrate water into the ground and to sustainably manage surface water in urban areas. More knowledge is needed to understand the urban water balance and the processes connected to water to prevent and counteract subsidence that can cause damage and unforeseen expenses.

Increased knowledge and understanding of the urban water cycle in the transitional zone between the built and natural environment is necessary. In the case of Bergen city centre, past research has shown that the subsidence to a large degree is driven by the depletion of water in the underlying organic-rich cultural layers (Harvold *et al.* 2015; Matthiesen *et al.* 2015). For a complete understanding of the urban water cycle, hydrological and hydrogeological studies should be included (Wakobe *et al.* 2018). Hence, we combine datasets for flood risk and subsidence to develop a risk assessment map for areas prone to damage. The case study is set in Bergen city centre (Figure 1), on the west coast of Norway. Bergen is a coastal city where the annual precipitation is high, 2,250 mm/year (NMI 2019). The city is therefore prone to water-related damage caused by pluvial flooding, storm surges and stormwater flooding.

The subsidence data are computed using satellite-based persistent scatterer interferometry (PSI; Crosetto *et al.* 2016). PSI has long been used to compute subsidence, especially related to groundwater depletion (e.g. Schmidt & Bürgmann 2003; Teatini *et al.* 2012). In this study, data from the Sentinel-1 satellites have been used as an input. Further, subsidence data have been correlated with an LiDAR DEM (Norwegian Mapping Authority 2009)-based urban flood model result.

Dual hazard analyses have been carried out by two different analysis methods using ArcGIS (ESRI 2018). In both methods, the resulting map is a grid, which is a common areal unit when synthesizing multiple variables (Carver 1991; Damoom *et al.* 2019). The first method is a simple grid overlay, recording the occurrence of input data within the grid cells. The second method uses Getis-Ord G* statistics (Getis & Ord 1992) commonly called 'hot spot analysis' (ESRI 2019), which automatically detect clusters of incident data within the bounding area of flood data. As an example, Lu *et al.* (2012) use the 'hot spot analysis' to detect slow-moving landslides from InSAR data. Geographical Information System (GIS)-based analysis for risk assessment is widely used to investigate various hazards, such as flooding (Albano *et al.* 2017, 2018; Lyu *et al.* 2018) and for multi-criteria decision-making analysis (Erbaş *et al.* 2018; Damoom *et al.* 2019). As pointed out by Damoom *et al.* (2019), when combining different datasets GIS allows the user to visualize, inquire, analyse and interpret the vast amount of (geological) data for a better understanding and problem-solving. Therefore, the risk assessment analysis presented in this paper aims to identify areas prone to the dual hazards of both flooding and subsidence. Dual hazard assessment maps, based on existing flooding and subsidence data, were executed using overlay and 'hot spot' analysis in the GIS. Results can be used as a tool to select areas that need mitigation and damage prevention measures, both for buildings and urban infrastructure. Risk assessment, shown in this case study, may be applied in urban (or rural) areas where data, such as subsidence and flooding, are available.

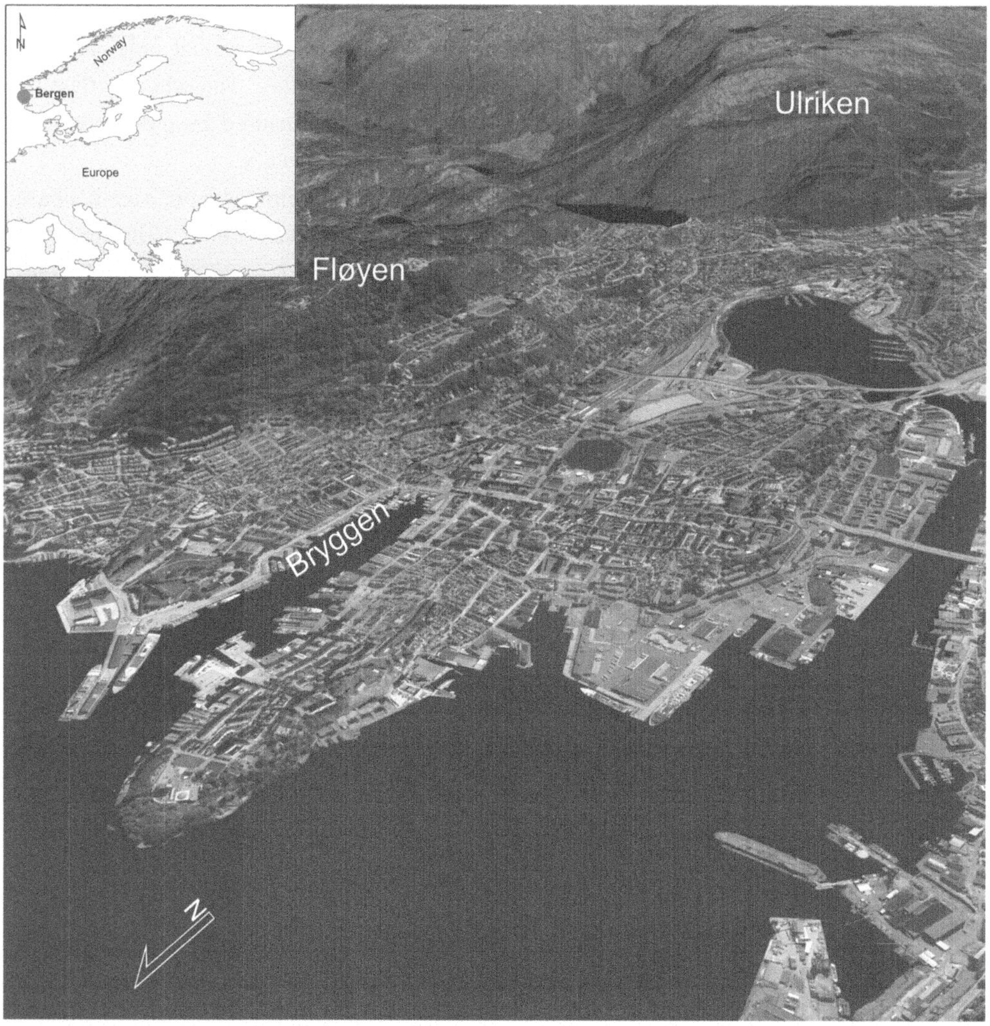

Figure 1 | Bergen city centre viewed towards the southeast with steep hillside and lower lying area along the shoreline (Google Earth, 2019). Please refer to the online version of this paper to see this figure in colour: http://dx.doi.org/10.2166/nh.2019.030.

STUDY AREA AND DATA

Bergen is the second-largest city in Norway, located on the west coast, with an area of 464 km^2 and a population of 278,556 (SSB 2017). The city has an annual average temperature of 8.6°C and an annual precipitation of 2,250 mm (NMI 2019). The climate is predicted to become wetter with more intense and frequent downpours, which will increase the pressure on surface water runoff and stormwater management (Hanssen-Bauer *et al.* 2017). The topography of Bergen city centre, as well as the surrounding areas, encompasses steep hillsides covered with forest vegetation on thin soil cover, down to flat-lying former shorelines with thicker natural sediments and anthropogenic layers. A 1 km relief goes from Fløyen (at 320 m a.s.l.) to Bryggen (at 1 m a.s.l.) (Figure 1). These natural conditions make surface runoff water abundant.

The study area has been constrained to the city centre, including the Medieval city and its surrounding area. In the city centre, the anthropogenic cultural heritage layers are thick with a rich organic content locally more than 10 m thick (Figure 2). The old shoreline from the 12th century (Hansen 1994) is shown in Figure 2. Since Bergen has close to no isostatic land uplift (Mangerud 2004), the progressing

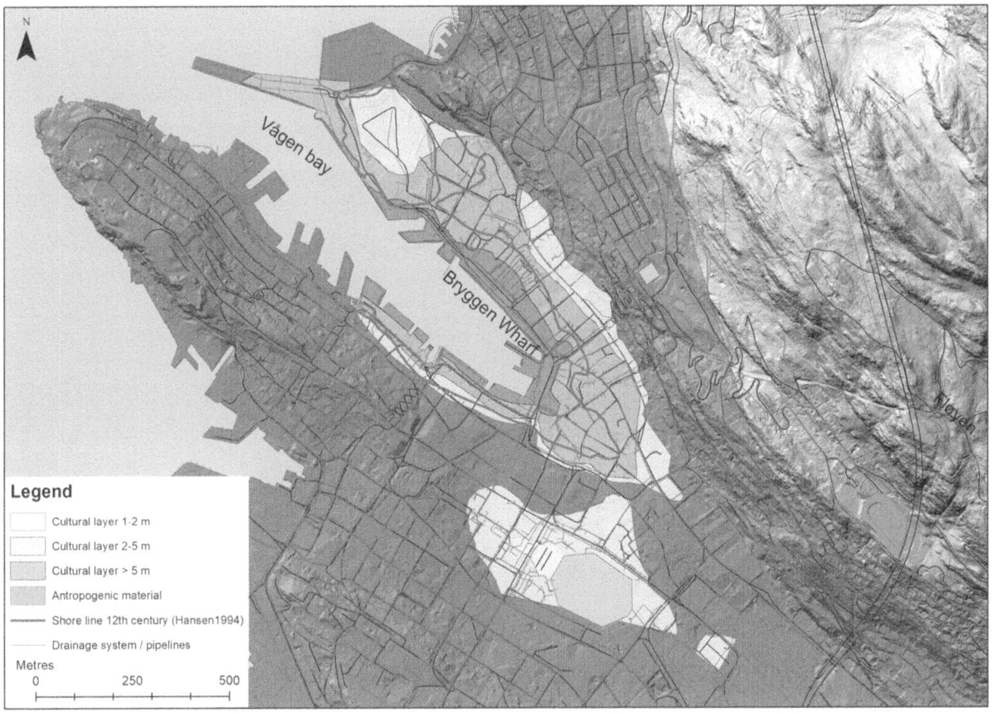

Figure 2 | In Bergen city centre, the subsurface consists of exposed bedrock in the hillside (light grey colour), anthropogenic material (dark grey colour) and up to 10 m of cultural layers (brown colour), on top of beach sand, clay and till before reaching bedrock below. (Directorate for Cultural Heritage, 2018, Norwegian Map Authority, 2018). Please refer to the online version of this paper to see this figure in colour: http://dx.doi.org/10.2166/nh.2019.030.

shoreline of today is due to filling of the anthropogenic material such as waste into the bay area, Vågen. These layers are more prone to destruction due to lack of infiltration of surface water; therefore, the Bryggen project was initiated in 2010 to save the UNESCO World Heritage site of the Hanseatic League Wharf (Ersland 2015; Rytter & Schonhowd 2015a). Rytter & Schonhowd (2015a) document the connection between soil moisture, groundwater level and the decay or preservation of organic anthropogenic material. The lack of soil moisture and very low groundwater levels can lead to the higher oxygen concentration in the organic matter and acceleration of disintegration. The organic layers then collapse and compact (Matthiesen *et al.* 2015), resulting in subsidence of the ground and damage to buildings and infrastructure (Jensen 2015; Rytter & Schonhowd 2015b). Bryggen is an example where measures have been taken by implementing SuDS to infiltrate surface water into the subsurface to increase soil moisture and groundwater level and thereby preserve the cultural layers and stabilize the ground (de Beer *et al.* 2012; Boogaard 2015; de Beer & Seither 2015).

Drainage system in Bergen city

To handle the surface water and stormwater, Bergen city has a drainage system with the purpose of transporting water effectively out of the city. In the greater parts of the city, especially in the inner centre, the stormwater is brought together with the wastewater from the industry and household (Figure 2; Bergen Kommune 2006). When intense rainfalls occur, the capacity of the drainage system is strained, which may cause the emission of wastewater. Since the relief in the city centre is steep (Figure 1) and the surface has low permeability, flooding arises when large and intense rainfalls occur in short time spans. Due to climate change, events with downpour will be more intense and frequent. This, in addition to predicted sea-level rise, will give more frequent and intense flooding where there are topographic depressions (Hanssen-Bauer *et al.* 2017), as seen in Figure 3.

For this study, we included a dataset of the pipelines for wastewater and sewage. It should be noted that the sewage system may be a combined stormwater and sewage, or a

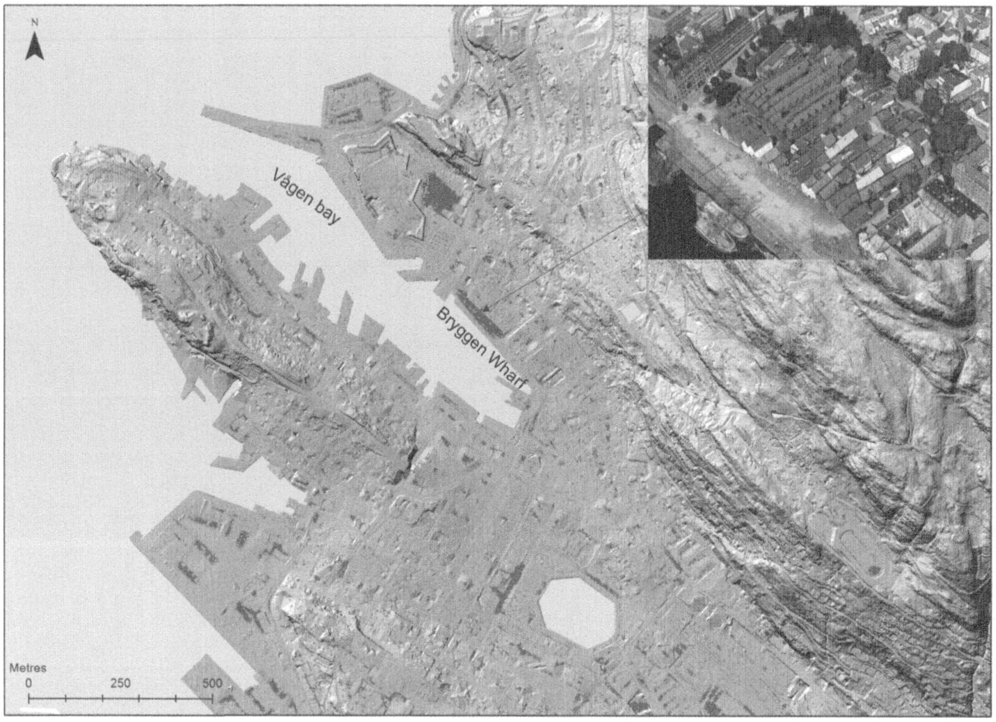

Figure 3 | Pluvial flood model result of Bergen city based on DEM and rainfall input, where terrain and depressions control the flow path and accumulation of surface water. Increase in colour intensity with the increasing surface water depth. The inserted photo shows the area in front of Bryggen Wharf prone to pluvial flooding. Please refer to the online version of this paper to see this figure in colour: http://dx.doi.org/10.2166/nh.2019.030.

separate system: these are not differentiated in the dataset (Bergen Kommune 2006).

Flood modelling

Pluvial, urban flooding has received increased attention over the last decade (Mignot *et al.* 2019), due to the costly damage on infrastructure and society (Miller & Hutchins 2017; Sörensen & Mobini 2017). There are many tools for urban stormwater flood modelling as pointed out by Balstrøm & Crawford (2018), which have been improved after the July 2011 Copenhagen event with close to a 100 mm/h rainfall (Miller & Hutchins 2017; Sörensen & Mobini 2017; Mignot *et al.* 2019). The flood modelling itself is not the scope of this work but the dual hazard of flooding and subsidence. The flood map was created as a case study of Bergen in the INXCES project described in Boogaard *et al.* (2017a, 2017b), and the results are further used for analysis in this study. The urban flood modelling was created using the Calamity Levels of Urban Drainage Systems (CLOUDS by Tauw bv) method with the aim of modelling and simulating water flow and water accumulation (Kluck *et al.* 2010; Boogaard *et al.* 2017a, 2017b, 2018). The simulation was run with a precipitation of >60 mm/h, where 20 mm/h is estimated to run in the sewer system and 40 mm/h on the surface. This represents an extreme storm or a 100-year event (Kluck *et al.* 2010, 2015). With this assumption, the digital elevation model (DEM; Norwegian Mapping Authority 2009) and rainfall distribution serve as the main input. The flood simulation was done to increase the understanding of which urban areas are most prone to flooding as well as indicating runoff flow paths for surface water (Figure 3). The Bryggen Project is a best management practice that demonstrates the linkage between infiltration of surface water, recharge of groundwater, preserving cultural layers and preventing subsidence (de Beer *et al.* 2012; Boogaard 2015; de Beer & Seither 2015; Matthiesen *et al.* 2015; Rytter & Schonhowd 2015a). This flood simulation indicates areas where infiltration of surface water will be most advantageous with regard to reducing flooding as well as subsidence. This can further be used to plan floodways for the city.

The resulting map shows stormwater accumulation, where the darkest blue colour indicates a greater water depth (Figure 3) (the colour figure can be viewed online). The DEM was created from LiDAR data produced from the FKB-Laser (Felles KartdataBase/common map database) dataset consisting of 1 point per m^2 (Norwegian Map Authority/Kartverket 2009). A detailed description of method, calculations and results from the flood modelling is presented in Boogaard *et al.* (2017a) and Kluck *et al.* (2010). For a complete comprehension of the urban water balance, hydrological and hydrogeological studies should be included (Wakode *et al.* 2018).

Present-day storm surge

In November 2018, the Norwegian Mapping Authority launched an open access web service with models of current and future (2090) sea-level rise and storm surges. The data, map tool and services are aimed at the planning of coastal areas (DSB 2017). The storm surge height intervals are mean high water, 20-year, 200-year and 1000-year return periods. One of the Mapping Authority's datasets entitled '200-year storm surge' (Figure 4) shows sea level under these extreme conditions. In Bergen, there are small differences in sea-level heights for the different return periods of storm surges (https://www.kartverket.no/sehavniva/). The dataset for present-day 200-year storm surge was chosen as the most relevant occurrence for further analysis and was incorporated into the dataset of pluvial flood for further use (Figure 4). Some of the pluvial flooded areas (Figure 3) coincide with the storm surge flooded areas.

Subsidence data

The subsidence data used in this study were produced by the Norwegian Ground Motion Service (Figure 5; www.insar.no). Using radar images from the Copernicus Programme's Sentinel-1 satellites, the service provides over two billion deformation measurements over the entire Norwegian mainland. At each point, both the average velocity (along the satellite-to-ground line-of-site) and a cumulative deformation time series are provided. The Sentinel-1 satellites provide full coverage of Europe every 6 days. The wide acquisition swath (250 km), along with Norway's northern

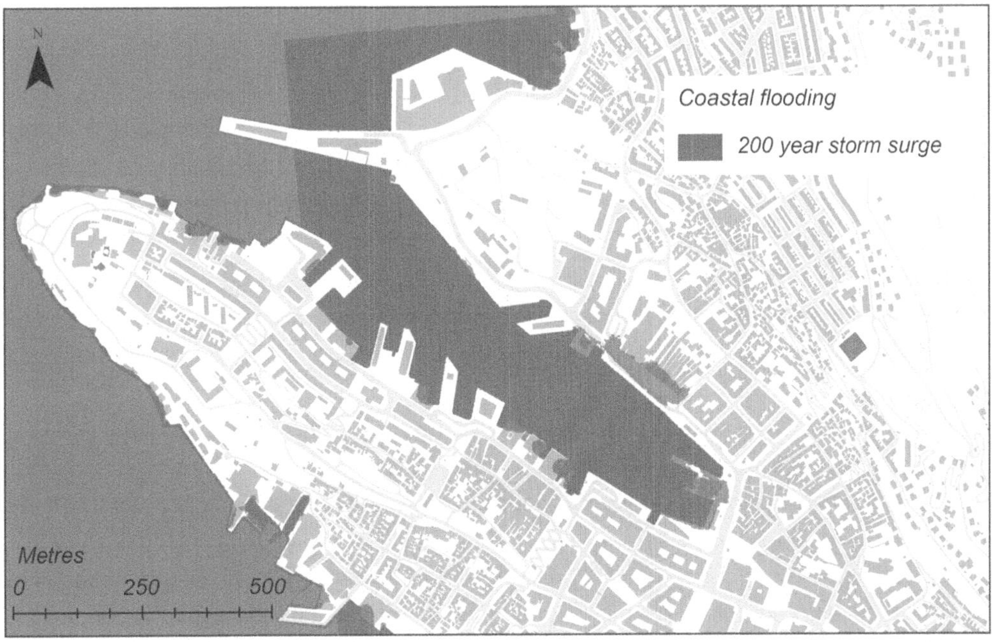

Figure 4 | Areas prone to coastal flooding during a 200-year storm surge are indicated with blue areas on land. Data from the Norwegian Map Authority (2018) (https://www.kartverket.no/sehavniva/). Please refer to the online version of this paper to see this figure in colour: http://dx.doi.org/10.2166/nh.2019.030.

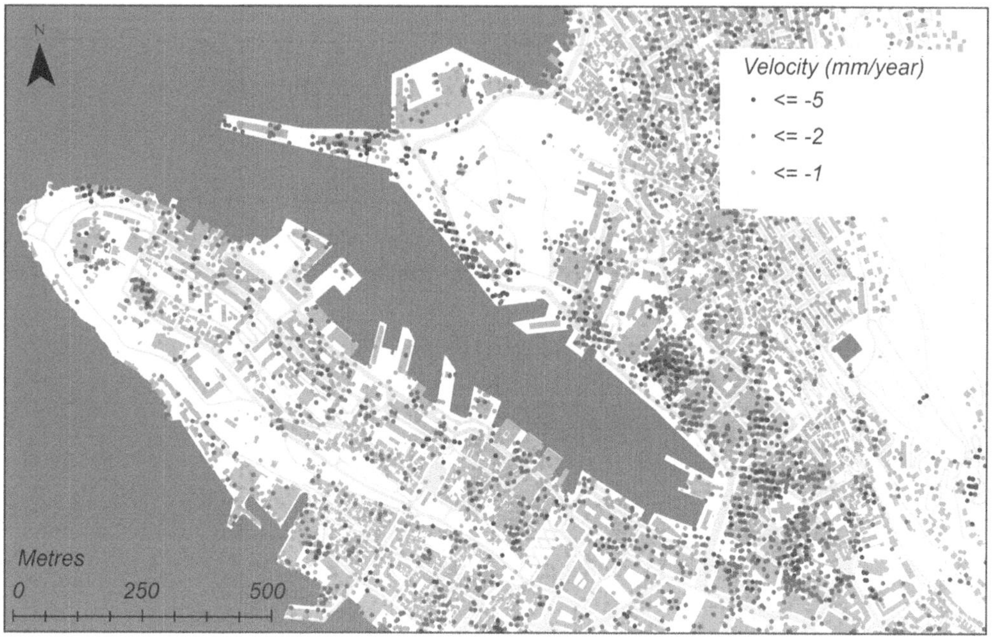

Figure 5 | PSI data from Sentinel-1 for the time period 2015–2018 collective ground movement, subsidence (vertical velocity) in mm/year. Data from the Norwegian Map Authority (2018). Please refer to the online version of this paper to see this figure in colour: http://dx.doi.org/10.2166/nh.2019.030.

latitude, results in multiple overlapping datasets for each area on the ground. For this study, four independent datasets were used, two from ascending (north-going) orbits and two from descending (south-going) orbits. The input data cover the time period 2015–2018, where two datasets are from 2015 to 2018, while two datasets are from 2016 to 2018 (Figure 5). Only data from June to October were used to reduce the possible effects of snow cover. The PSI technique does not return any data from vegetated areas. In the built environment, datapoints commonly represent buildings and other surface constructions.

One advantage of multiple, independently processed PSI datasets is that they can be compared with each other as a basic quality control step. In our study, the datasets were self-consistent. For a smaller area, at the site of the Hanseatic Wharf 'Bryggen', the PSI data have been controlled by comparison with ground-based monitoring of movement (Jensen 2004, 2015; Haukedal 2017). These studies show that both measuring techniques reveal similar patterns of movement and the order of subsidence within the same time period. However, ground-based measurements are time-consuming and costly compared to satellite data collection.

For this study, a threshold for the PSI data was set to −1 mm, only negative vertical movement, subsidence, from −1 mm and larger was included. All data with values 0 mm or more, positive (+) vertical movement was discarded.

METHODOLOGY – RISK ASSESSMENT APPROACH

The Geographical Information System tools such as ArcGIS and ArcGIS Pro (ESRI 2018) were used for the analysis in this study, with the aim of detecting areas with a risk of both subsidence and flooding. To prepare the datasets for analysis, the results from the flood model were georeferenced and vectorized and clipped against the shore. The original flood model consisted of many small and scattered polygons. Since the focus was on areas with severe flood problems, flood polygons spaced closer than 3 m were aggregated, while the areas smaller than 10 m^2 were removed. Then, the results from the pluvial flooding were merged with the 200-year storm surge data. Only PSI points with more than 1 mm/year subsidence were used (Figure 5). The uncertainties connected with these datasets will be discussed later.

The first and simplest overlay is a plain visual overlay of the input data, showing flood data (blue areas in Figure 6(a)) with subsidence data (red points in Figure 6(b)) on top (Figure 6(c)) (the colour figures can be viewed in the

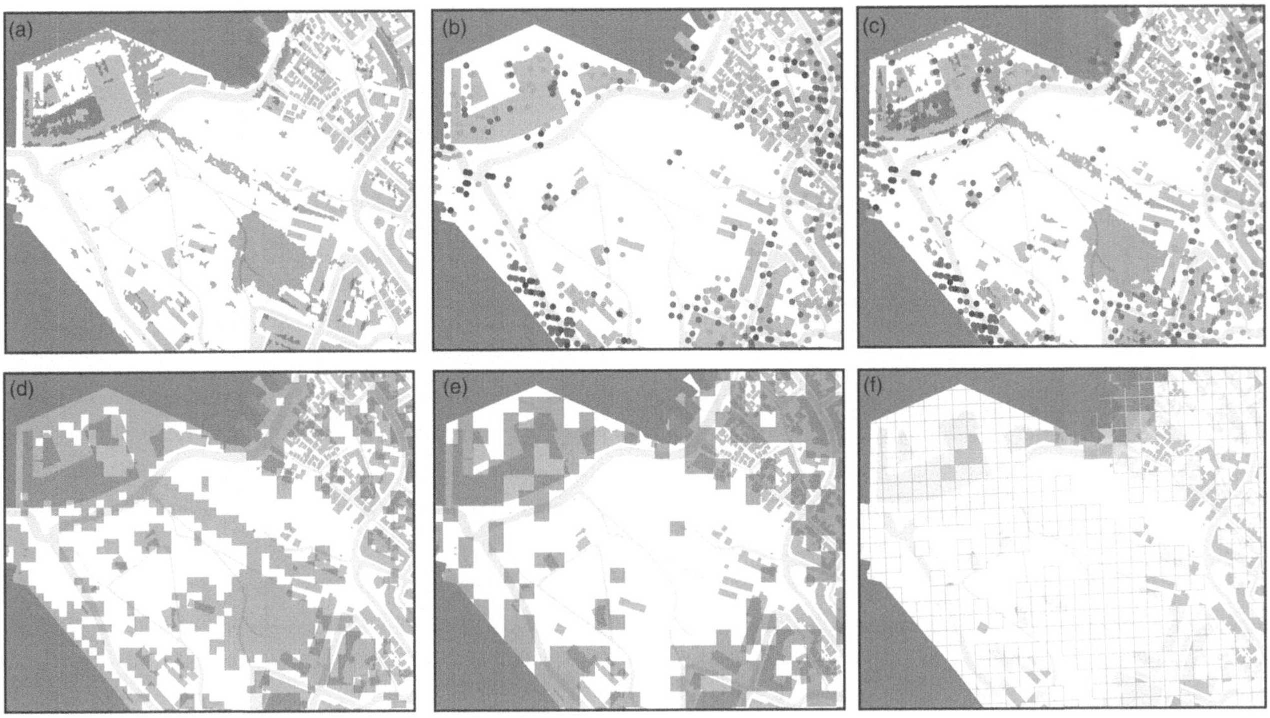

Figure 6 | Top row: the datasets used in the analysis: (a) flooding, (b) subsidence and (c) the combination of the two datasets. Bottom row: results from methods. (d) Method 1 with grid cells with 10 × 10 m, (e) method 1 with 20 × 20-m grid cells and (f) method 2, the 'hot spot analysis' of subsidence within the flooded area. This method uses 20 × 20-m grid and the three different colours displaying 90%, 95% and 99% confidence levels. Please refer to the online version of this paper to see this figure in colour: http://dx.doi.org/10.2166/nh.2019.030.

online version of the article). A visual overlay is useful both to evaluate results from automatic tools and as a complementary map for detailed insights. For planning purposes, pre-selecting areas for action lead to more effective decision-making (Campbell *et al.* 2017; Hooimeijer *et al.* 2017; Hanssen 2018). This work uses grids to synthesize the input data. Grid maps reduce the degree of detail and are expected to give the impression of data uncertainty because it clearly does not follow the pattern of flooding nor the built infrastructure. Two methods were used (Figure 6): The first method does not take the spatial clustering of subsidence into account (Figures 6(d) and 6(e)). The second method analyses the clustering of subsidence within areas prone to flooding (Figure 6(f)).

Description of the simple grid overlay method (1)

In the first method, grids of different sizes are created followed by a selection of grid cells that cover areas with a risk of both flooding (Figure 6(a)) and subsidence (Figure 6(b)). See Figures 6(d) and 6(e) for selected areas, respectively, for grids of 10 × 10 m and 20 × 20 m. The method followed two manual operations:

(I) Two different grids were made with grid size set to 10 × 10 m (Figure 6(d)) and to 20 × 20 m (Figure 6(e)). The flood data map extent was used as the template extent. (II) Grid cells containing both flood and subsidence data are given the colour orange in the map, as shown in Figures 6(d) and 6(e).

Description of the 'hot spot analysis' with aggregated flood areas method (2)

This method uses the optimized 'hot spot analysis' tool to create a grid showing hot spots of subsidence data within areas with a risk of flooding (Figure 6(f)). This tool uses the Getis-Ord Gi* statistic to identify statistically significant hot spots (ESRI 2019). For this method, we went through the following parameters: the main input was the subsidence data and grid cells of 20 × 20 m were selected. Aggregation was selected to count incidents of subsidence within the grid cells within areas prone to flooding. The result was a map with grid cells showing statistically significant hot spots of subsidence that also are at risk of flooding, as shown in Figure 6. A visual comparison of the results with the cartographic overlay as shown in Figures 6(c) and 7 was done to ensure that the areas with the highest values of subsidence were represented.

RESULTS AND DISCUSSION

The areas identified to be at dual risk in this study could further be targeted for mitigation measures that allow surface water to infiltrate the subsurface. Firstly, such measures would help maintain the anoxic conditions necessary to impede the decay of the rich organic layers. Secondly, mitigation measures could help stabilize the groundwater levels and assist in preventing further subsidence. Participants of the Bryggen Project demonstrated that the groundwater levels could be stabilized by introducing SuDS for retaining, storing and further infiltrating surface water (de Beer *et al.* 2012; Boogaard 2015; de Beer & Seither 2015; Matthiesen *et al.* 2015; Rytter & Schonhowd 2015b; Boogaard *et al.* 2016). Large areas of impermeable surface in the city centre also contributed to the risk of flooding. Natural water management practices, like the implementation of SuDS, help increase the infiltration of floodwater to subsurface soils and groundwater. This study gives an example from Bergen city but is relevant for cities having similar challenges related to flooding and subsidence.

Datasets and selected methods for analysis

A visual analysis of the input data reveals an image of a city widely affected by subsidence and flooding after heavy rainfall or storm surges, as shown in Figure 7. To make visual analysis easier, the PSI data are shown with points of increasingly darker red for higher degrees of subsidence (the colour figures can be viewed in the online version of the article). The flooded areas are shown in blue. Areas most prone to flooding and subsidence become prominent in this visualization (Figure 7).

Subsidence data

It should be noted that PSI datapoints may represent points on the ground or points on the city infrastructure,

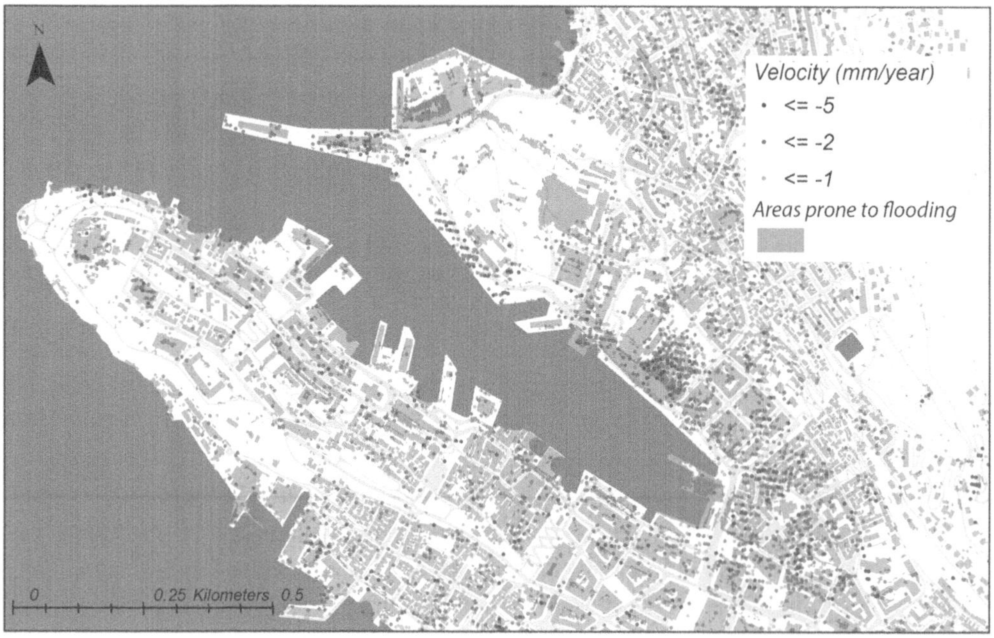

Figure 7 | The PSI data indicates that subsidence is shown in red and the flooded areas in blue. With an overlay of the two datasets, the map shows a city widely affected by subsidence as for areas prone to flooding. Please refer to the online version of this paper to see this figure in colour: http://dx.doi.org/10.2166/nh.2019.030.

such as buildings. Any hard object on the surface may reflect a signal. As such, there is always the possibility that individual points are measuring the deformation of the city infrastructure or in the building itself, and not ground subsidence. Additionally, the PSI technique does not return any measurement in vegetated areas, such as yards or parks. Nonetheless, more than 300,000 datapoints were used in this study providing orders of magnitude more information than could have been obtained using traditional surveying techniques. Although there are many historic buildings in the area, most have been rehabilitated in the last decade and we do not expect that building deformation is a significant part of what is measured. Therefore, we have great confidence that PSI data are suitable for the risk assessment. In this study, all PSI points with more than 1 mm subsidence are included. The exact value of vertical velocity is not used in either of the analyses, only the presence in the simple grid overlay (method 1, Figure 8) and the cluster of points in the 'hot spot analysis' (method 2, Figure 9). For method 2, a visual control of the result shows that areas of high value are also selected as hot spot areas.

Flood data

Results from the urban flood modelling, used in this study, emphasize the areas prone to flooding (Boogaard *et al.* 2017a). The flood modelling is based on the DEM and on the rainfall distribution where depressions in the terrain will control the flooded areas. Manmade constructions, including roads, will create sinks where the flooding will occur (Kluck *et al.* 2010, 2015; Boogaard et al. 2017a, 2017b; Balstrøm & Crawford 2018). For the flood results presented here, this is regarded as inevitable because the study is in an urban and built environment.

Planners are interested in surface water flood modelling and simulation at a coarser and more overall level (Balstrøm & Crawford 2018) for the purpose of prioritizing and decision-making (Campbell *et al.* 2017; Hanssen 2018). For a complete flood risk assessment analysis, hydrological and hydrogeological studies (Wakode *et al.* 2018), an updated flood model, based on an updated DEM, topographic data and flow parameters should be included. The flood risk due to storm surge is based on the estimated highest level of storm surge at present day (Norwegian Mapping Authority; www.Kartverket.no). The storm surges are

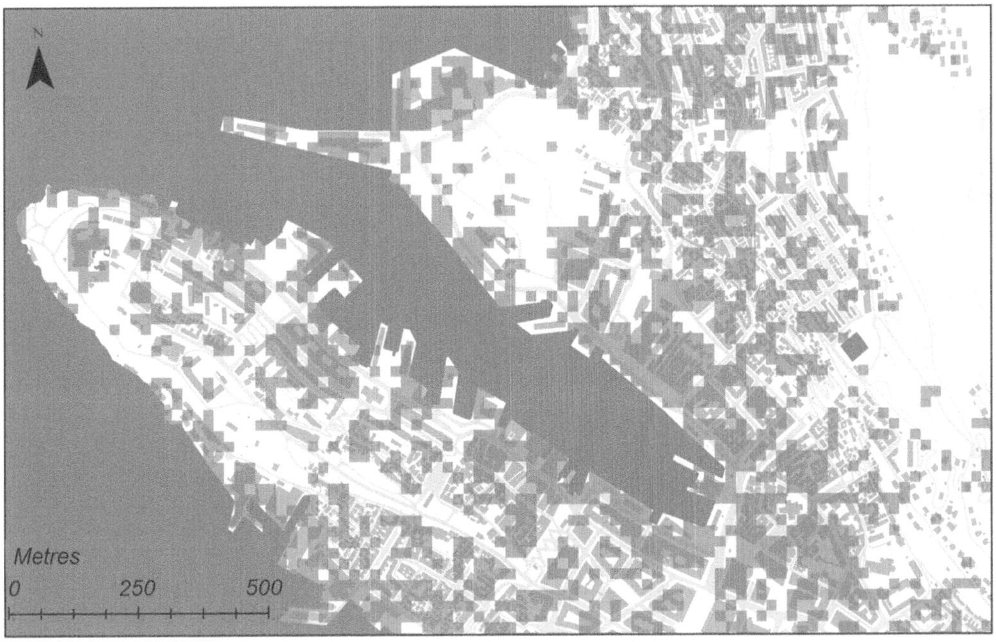

Figure 8 | Simple overlay analysis with 20 × 20-m grid shows where both subsidence and flooding occur. Please refer to the online version of this paper to see this figure in colour: http://dx.doi.org/10.2166/nh.2019.030.

modelled with 20-year, 200-year and 1000-year intervals (Map Authority). For Bergen city centre, the differences are minor. The 100-year design precipitation (Kluck *et al.* 2010, 2015) for the pluvial flooding is therefore combined with a 200-year storm surge, as an extreme event. In further steps for risk assessment, this dataset should be updated and include the worst-case scenario of sea-level rise (IPCC 2014).

Simple grid overlay – method 1

For the simple grid overlay with grid sizes of 10 × 10 m and 20 × 20 m, the result is numerous small areas as shown in Figure 8. It is clearly illustrated in the case of the city centre that a simple grid overlay method gives minimal guidance for authorities as to which areas should be prioritized for dual hazard mitigation. Due to the characteristics of the two datasets; flooded areas in streets and PSI data on buildings and grid cells of 10 × 10 m and smaller give a result of scattered patches and no area of significance. However, when the grid cells are 20 × 20 m, areas prone to both flooding and subsidence are distinguished, as shown in Figure 8.

'Hot spot analysis' - method 2

The hot spot analysis, method 2, does the narrowest selection of areas, using the aggregated flood data and a count of subsidence hot spots within each 20 × 20 m grid cell (Figure 9). The results show that within our study area, there are several areas of significance. For a decision-making process, it would be easier to prioritize areas for mitigation using the 'hot spot analysis' for risk assessment mapping, as shown in Figure 9.

Risk assessment map combined with the existing drainage system

As an example of usability, the risk assessment maps from the simple overlay analysis (method 1, 20 × 20 m grid cells) and the 'hot spot analysis' (method 2) have been combined with the existing drainage system. A 'near-analysis' with 3 m radii of areas in dual hazard and pipelines intersect shows areas where the drainage system is under great pressure when heavy and rapid rainfall or a storm surge occurs (Figure 10). This is standard procedure within water management (Marsalek & Chocat 2002; Marsalek *et al.* 2006). However, this study shows the connected drainage pipes

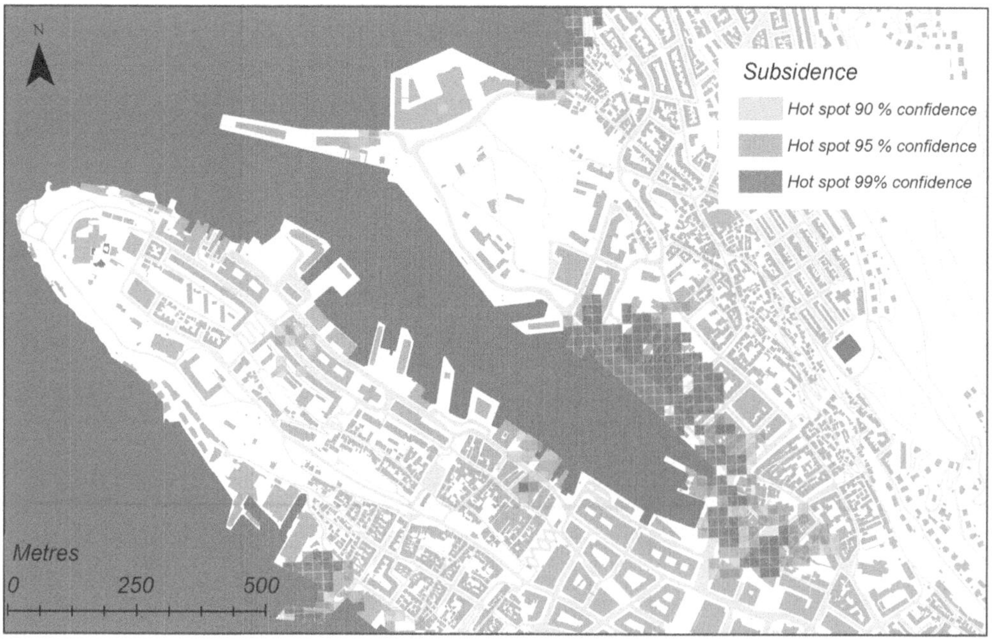

Figure 9 | 'Hot spot analysis' where clusters of subsidence are within areas of pluvial or coastal flooding. The grid cells are 20 × 20 m. Please refer to the online version of this paper to see this figure in colour: http://dx.doi.org/10.2166/nh.2019.030.

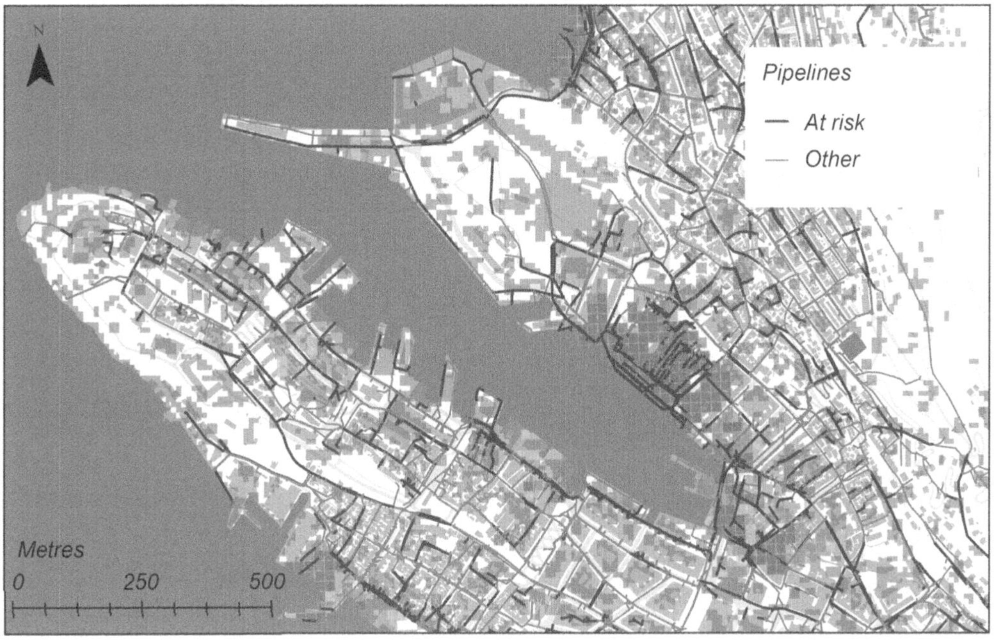

Figure 10 | A 'near-analysis' of pipelines shows all pipes affected by both subsidence and risk of flooding within 3 m distance (red lines). The results from the simple overlay analysis (method 1) with 20 × 20-m grid cells and the 'hot spot analysis' (method 2) are included in the map. Please refer to the online version of this paper to see this figure in colour: http://dx.doi.org/10.2166/nh.2019.030.

and manholes that in addition to high water pressure and excess surface water are prone to ground subsidence that may cause damage and disconnect the pipes (Figure 10). In these areas, it is expected that the drainage system has a greater need for maintenance and thereby costs.

Figure 11 compares all methods, where (A) displays the raw data where blue colour shows flooding and red colour shows subsidence (the colour figures can be viewed in the online version of the article). This visualization indicates that the larger parts of the city are influenced by flooding or subsidence, or both. Comparing the two methods: (1) simple grid overlay and (2) 'hot spot analysis' (Figure 11), the best choice of the method depends on the end-use. Method 1 uses input data raw and has no regard to the size of areas flooded or the density or degree of subsidence. Consequently, the result for Bergen marks areas on almost all buildings in the study area (Figures 8 and 11(b)). When using small grid sizes and without consideration of nearby objects, there is a risk of overlooking relevant areas. There is no prioritizing, and one can argue whether this map result is of any benefit to Bergen's decision-makers other than seeing that there are large areas of dual hazard. It may also contribute to a loss of information due to the cartographic overlay of the input dataset (Figure 11(a)). Nonetheless, the result does suggest that there is a need for general guidelines for city management and building owners. At this level of detail, and if the target user group was property owners, the method can focus on buildings that are prone to flood and subsidence. A 'near-analysis' would possibly be a better alternative as exemplified with pipelines in Figure 10. The results from the 'hot spot analysis' (method 2) are more selective and areas are clearly prioritized (Figure 11(c)). For scientific research on the relationship between flooding and subsidence, or for the municipality to select areas for greater follow-up, this method gives significant results for the clearest selection of areas (Figures 9 and 11).

Risk assessment as a tool for end-users

Subsidence in urban areas is often related to water. A lack of water in the subsurface may lead to compactions of sediments and where organic matter is present, decay and decomposition (Chaussard *et al.* 2014; de Beer & Seither 2015; Matthiesen *et al.* 2015; Castellazzi *et al.* 2016; Motagh *et al.* 2017). Excess water causes flooding and increased erosion (Dixon *et al.* 2006; Miller *et al.* 2008; Yin *et al.* 2016).

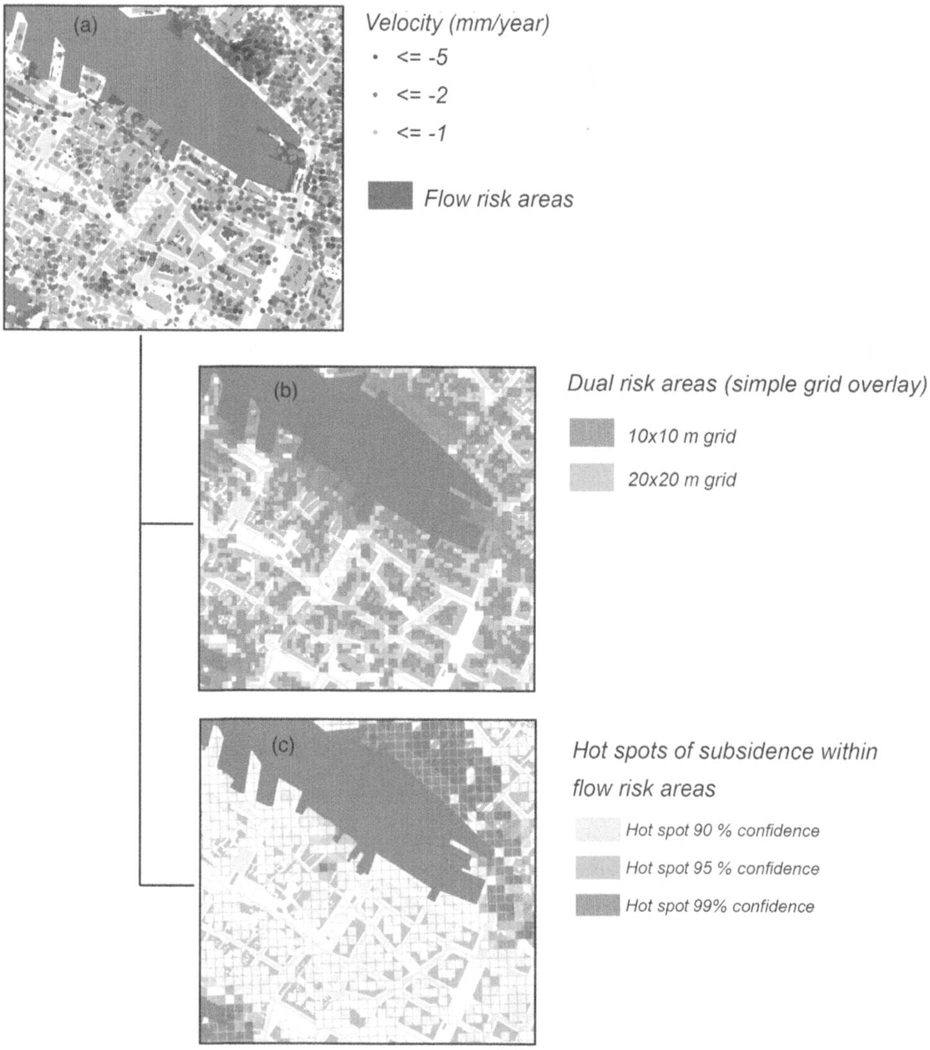

Figure 11 | Comparing the methods. (a) The input data are shown in blue for flooded areas and red for subsidence by PSI data. (b) In method 1, the 10 × 10 (dark orange colour) and 20 × 20 (light orange colour) metre grid cells are all containing both flood risk and subsidence. (c) In method 2, fewer areas are selected based on a 'hot spot analysis' on subsidence bounded by the existence of aggregated flood data. The colour nuance reflects 99%, 95% and 90% confidence levels as displayed in Figure 9. Please refer to the online version of this paper to see this figure in colour: http://dx.doi.org/10.2166/nh.2019.030.

The subsurface of any city is complex, and in Bergen, it can be roughly divided into three layers: natural ground consisting of bedrock and sediments on top, cultural layers consisting of domestic waste, with up to 100% organic matter (Matthiesen *et al.* 2015; Rytter & Schonhowd 2015b) and anthropogenic materials, such as agglomerate, asphalt and material for drainage. The subsidence occurring is not constrained by geological structures and cannot be explained by geological processes alone. However, water, both surface water and groundwater, plays an important part in the process.

Pregnolato *et al.* (2017), in their risk assessment of roads in Newcastle, UK, show that roads are prone to flooding during heavy rainfall. Similarly, the risk assessment presented in this study can help the municipality prioritize areas for mitigation or that need on-going surveillance. A current discussion in Norway is how to implement climate adaptation into best management practice for municipalities (Hanssen 2018). Hanssen (2018) shows how well flood risk maps function to translate natural science information into local planning and decision-making. This shows that maps are credible and essential tools, but that they need to be brought to the table by planners and interpreted in a local context. Hanssen (2018) conclude that local climate adaption is dependent on well-functioning interaction between multiple levels as well as disciplines and emphasis on strengthening the role of the government agencies as 'knowledge translators' ('kunnskapsoversettere'; Hanssen 2018). The risk assessment map methodology presented in this study aims to translate knowledge into maps to assist the end-user to select areas for implementation of, for example, SuDS by identifying areas prone to the dual hazard of flooding and subsidence. Resilience of the built environments has not been well studied (Thornbush *et al.* 2013), and results from this study may help the Bergen Municipality to plan mitigation and further adaptation to prevent areas of flooding, by increasing infiltration of surface water and decreasing flooding, as well as the processes causing subsidence. Managing stormwater is not just important for protecting water resources and aquatic ecology but also to restore urban water cycle processes that are critical to the health of urban watersheds. These include infiltration and groundwater recharge, evapotranspiration and chemical/biological transformations, especially due to more frequent and intense rainfall and flooding (UN-Water 2018).

CONCLUSIONS

There is a link between areas that suffer from subsidence and areas with an excess or shortage of water. The aim of this study was to locate areas in Bergen city centre that are prone to the dual hazard of subsidence and flooding. This was achieved by processing existing data and maps that identify areas prone to PSI data for risk of subsidence, a flood model map and a storm surge map for areas prone to flooding.

We have demonstrated that a 'hot spot analysis', for the subsidence data within areas prone to flooding, provides an effective means of selecting areas for further field evaluation. Data for climate adaptation analyses are increasing and open access. The method can easily be repeated with updated PSI and flood data. The areas selected are constrained and could serve as a starting point in prioritizing areas by the municipality for detailed hydrological and hydrogeological studies of the urban water cycle and further implementation of water management solutions, like SuDS.

The subsurface in cities is complex due to a mixture of natural and built environments. The processes causing subsidence are not easily understood but are commonly related to water. Increasing infiltration of surface water may prevent the processes causing subsidence. Managing stormwater in this way is not only important for protecting water resources and the aquatic environment – it can help restore and maintain urban water cycle processes critical to making cities resilient to the effects of climate change.

Further work

The increased availability of data, both large datasets and timeseries, makes analyses, such as the risk assessment presented here, much more achievable. The Copernicus program is revolutionary in that it promises this type of data for decades to come, free and open. Risk assessment similar to that conducted in this study is relevant for all cities that are prone to coastal and/or pluvial flooding or possible the combination of flooding and subsidence. The www.InSAR.no service is an open access portal, displaying data used in this study, and is an example of possibilities with the upcoming EU Ground Motion Service.

The latest available PSI data and a new and updated flood model based on the latest and most detailed DEM and topographic data should be used before selecting areas in a potential follow-up of this study. This risk assessment should be also followed up by hydrological and hydrogeological field investigations to evaluate the results and to find the best management practices for the given location and problem.

This study will be expanded to categorize PSI data indicating subsidence by trends in timeseries and combining them with other datasets. This would increase the knowledge of the subsurface processes and the effects of interventions, and thereby ultimately identify effective actions to decrease effects related to the urban water cycle.

Further, end-users should be involved in the development of risk assessment maps, for example in the evaluation of the usability of prototypes, like the ones presented here. Choosing an adequate method for risk assessment with the end-user tasks in focus is important and will give more applicable results. Trying out multiple methods for analysis and visual analysis for quality control of map results was emphasized in this study and is strongly recommended in further studies.

ACKNOWLEDGEMENTS

This research is supported through the Water JPI funded the INXCES research project 'INnovation for eXtreme Climatic EventS' www.inxces.eu. Sentinel-1A/B images are provided by the European Space Agency (ESA). Thanks to the Norwegian Ground Motion Services for open and accessible data, www.insar.no. We greatly appreciate the help of John Dehls (NGU). We thank the Bergen Municipality, Tauw and Hogeschool van Amsterdam, the Hanze University of Applied Sciences, Groningen, for support for this work. Thanks to the Norwegian Hydrology Council for the NHC2018 conference in Bergen and this special conference issue. Great appreciation for the support of Anne Liinamaa-Dehls at the Geological Survey of Norway (NGU) during proof-reading and editing of this article. Great appreciation to two anonymous reviewers and the special issue editor of *Journal of Hydrology Research*, Kolbjørn Egeland, for improving this paper.

REFERENCES

Albano, R., Mancusi, L. & Abbate, A. 2017 Improving flood risk analysis for effectively supporting the implementation of flood risk management plans: the case study of 'Serio' valley. *Environmental Science & Policy* **75**, 158–172.

Albano, R., Sole, A., Adamowski, J., Perrone, A. & Inam, A. 2018 Using FloodRisk GIS freeware for uncertainty analysis of direct economic flood damages in Italy. *International Journal of Applied Earth Observation and Geoinformation* **73**, 220–229. https://doi.org/10.1016/j.jag.2018.06.019.

Balstrøm, T. & Crawford, D. 2018 Arc-Malstrøm: A 1D hydrologic screening method for stormwater assessments based on geometric networks. *Computers and Geosciences* **116**, 64–73. https://doi.org/10.1016/j.cageo.2018.04.010.

Berardino, P., Costantini, M., Franceschetti, G., Iodice, A., Pietranera, L. & Rizzo, V. 2003 Use of differential SAR interferometry in monitoring and modelling large slope instability at Maratea (Basilicata, Italy). *Engineering Geology* **68**, 31–51. doi:10.1016/S0013-7952(02)00197-7.

Bergen kommune 2006 Hovedplan for avløp og vannmiljø 2005-2015. Bystyrets Kontor (ed.). Bergen, Norge: Bergen kommune (In Norwegian).

Bolund, P. & Hunhammar, S. 1999 Ecosystem services in urban areas. *Ecological Economics* **29**, 293–301.

Boogaard, F. C. 2015 Stormwater quality and sustainable urban drainage management. In: *Monitoring, Mitigation, Management: The Groundwater Project – Safeguarding the World Heritage Site of Bryggen in Bergen* (J. Rytter & I. Schonhowd, eds). Riksantikvaren, Bergen, pp. 136–149.

Boogaard, F. C., Wentink, R., Vorenhout, M. & de Beer, J. 2016 Implementation of sustainable urban drainage systems to preserve cultural heritage – pilot Motte Montferland. *Journal of Conservation and Management of Archaeological Sites* **18** (1–2), 328–341. https://doi.org/10.1080/13505033.2016.1182767.

Boogaard, F. C., Vojinovic, Z., Chen, Y.-C., Kluck, J. & Lin, T.-P. 2017a High Resolution Decision Maps for Urban Planning: A Combined Analysis of Urban Flooding and Thermal Stress Potential in Asia and Europe. In *MATEC Web of Conferences*. 103, 04012. ISCEE 2016. doi:10.1051/matecconf/201710304012.

Boogaard, F. C., Kluck, J., Bosscher, M. & Schoof, G. 2017b Flood model Bergen Norway and the need for (sub-)surface INnovations for eXtreme Climatic EventS (INXCES). *Procedia Engineering* **209**, 56–60. doi:10.1016/j.proeng.2017.11.130.

Boogaard, F. C., Krol, D., Kluck, J. & Bosscher, M. 2018 Multi-Level Climate Effect Modeling, best management practices From private property to national level, Cirre 2018. In: *Conference of Interdisciplinary Research on Real Estate*, September 2018, Groningen.

Campbell, D., de Beer, J., Mielby, S., van Campenhout, I., van der Meulen, M., Erikkson, I., Venvik Ganerod, G., Lawrence, D., Bacic, M., Donald, A., Gogu, C. R. & Jelenek, J. 2017 Transforming the relationship between geoscientists and urban decision-makers. European COST Sub-Urban Action (TU1206). *Procedia Engineering* **209**, 4–11. doi:10.1016/j.proeng.2017.11.124.

Carver, S. J. 1991 Integrating multi-criteria evaluation with geographical information systems. *International Journal of Geographical Information Systems* **5** (3), 321–339.

Castellazzi, P., Arroyo-Domínguez, N., Martel, R., Calderhead, A. I., Normand, J. C. L., Gárfias, J. & Rivera, A. 2016 Land subsidence in major cities of Central Mexico: interpreting InSAR-derived land subsidence mapping with hydrological data. *International Journal of Applied Earth Observation and Geoinformation* **47**, 102–111. http://dx.doi.org/10.1016/jag2015.12.002.

Chaussard, E., Wdowinski, S., Cabral-Cano, E. & Amelung, F. 2014 Land subsidence in central Mexico detected by ALOS InSAR time-series. *Remote Sensing of Environment* **140**, 94–106. https://doi.org/10.1016/j.rse.2013.08.038.

Crosetto, M., Monserrat, O., Cuevas-González, M., Devanthéry, N. & Crippa, B. 2016 Persistent Scatterer Interferometry: A review. *Journal of Photogrammetry and Remote Sensing* **115**, May 2016, 78–89.

Damoom, M. M., Hashim, S., Aljohani, M. S., Saleh, M. A. & Xoubi, N. 2019 Potential areas for nuclear power plant siting in Saudi Arabia: GIS-based multi-criteria decision making analysis. *Progress in Nuclear Energy* **110**, 110–120. https://doi.org/10.1016/j.pnucene.2018.09.018.

de Beer, J. D. & Seither, A. 2015 Groundwater balance. In: *Monitoring, Mitigation, Management: The Groundwater*

Project – Safeguarding the World Heritage Site of Bryggen in Bergen (J. Rytter & I. Schonhowd, eds). Riksantikvaren, Bergen, Norway, pp. 106–123.

de Beer, J. D., Matthiesen, H. & Christensson, A. 2012 Quantification and visualization of in situ degradation at the World Heritage Site Bryggen in Bergen, Norway. *Conservation and Management of Archaeological Sites* **14**, 215–227.

Directorate for Cultural Heritage (Riksantikvaren) 2018 Cultural layers vector data. https://www.riksantikvaren.no/.

Dixon, T. H., Amelung, F., Ferretti, A., Novali, F., Rocca, F., Dokka, R., Sella, G., Kim, S.-W., Widowinski, S. & Whitman, D. 2006 Subsidence and flooding in New Orleans. *Nature* **441** (7093), 587–588. https://doi.org/10.1038/441587a.

Erbaş, M., Kabak, M., Özceylan, E. & Çetinkaya, C. 2018 Optimal siting of electric vehicle charging stations: a GIS-based fuzzy Multi-Criteria Decision Analysis. *Energy* **163**, 1017–1031. https://doi.org/10.1016/j.energy.2018.08.140.

Eriksen, H. Ø., Lauknes, T. R., Larsen, Y., Corner, G. D., Bergh, S. G., Dehls, J. & Kierulf, H. P. 2017 Visualizing and interpreting surface displacement patterns on unstable slopes using multi-geometry satellite SAR interferometry (2D InSAR). *Remote Sensing of Environment* **191**, 297–312. https://doi.org/10.1016/j.rse.2016.12.024.

Ersland, G. A. 2015 The history of Bryggen until C. 1900. In: *Monitoring, Mitigation, Management: The Groundwater Project – Safeguarding the World Heritage Site of Bryggen in Bergen* (J. Rytter & I. Schonhowd, eds). Riksantikvaren, Bergen, pp. 12–23.

ESRI 2018 *ArcPro 2.3.1*. Environmental System Research Institute, Redlands, CA.

ESRI 2019 http://desktop.arcgis.com/en/arcmap/10.3/tools/spatial-statistics-toolbox/hot-spot-analysis.htm

Fletcher, T. D., Andrieu, H. & Hamel, P. 2013 Understanding, management and modelling of urban hydrology and its consequences for receiving waters: a state of the art. *Advances in Water Resources* **51**, 261–279. https://doi.org/10.1016/j.advwatres.2012.09.001.

Getis, A. & Ord, J. K. 1992 The analysis of spatial association by use of distance statistics. *Geographical Analysis* **24** (3), 189–206. doi:10.1111/j.1538-4632.1992.tb00261.x.

Google Earth 2019 Bergen City 60°24'11"N, 5°19'23"W, 3D map, viewed 30 April 2019. https://earth.google.com/web/.

Hansen, G. 1994 Den overordnede bebyggelsestopografi omkring 1190 i Bergen: belyst ut fra arkæologiske, naturtopografiske og skriftlige kilder. Riksantikvaren, Utgravningskontoret for Bergen (In Norwegian).

Hanssen, G. S. 2018 Planlegging for risikosamfunnet: Hvordan fungerer flomsonekart med klimapåslag som kunnskapsoversettelse? Kart og Plan 1-2018, pp. 17-36 (In Norwegian).

Hanssen-Bauer, I., Førland, E. J., Haddeland, I., Hisdal, H., Lawrence, D., Mayer, S., Nesje, A., Nilsen, J. E. Ø., Sandven, S., Sandø, A. B., Sorteberg, A. & Ådlandsvik, B. 2017 *Climate in Norway 2100 – A Knowledge Base for Climate Adaptation*. NCCS Report no. 1/2017. M-741/2017. ISSN nr. 2387-3027.

Harvold, K., Larsen, K., de Beer, J., Boogaard, F., Vandrup Martens, V., Matthiesen, H., Muthanna, T. M., Seither, A., Skogheim, R. & Vorenhout, M. 2015 (Urban WATCH), ISBN: 978-82-92935-13-2, CIENS Report 2015:1

Haughton, G. & Hunter, C. 1994 *Sustainable Cities*. UK Jessica Kingsley Publishers, London, pp. 357.

Haukedal, J. 2017 *Setningsutvikling i Bergen sentrum: Urbanhydrogeologi, InSAR, setning og kulturlag*. Master Thesis, Natural Geography at the University of Bergen. (In Norwegian).

Hooimeijer, F. L., Lafleura, F. & Trinh, T. T. 2017 Drawing the subsurface: an integrative design approach. *Procedia Engineering* **209**, 61–74. doi:10.1016/j.proeng.2017.11.131.

IPCC 2014 Climate change 2014: impacts, adaptation, and vulnerability. Part A: Global and sectoral aspects. In: *Contribution of Working Group II to the Fifth Assessment Report of the Intergovernmental Panel on Climate Change* (C. B. Field, V. R. Barros, D. J. Dokken, K. J. Mach, M. D. Mastrandrea, T. E. Bilir, M. Chatterjee, K. L. Ebi, Y. O. Estrada, R. C. Genova, B. Girma, E. S. Kissel, A. N. Levy, S. MacCracken, P. R. Mastrandrea & L. L. White, eds). Cambridge University Press, Cambridge, UK and New York, NY, USA, pp. 1–1132.

Jensen, J. A. 2004 *Vurdering og tolkninger av setningsmålinger*. Report No. 400425-6. Multiconsult avd. NOTEBY (In Norwegian).

Jensen, J. A. 2015 Subsidence at the Bryggen site. In: *Monitoring, Mitigation, Management: The Groundwater Project – Safeguarding the World Heritage Site of Bryggen in Bergen* (J. Rytter & I. Schonhowd, eds). Riksantikvaren, Bergen, pp. 152–164.

Kluck, J., Claessen, E. G., Blok, G. M. & Boogaard, F. C. 2010 Modelling and mapping of urban storm water flooding, communication and prioritizing actions through mapping urban flood resilience. *NOVATECH 2010*.

Kluck, J., Boogaard, F. C., Goedbloed, D. & Claassen, M. 2015 Storm Water Flooding Amsterdam, from a quick Scan analyses to an action plan. In *International Waterweek 2015*, 3 November 2015, Amsterdam.

Lauknes, T. R., Piyush Shanker, A., Dehls, J. F., Zebker, H. A., Henderson, I. H. C. & Larsen, Y. 2010 Detailed rockslide mapping in northern Norway with small baseline and persistent scatterer interferometric SAR time series methods. *Remote Sensing of Environment* **114**, 2097–2109. https://doi.org/10.1016/j.rse.2010.04.015.

Lu, P., Casagli, N., Catani, F. & Tofani, V. 2012 Persistent Scatterers Interferometry Hotspot and Cluster Analysis (PSI-HCA) for detection of extremely slow-moving landslides. *International Journal of Remote Sensing* **33** (2), 466–489. doi:10.1080/01431161.2010.536185.

Lyu, H. M., Sun, W. J., Shen, S. L. & Arulrajah, A. 2018 Flood risk assessment in metro systems of mega-cities using a GIS-based modelling approach. *Science of the Total Environment* **626**, 1012–1025. https://doi.org/10.1016/j.scitotenv.2018.01.138.

Mangerud, J. 2004 Ice sheet limits in Norway and on the Norwegian continental shelf. In: *Quaternary Glaciations – Extent and Chronology: Europe*, Vol. 1 (J. Ehlers & P. Gibbard, eds). Elsevier, Amsterdam, pp. 271–294.

Marsalek, J. & Chocat, B. 2002 International report: stormwater management. *Water Science & Technology* **46** (6–7), 1–17. https://doi.org/10.2166/wst.2002.0657.

Marsalek, J., Jiménez-Cisneros, B. E., Malmquist, P.-A., Karamouz, M., Goldenfum, J. & Chocat, B. 2006 *Urban Water Cycle Process and Interactions*. UNESCO Publishing and Taylor and Francis Group, Paris.

Matthiesen, H., Hollesen, J. & Gregory, D. 2015 Preservation Conditions and Decay Rates. In: *Monitoring, Mitigation, Management: The Groundwater Project – Safeguarding the World Heritage Site of Bryggen in Bergen* (J. Rytter & I. Schonhowd, eds). Riksantikvaren, Bergen, pp. 76–91.

Mignot, E., Li, X. & Dewals, B. 2019 Experimental modelling on urban flooding: a review. *Journal of Hydrology* **568**, 334–342. https://doi.org/10.1016/j.jhydrol.2018.11.001.

Miller, J. & Hutchins, M. 2017 The impacts of urbanisation and climate change on urban flooding and urban water quality: a review of the evidence concerning the United Kingdom. *Journal of Hydrology: Regional Studies* **12**, 345–362. http://dx.doi.org/10.1016/j.ejrh.2017.06.006.

Miller, R. L., Fram, M. S., Fujii, R. & Wheeler, G. 2008 Subsidence reversal in a re-established wetland in the Sacramento-San Joaquin Delta, California, USA. *San Francisco Estuary and Watershed Science* **6** (3). article 1.

Motagh, M., Shamshiri, R., Haghighi, M. H., Wetzel, H.-U., Akbari, B., Nahavandchi, H., Roessner, S. & Arabi, S. 2017 Quantifying groundwater exploration induced subsidence in the Rafsanjan plain, southeastern Iran, using InSAR time-series and in situ measurements. *Journal of Engineering Geology* **2018**, 134–151. http://dx.doi.org/10.1016/j.enggeo.2017.01.011.

Norwegian Directorate for Civil Protection (DSB) 2017 Integrating Sea Level Rise and Storm Surges in Local Planning. ISBN 978-82-7768-447-5 (PDF). Available from: https://www.dsb.no/globalassets/dokumenter/veiledere-handboker-og-informasjonsmateriell/veiledere/integrating-sea-level-rise-and-storm-surges-in-local-planning.pdf

Norwegian Ground Motion Center 2018 Interferometric synthetic-aperture radar data 2016-2018. Available from: http://www.insar.no

Norwegian Map Authority (Kartverket) 2009 Produktspesifikasjon FKB-Laser Versjon 2.0 2009-02-01, Produktspesifikasjon Nasjonal modell for høydedata fra laserskanning (FKB-Laser), *Statens kartverk* 2009.

Norwegian Map Authority (Kartverket) 2018 Topographical data, sea level and storm surges. Available from: https://www.kartverket.no/sehavniva/.

Norwegian Meteorological Institute (NMI) 2019 Available from: https://www.yr.no/place/Norway/Hordaland/Bergen/Bergen/statistics.html.

Pregnolato, M., Ford, A., Glenis, V., Wilkinson, S. & Dawson, R. 2017 Impact of climate change on disruption to urban transport networks from pluvial flooding. *Journal of Infrastructure Systems* **23** (4). doi:10.1061/(ASCE)IS.1943-555X.0000372.

Rytter, J. & Schonhowd, I. (eds) 2015a *Monitoring, Mitigation, Management: The Groundwater Project – Safeguarding the World Heritage Site of Bryggen in Bergen*. Riksantikvaren, Bergen, pp. 213. ISBN: 978-82-7574-085-2. Available from: http://prosjektbryggen.no/publikasjoner/.

Rytter, J. & Schonhowd, I. 2015b Operation groundwater rescue. In: *Monitoring, Mitigation, Management: The Groundwater Project – Safeguarding the World Heritage Site of Bryggen in Bergen* (J. Rytter & I. Schonhowd, eds). Riksantikvaren, Bergen, pp. 50–59.

Schmidt, D. A. & Bürgmann, R. 2003 Time-dependent land uplift and subsidence in the Santa Clara valley, California, from a large interferometric synthetic aperture radar data set. *Journal of Geophysical Research September 2003*. https://doi.org/10.1029/2002JB002267.

Sörensen, J. & Mobini, S. 2017 Pluvial, urban flood mechanisms and characteristics – assessment based on insurance claims. *Journal of Hydrology* **555**, 51–67. https://doi.org/10.1016/j.jhydrol.2017.09.039.

Statistics Norway (SSB) 2017 Available from: https://www.ssb.no/befolkning/statistikker/folkemengde/aar-per-1-januar/#relatert-tabell-1.

Teatini, P., Tosi, L., Strozzi, T., Carbognin, L., Cecconi, G., Rosselli, R. & Libardo, S. 2012 Resolving land subsidence within the Venice Lagoon by persistent scatterer SAR interferometry, Physics and Chemistry of the Earth, Parts A/B/C, **40–41**, 72–79. ISSN 1474-7065.

Thornbush, M., Golubchikov, O., Golubchikov, O. & Bouzarovski, S. 2013 Sustainable cities targeted by combined mitigation-adaptation efforts for future-proofing. *Sustainable Cities and Society* **9**, 1–9. http://dx.doi.org/10.1016/j.scs.2013.01.003.

United Nations, Department of Economic and Social Affairs, Population Division 2016 The World's Cities in 2016 - Data Booklet (ST/ESA/ SER.A/392). Available from: https://www.un.org/en/development/desa/population/publications/pdf/urbanization/the_worlds_cities_in_2016_data_booklet.pdf

UN-Water 2018 *United Nations World Water Assessment Programme, 2018*. The United Nations World Water Development Report 2018: Nature-Based Solutions for Water. UNESCO, Paris. Available from: http://www.unesco.org/new/en/natural-sciences/environment/water/wwap/wwdr/2018-nature-based-solutions/.

Wakode, H. B., Baier, K., Jha, R. & Azzam, R. 2018 Impact of urbanization on groundwater recharge and urban water balance for the city of Hyderabad, India. *International Soil and Water Conservation Research* **6**, 51–62. https://doi.org/10.1016/j.iswcr.2017.10.003.

Yin, J., Yu, D. & Wilby, R. 2016 Modelling the impact of land subsidence on urban pluvial flooding: a case study of downtown Shanghai, China. *Science of the Total Environment* **544**, 744–753. http://dx.doi.org/10.1016/j.scitotenv.2015.11.159.

First received 1 March 2019; accepted in revised form 2 December 2019. Available online 31 December 2019

Limitations in using runoff coefficients for green and gray roof design

Lotte Askeland Schärer, Jan Ove Busklein, Edvard Sivertsen and Tone M. Muthanna

ABSTRACT

Climate change combined with urbanization increases the performance demand on urban drainage systems. Green roofs are one of the most used green infrastructure measures to alleviate the pressure on the urban drainage system through the detention and retention of runoff. The rational method with the runoff coefficient (*C*) is one of the most commonly used design tools for stormwater design in Norway. This method relies on a runoff coefficient being available for green roofs, which is typically not the case. This paper compares laboratory and experimental field studies to investigate runoff coefficients from different types of detention-based roofs. The methodology described in the German 'FLL Guideline', one of the world's most commonly used green roof standards, was used to measure the runoff coefficients for the different components making up a typical green roof. The contribution from each layer is reflected in the runoff coefficients. The runoff coefficients from the field experiments were calculated using observed precipitation and runoff from existing green roofs in Oslo, Trondheim, Sandnes, and Bergen, Norway. Events that had a cumulative precipitation comparable to the laboratory events, but longer durations, were selected. These events gave significantly lower and varying runoff coefficients, clearly demonstrating the limitation of choosing a suitable runoff coefficient for a given roof. However, laboratory experiments are important in understanding the underlying flow processes in the different layers in a detention-based roof.

Key words | detention, green roof design, laboratory scale, rational method, runoff coefficients

Lotte Askeland Schärer
Tone M. Muthanna (corresponding author)
Department of Civil and Environmental Engineering,
The Norwegian University of Science and Technology (NTNU),
S.P Andersens vei 5, 7491 Trondheim,
Norway
E-mail: *tone.muthanna@ntnu.no*

Jan Ove Busklein
Edvard Sivertsen
SINTEF Building and Infrastructure,
Høgskoleringen 7b, 7034 Trondheim,
Norway

INTRODUCTION

An increased performance demand on the urban drainage system from climate change and urbanization is a worldwide challenge. Climate change leads to a change in rainfall frequency, a general increase in the intensity and frequency of extreme events (Intergovernmental Panel on Climate Change [IPCC] 2013; Hanssen-Bauer *et al.* 2015). Combined with urbanization, damaging rain-induced flood events will increase in frequency (Norges Offentlige Utredning [NOU] 2015, p. 30). In Norway, a three-step approach to stormwater management has nationally been adopted. Step one: infiltration of all small events; step two: detention of medium events; and step three: ensure safe flood ways for the larger events. The first two steps are mainly about reducing the impermeable surface area, and increasing infiltration and evapotranspiration. Rooftops typically make up as much as 40–50% of the paved surfaces in cities, which

doi: 10.2166/nh.2020.049

make detention-based roofing a promising solution (Stovin *et al.* 2012; Berretta *et al.* 2014; Sobczyk & Mrowieck 2016; Hamouz *et al.* 2018).

Rooftop detention can be accomplished through different solutions where green roofs are the most common. Green roofs are made to collect, store, and retain precipitation through evapotranspiration and detention in the substrate. By converting impermeable roofs to something more akin to natural landscape, one can achieve a significantly reduced and delayed peak runoff (WEF 2012, p. 326). The typical buildup of a green roof consists of plants, substrate, root barrier, drainage layer, and an impermeable membrane. For the vegetation sedums, plants are commonly used. The robustness of these plants requires little maintenance and less soil. These plants are robust and require little maintenance and little soil. These types of green roofs are called extensive green roofs and are characterized by their thin profile thickness of less than 100–150 mm (WEF 2012, p. 326; Berretta *et al.* 2014). However, detention can also be achieved through various non-vegetated detention substrates and media. These types of roofs commonly use an extruded clay aggregate layer to achieve detention caused by the porous media the water flows through. In addition, a top layer of pavers is needed to keep the detention layer in place (Andenæs *et al.* 2018).

Retention of water through evapotranspiration and detention through temporary storage and peak flow delay in the substrate and drainage layers are the two most commonly studied hydrologic functions of green roofs, according to a review article by Andenæs *et al.* (2018). Detention-based studies investigate detention performance with focus on peak flow reductions for single events, whereas the retention-based studies investigate water retention in the form of evapotranspiration over a longer period of time. A study by Johannessen *et al.* (2017) investigated the green roof performance potential in cold and wet regions. The evapotranspiration was found to be a limiting factor for the green roof retention capacity, with almost negligible values in the winter. Hamouz *et al.* (2018) presented an extruded clay aggregate-based detention layer overlaid with lightweight concrete pavers to keep the extruded clay in place (wind protection). The retention on this roof was found to be lower than a typical green roof, as this system does not offer any transpiration, and evaporation can only occur in the slits between the pavers. Though the retention was less in the extruded clay aggregate roof system, it showed very promising detention capacity. Stovin *et al.* (2015) performed an outdoor study in Sheffield, UK, based on nine test beds with different substrates and vegetation. Rainfall- and runoff data over a 4-year period were collected. This study provides both lower retention and detention on the non-vegetated test beds, as well as for the large-pored and permeable substrate. Johannessen *et al.* (2018) studied retention and detention performances for extensive green roofs in different Norwegian locations. In order to investigate detention metrics, it was necessary to identify single events in the continuous time series. This was particularly challenging in a coastal climate with a more or less continuous stream of low-pressure weather systems from the Atlantic. This resulted in large variability in metrics, even with 3–8 years of collected field data. This variability showcases the need for geographical site-specific design of green roofs. However, there is a need to understand the water detention in the various components and layers of green and gray roofs in order to improve performance prediction for use in design. Currently, this information is to a large extent unknown and not available. Further, with more knowledge of the performance of each layer, it will be possible to optimize layer composition for different climatic zones, as well as meeting local discharge regulations and building restrictions.

One possible and commonly used metric to capture the detention performance is the runoff coefficient from the rational method (Kuichling 1889). The rational method is one of the most commonly used design tools for urban runoff calculations, where the runoff is found as a function of the area times the rainfall intensity times a runoff coefficient. The runoff coefficient is given as the relationship between precipitation and runoff. It can be calculated either by the ratio between the intensities of the peaks or the volumes. This ratio is typically applied at the outlet of a watershed. In order to improve design calculations, more knowledge is needed on the runoff coefficient for detention-based roofs, broken down to the individual layers in the roof systems. Hence, the objective of this paper is to compare laboratory and experimental field installations to investigate runoff coefficients for different layered roofs with focus on the detention. Further, it is discussed to what extent the use of runoff coefficients from detention-based roofs is an appropriate tool. More specifically, we wanted to answer the following research questions:

1. What are the runoff coefficients of different types of detention-based roof systems?
2. How does laboratory measured runoff coefficients compare with field observations?
3. How appropriate is the use of runoff coefficients for detention-based roof design?

STUDY AREA AND DATA

This study is based on data from a set of laboratory experiments and four field test roofs at different locations in Norway. The green roofs are located in the cities of Oslo, Trondheim, Sandnes and Bergen in Norway. Three of the four locations are characterized by a coastal climate, classified as temperate oceanic climates (Cfb) in the Köppen-Geiger classification, while Oslo, located in eastern Norway, is classified with warm summers and a humid continental climate (Dfb) in the Köppen-Geiger classification (Peel Finlayson & McMahon 2007). Intensity, Duration, and Frequency (IDF) curves, given by the Norwegian Center for Climate Services (NCCS, www.klimaservicesenter.no), show that the climate in Oslo differs most from the other locations, with more frequent, shorter, and more intense precipitation events compared to the other sites. The IDF curves for Bergen, Sandnes, and Trondheim show events of lower intensity with smaller differences in intensities between the different return periods.

The four roofs chosen for the study were constructed for field research, described by Johannessen *et al.* (2018). The roofs consist of different sections with test beds from three to five test beds, made up of varying commercial green roof solutions. This study focused on one of the four roof sections at each location, namely the roof consisting of a 10 mm felt mat underneath a layer of sedum, which is equivalent to R4 in Figure 1. The area of the roof in Oslo is 2 m × 4 m, with a slope of 5.5%. In Trondheim, the area is 7.5 m × 2 m, and in Bergen, the area is 4.9 m × 1.6 m, both with a slope of 16%. The area in Sandnes is 5.4 m × 1.6 m, with a steeper slope of 27%. Climatic data from a period of 3 years are collected in Trondheim, Sandnes, and Bergen. For the roof in Oslo, data are collected over a period of 8 years (Johannessen *et al.* 2018).

METHODS

For the laboratory part of the investigations, the German standardized method (FLL 2008), which has standardized the procedure for investigating the runoff coefficients of green roofs, was used. In order to understand the behavior of each individual layer making up the different roof configurations, the runoff coefficients for individual layers were tested first, and subsequently, the different roof configurations were tested as a complete solution. In order to relate the results from the laboratory to field observations from four different locations in Norway were compared to the runoff coefficients measured in the laboratory.

Laboratory measurements

The method used to determine the runoff coefficient for the different roof layers and the combination of layers is in this study based on the 2008 edition of German FLL's guidelines for planning construction and the maintenance of green roofing (FLL 2008, p. 100). There is no national guideline available in Norway. The FLL standards have previously been used in Norway by Busklein *et al.* (2014). Using the FLL standards in the current study enables an easy comparison to previously conducted studies and thereby facilitated the discussion of the results

The materials tested in this study are typical components of green or extruded clay aggregate-based roof solutions. To find the runoff coefficient (C), for the different roof configurations, in total 10 single layers or combinations of the layers were tested. Configurations for each experiment (run, R) are shown in Figure 1. In addition to these ten runs, a reference roof test was added, which tested the plain roofing material without any additions. This was used to compare the results to standard black roof, denoted reference roof from hereon.

The FLL guideline requires that the test roof should be constructed with a 2% drainage gradient, a width of 1 m, and be placed inside a wind- and rain-protected testing hall. The method specifies a block rain of 27 mm over the

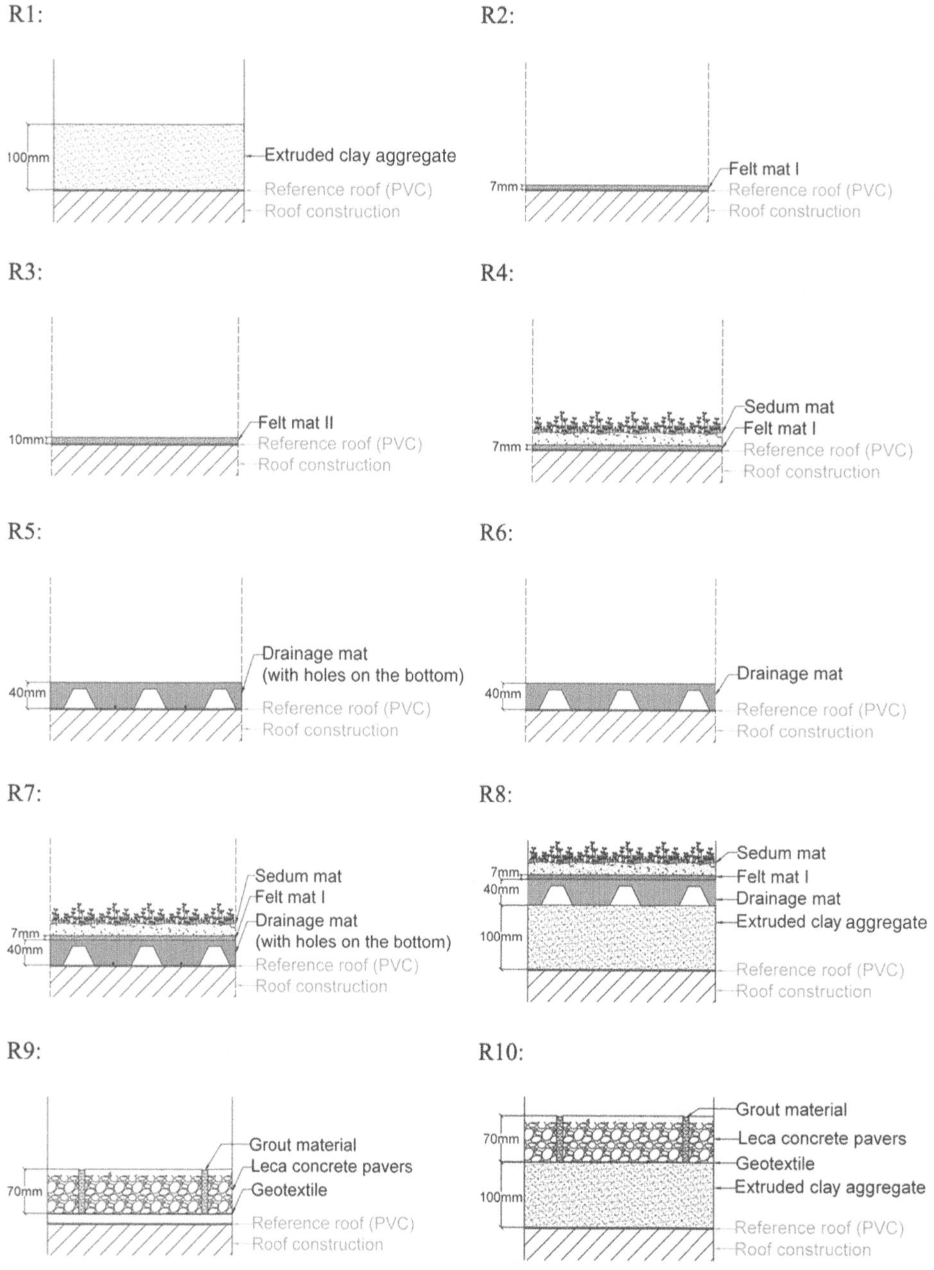

Figure 1 | Section drawings for the composition of the different runs; R#.

duration of 15 min. This is a high intensity event for the locations in this study, which exceeds the 100-year event for all the locations in the study. As green roofs are predominantly designed to handle the smaller events, described as a step 1 solution in the three-step approach to stormwater management in Norway (Lindholm *et al.* 2008), it was decided to include a more relevant and moderate event for comparison between the field and laboratory results. The four locations are located in different climatic classification regions, as described in the case study section; however, a 11.4 mm event over 15 min was chosen to represent a more typical event which should be handled by green roofs. This represents a 5-year event in Bergen, a 10-year event in Sandnes, and between a 2- and 5-year event in

Oslo (10.3 mm/15 min for 2 years and 14.1 mm/15 min for 5 years), while, in Trondheim, it is equivalent to a 50-year return event. Trondheim observes significantly lower intensity events compared to the three other locations. The runs using this event were tested on the two complete roof configurations denoted in R8 and R10 in Figure 1.

Prior to the test, the roof material should be pre-wetted to saturation by continuous irrigation for 10 min beyond reaching a constant runoff rate. This is followed by a subsequent 24-h drainage time, after which field capacity is assumed. The method then prescribes three repetitions for each test with 24-h intervals. The runoff coefficient C is then given by the following equation:

$$C = \frac{R}{V} \tag{1}$$

where V is the total volume of water added in liters and R is the volume of runoff in liters at the time when the precipitation ends, in this case 15 min.

In this study, the area of the modeled roof was 2 m × 2 m. The precipitation was supplied using 16 nozzle tubes placed 80 cm above the roof construction. The system was calibrated to give a total amount of 27.4 mm in 14.67 min, which was considered accurate enough to the 15 min prescribed treatment time. The runoff was measured with a 0–100 mBAR PTX1400 pressure transducer in a collection tank at the downstream end with a two second time resolution (Figure 2). For R2, R3, R5, R6 and R9, the prescribed 24-h period between the pre-wetting and the test was omitted, as these were single-layer runs where a 24-h period would have completely dried them up. For these runs, field capacity was assumed at the end of runoff from the pre-wetting phase. The permeable pavement (R9) was lifted 1.5 cm by using steel rods since the water flows vertically through the joints but not horizontally through the concrete pavers.

Figure 2 | Illustration pictures from the laboratory: (a) reference roof; (b) example of built up, R7; and (c) running the simulation.

Darcy's and Manning's formulas are used to calculate the horizontal flow occurring in the different drainage layers. The permeable layers and the drainage layers can be described as a filter with flow across the filter. The flow may be described by Darcy's formula:

$$Q = \frac{K(h_{sf} + d)}{d} \times A_{sf} \tag{2}$$

Q (V/T) is the flow through the media, K (L/T) is the hydraulic conductivity, h_{sf} (L) is the depth of ponding over the filter media surface, d (L) is the thickness of the filter media, and A_{sf} (L^2) is the surface area of the filter media. When a free surface flow occurs, the flow can be described by Manning's formula as:

$$Q = \frac{1}{n} A R^{2/3} S^{1/2} \tag{3}$$

Q (V/T) is the flow, n is the Manning's roughness coefficient, A (L^2) is the cross-section of the flow, R (L) is the hydraulic radius, given as flow depth for wide 'channels', and S (L/L) is the slope.

Field measurements

Data from selected precipitation events at the four field locations were used for comparison with the laboratory results. The events were selected from continuous precipitation records at each location. A precipitation event was defined as precipitation after a minimum of 6 h of -antecedent dry weather period. From these, only events producing runoff were selected. In a final step, only events from May to October were selected in order to avoid data from snow-covered roofs, which may appear in Oslo and Trondheim. The outcome of the selection procedure is displayed in Table 1.

Table 1 | The selection of events from the total number of single events in the continuous data series from the field observed green roofs

Observation period		Total number of observed precipitation events	Events resulting in runoff	Events from May to October (excluding winter)	Events with a duration less than 1,440 min (1 day)
Bergen	01.01.15–21.08.17	122	47	35	26
Oslo	02.09.09–06.12.17	655	263	192	179
Trondheim	01.01.15–18.12.17	201	47	29	20
Sandnes	22.04.15–21.10.17	158	72	47	38

The remaining events, ranging from 26 in Trondheim to 179 in Oslo, were plotted in IDF curves for each respective area (NCCS 2018) (see Figure 3). From the events displayed in Figure 3, we selected a small subset where cumulative precipitation is similar to the laboratory events, and the duration is as short as possible. To get events with similar cumulative precipitation as in the laboratory experiments, it was necessary to select events with duration up to 450 min. Figure 3 shows the IDF curves together with the initial set of events, the final set of selected events, and the simulated laboratory events.

Precipitation and runoff data from each event were used to calculate the runoff coefficients for the roofs based on the same definition as in Equation (1), which means that the

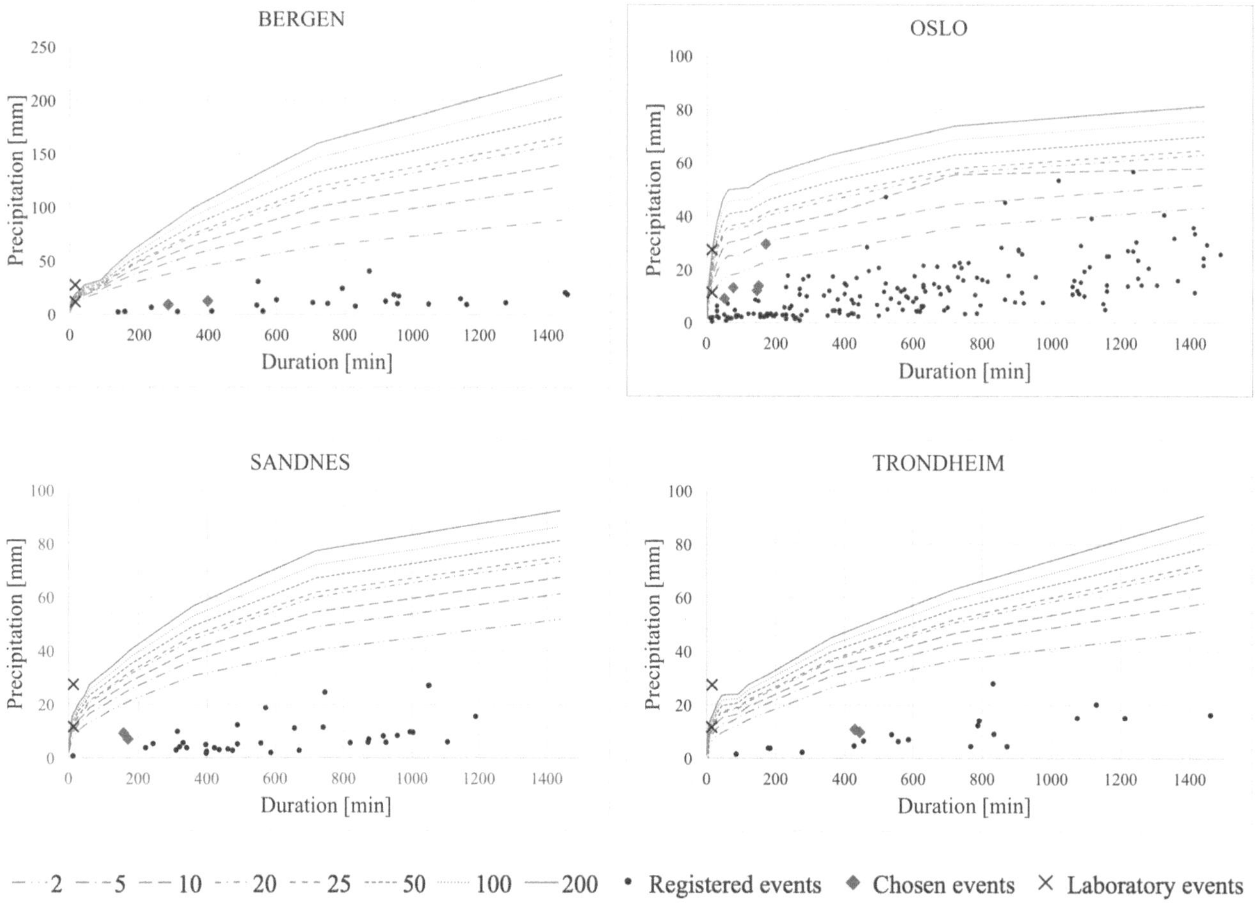

Figure 3 | IDF curves and precipitation events from the field measurements, where X for laboratory events includes both the FLL prescribed precipitation event of 27 mm/15 min and the 11.4 mm/15 min selected based on the field locations.

runoff coefficients for the field cases were calculated as the runoff volume over the total precipitation volume for the time span covered by the precipitation event. This makes the duration different for all the events as it is based on the actual duration of the event. In the events where the precipitation starts off very small, almost negligible, the events were set to start when the precipitation exceeded 0.5 mm.

RESULTS AND DISCUSSION

In this section, the results from the laboratory and field measurements are presented and compared.

Laboratory measured runoff coefficients

The laboratory conducted tests showed small variance between the three repetitions for each run (denoted R1 through R10 in Figure 1), which indicates that the set-up had a satisfactory reproducibility, with a standard deviation of 0.02 for calculated runoff coefficients. For further analysis, a simple average of the three repetitions is used. Average runoff and intensity curves from each of the runs exposed to 27.4 mm precipitation are presented in Figure 4.

For several runs, a free water surface above the layer being tested occurred (R1, R2, R3, and R4). This results in that runoff flows as overland flow on the surface of the layer and directly into the collection tank, which would affect the runoff coefficient calculation. The drainage board with extra drainage holes at the bottom (R5) was nearly empty through the irrigation, indicating that the holes were not serving to detain the runoff. The holes were made to function as a slow draining of the storage volume in these drainage boards. The cups on the drainage board without these extra drainage holes (R6) were full at the start of the run because of the pre-wetting in the procedure. This

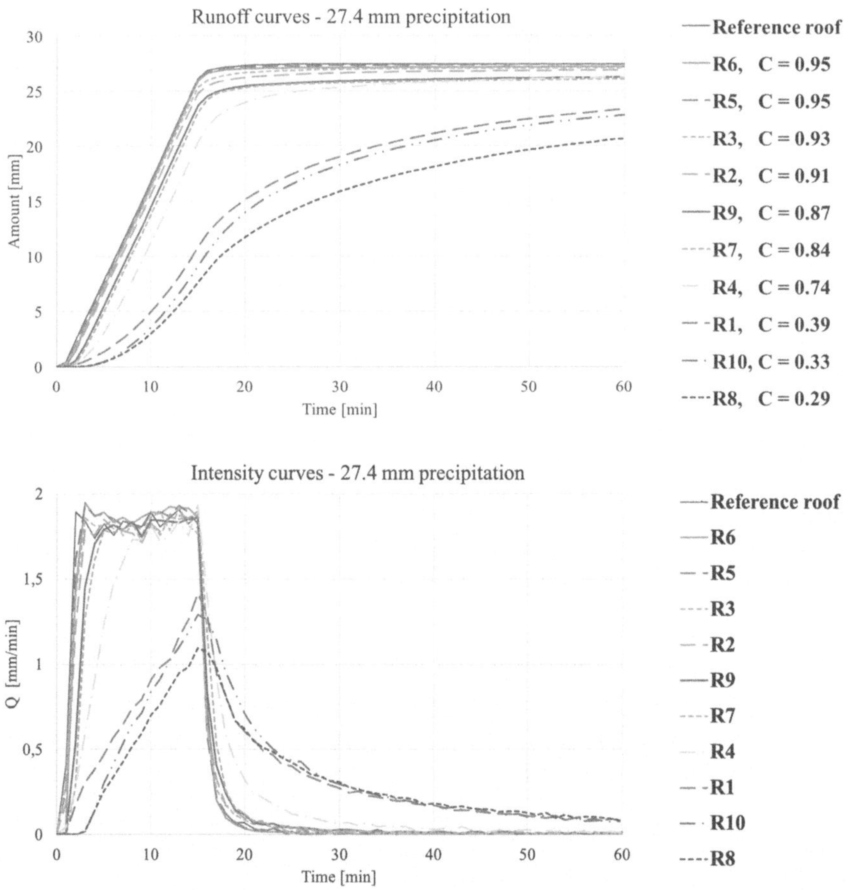

Figure 4 | Average runoff curves, intensity curves, and runoff coefficients, *C*, for each run.

caused the initial storage to be full at the onset of precipitation. A general deterioration of the sedum was observed as the experiments progressed through different configurations, as the sedum mats were reused for several runs, which resulted in multiple moving operations of the sedum. This was accounted for by the fragile mats that serve as a placeholder for the substrate and the sedum, which really is not made to be moved between the different configurations.

It can be seen from the curves represented by the single layers (R1, R2, R3, R5, and R6) that the 10 cm layer of an extruded clay aggregate medium (R1) stands out as the single-layer component with the highest detention capacity both with respect to volume and peak flow reduction. It had a considerably lower runoff coefficient than the other four individually tested layers, which all appear to be more similar to the reference roof. The two felt materials tested, from the different producers (R2 and R3), have only minor differences in the hydraulic behavior. It was somewhat unexpected that the thinner of the two mats had the greatest detention and the smallest runoff coefficient. This may be due to a more tightly packed material, with less pore volume inside for the thicker mat. The drainage boards with and without the extra drainage holes on the bottom (R5 and R6) resulted in the same runoff coefficient, indicating that the extra drainage holes of 3 mm does not increase the detention. Smaller drainage holes may have an increasing effect on the detention, but at the same time smaller holes are more vulnerable to clogging. Since the cold and wet coastal climate leads to lower evapotranspiration (Johannessen *et al.* 2017), water stored in the layers of the roof may never evaporate. This reduces the retention capacity and the layer only works as a 'one time retention volume'. In this detention-based testing method, the drainage board without the extra holes used in the laboratory was always full while testing, and the runoff is a function of the surface friction and rainfall intensity. This leads to a high runoff coefficient comparable to the reference roof, as there is a higher friction in the roofing reference than the slick plastic surface of the drainage boards.

Green roofs are represented by three different combinations of layers tested with just the sedum, R4; sedum and felt mat, R7; sedum, felt mat over-the-drainage mat with extra drainage holes and R8; sedum, felt mat, drainage board and an extruded clay aggregate medium. The combination in R4 gives a lower runoff coefficient than the combination in R7, although R7 is thicker consisting of one more layer. The difference is most likely explained by the vertical movement of the water through the sedum and felt layers, followed by horizontal flow movement through the drainage board, which gave a low detention performance when it was tested alone. In R4, with the combination only consisting of a sedum mat and a felt mat, the water flows laterally through the layers. R8, which includes an extruded clay aggregate medium, gives the lowest coefficient of runoff. It also gives a substantially lower runoff coefficient than the extruded clay aggregate layer alone. This showcases the importance of understanding the interactions between the layers in the design phase. Here, it is possible that the horizontal flow occurs in both the sedum layer and the extruded clay aggregate media layer. It is unlikely that a free surface flow will occur on the extruded clay aggregate surface, as the vertical infiltration rate is much higher than the maximum intensity of 27 mm/15 mm, which is a rather high intensity event.

The non-vegetated roof was made up of the extruded clay aggregate with concrete pavers on top (R10 = R1 + R9). The runoff coefficient of the combined system was $C = 0.33$. This was to a large degree influenced by the 10 cm extruded clay aggregate layer with a runoff coefficient equal to 0.39. The concrete pavers covering the extruded clay layer made the runoff flow laterally over the pavers, entering the media in the cracks between the pavers before it flows laterally in the extruded clay aggregate-based layer. The test of the concrete pavers alone (R9) gave a runoff coefficient equal to 0.89, which, to a lesser extent, contributes to the detention capacity of the combined system (R10). Since the flow directions are the same for the layers in the combined system as for the single layers, multiplying the individually obtained runoff coefficients gives a runoff coefficient equal to 0.34 for the combined roof system R10. This is in good agreement with the value obtained directly for the combined system and within the standard deviation of the method. The runoff coefficients found in this study can be seen as detention based. Of the measurements in the laboratory, the extruded clay aggregate medium-based systems (R8 and R10) gave the lowest runoff coefficients.

The applied standard with 27 mm in 15 minutes is an extreme event in the study locations in this study. A

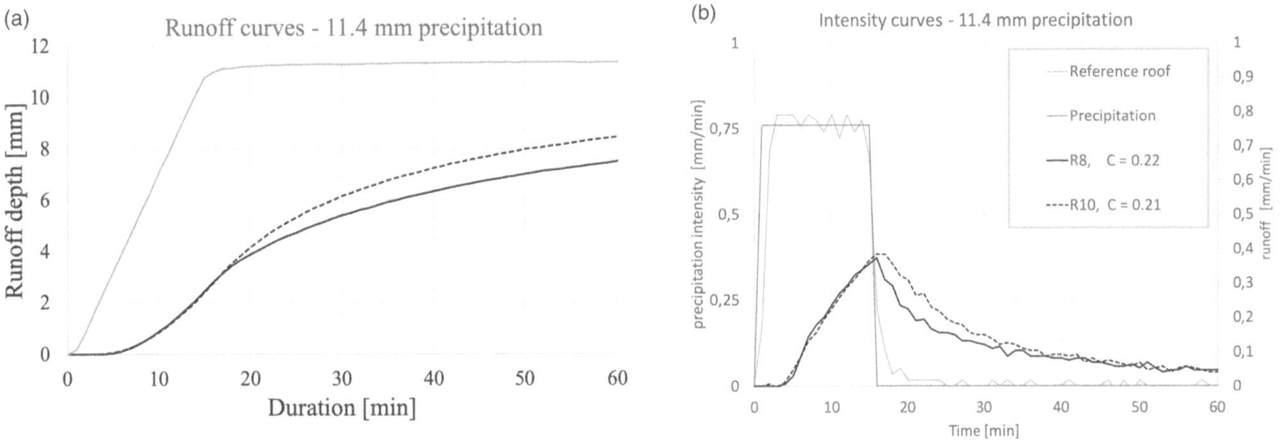

Figure 5 | Runoff curves (a), intensity curves and runoff coefficients (b) for the reduced precipitation event on the extruded clay aggregate-based roof systems; R8 and R10.

simulated rain event of 11.4 mm with a duration of 15 minutes was applied on the R8 and R10 roof systems, in order to obtain a rainfall event more suitable to the Norwegian climate. It can be seen that a reduction in applied precipitation depth leads to a reduced runoff coefficient (Figure 5).

Both tests gave similar runoff coefficient for the vegetated (R8) and non-vegetated roof (R10) systems, 0.22 and 0.21, respectively. For the higher intensity event, the 27 mm event, the non-vegetate roof had a slightly better detention, 0.29 versus 0.34 for the runoff coefficient. The change for the lower intensity event could be a result of a poorer vegetation mat towards the end of the experiments and the previously mentioned progressive deterioration of the vegetation mat due to all the handling configuring the different combinations. However, it can be concluded that the runoff coefficient increases with increasing intensity.

Laboratory experiments are important in understanding the underlying flow processes in the different layers in a detention-based roof. As interpreted from the laboratory experiments performed in this study, the runoff coefficients are mostly governed by the layers where horizontal flow occurs. In the runs where horizontal flow occurs through the porous media, as for the extruded clay aggregate (R8 and R10) and the sedum (R4), the flow is governed by Darcy's equation. This means that the flow through the media, among other things, is based on the hydraulic conductivity and the particle size distribution of the porous media. When the water flows across the drainage board (R7), overland flow may occur, which is governed by Manning's equation. Here, the shorter detention time may be explained by the friction and the slope of the roof.

Runoff coefficients based on field data

The selected events (cf. Figure 3) are presented as cumulative precipitation and runoff in Figure 6. These observations are considered most comparable to the events simulated in the laboratory based on total precipitation.

The graphs in Figure 6 indicate a varying performance response of a similar layered roof at the four locations (i.e. sedum spices and a felt mat of 10 mm). Due to varying intensities within the events, it can be observed that the curves from the field measurements are less smooth than those from the laboratory measurements. The detention performance varies between the events, whereas lag times vary from 1 to 351 min. The ratio between accumulated precipitation and runoff at the end of each precipitation event results in detention-based runoff coefficients varying between 0.023 and 0.41. Compared to the laboratory measured runoff coefficient for the same layered roof (R4), these field observations give a significantly lower value than the measured runoff coefficient of 0.74 for the 27 mm event.

In the field, the state of the roof at the onset of each precipitation event will vary, which could affect the performance. The moisture in the roof will vary for the field events, while it is constant at field capacity for the laboratory

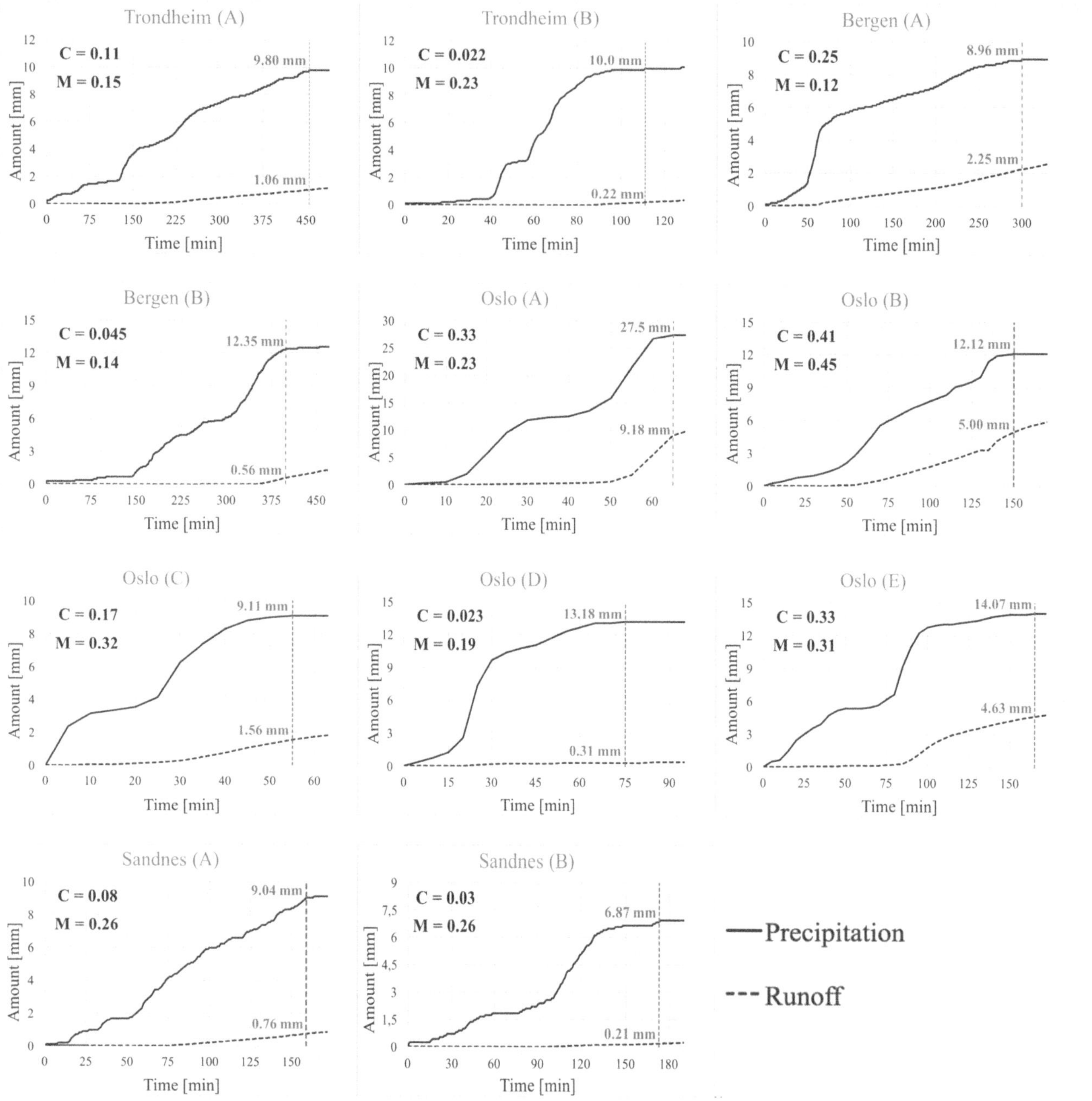

Figure 6 | Precipitation and runoff curves for the chosen field events, runoff coefficient (*C*) and moisture (*M*) at the start of precipitation. The blue represents the values used in the calculation of the runoff coefficient. Please refer to the online version of this paper to see this figure in colour: http://dx.doi.org/10.2166/nh.2020.049.

experiments. Lower soil moisture levels will make the roof capable of storing more water, which will lead to an increase in lag time and a decrease in runoff coefficient, *C*. This differs from the laboratory measurements, where the roof was at field capacity at the onset of precipitation. The events were selected based on the amount of precipitation. In all the chosen field events, this occurred over a longer duration than in the laboratory. Longer duration results in the lower average intensity of precipitation in the field than in the laboratory.

Runoff coefficients as a variable in designing detention-based roofs

In urban areas, where roofs are a large part of the impervious surfaces, the runoff from roofs is an important factor when dimensioning urban stormwater structures (Stovin *et al.* 2012; Berretta *et al.* 2014; Hamouz *et al.* 2018). As more detention-based roofs are established in the cities, the importance of including the effect of these in the runoff calculations increases (Sobczyk & Mrowiec 2016). The variations in the runoff coefficients calculated in this study prove the challenge of using a suitable value for a given roof.

Results from the laboratory measurements gave a variation in the runoff coefficient depending on the materials and compositions of the layers. It also gave a variation depending on the intensity of the added event which confirms the results from other studies (Stovin *et al.* 2015; Johannessen *et al.* 2018; Hamouz *et al.* 2018). The laboratory test method gave runoff coefficients with small standard deviations and high reproducibility, which indicates that the testing method was robust and reliable. In addition, the laboratory analysis improves the understanding of how water moves through the layers, making it easier to compare the layers and evaluate the contribution of each layer. However, the discrepancy to the field observations raises an important concern in using the laboratory measured runoff coefficients for design. The laboratory experiments were conducted at field capacity, which will be a conservative approach, still the resulting runoff coefficient was higher than the field comparisons for all the roofs.

The challenge with a standardized test method is the results' suitability to the location they may be used. The laboratory measurements are conducted in conditions which may not be realistic for a given location. The lack of suitability is especially an issue for detention-based roofs which are established to handle small-to-medium events, as defined in the three-step approach (NOU 2015, p. 67), and not the larger events with rare recurrences. Hence, these laboratory measurements, with 27 mm in 15 min, are more suitable for downstream stormwater calculations dimensioned for larger events and the performance of step 1 solutions in extreme events.

Measurements from the four roofs in the field resulted in significantly lower runoff coefficients than the values obtained from the laboratory test. There are many variables that affect the runoff peaks such as soil moisture content, intensities, and physical roof design. The field data record of 4–8 years can be considered substantial; however, it was still difficult to find events that could be compared with the laboratory experiments as there are many variables that may influence the performance. This clearly demonstrates that this approach is not well suited to capture the performance of detention-based roofs, such as green roofs. As an alternative to the typical event-based metrics and runoff coefficient focus on urban stormwater management is used for evaluating detention performance, Johannessen *et al.* (2018) presents flow duration curves based on time series as an alternative approach. Flow duration curves give valuable information on the runoff pattern from the roofs, which can be used in relation to local requirements. A volume of storage-based approach would also complement a pure peak flow approach that today is still very commonly used. Both these methods would shift the focus from a pure peak flow focus to a total water management focus, where detention-based roofs are part of a series of solutions.

CONCLUSION

In this study, the runoff coefficients for different layered detention-based roofs have been investigated. The results highlight the complexity of using a runoff coefficient approach to design of these roofs. The laboratory measurements gave a varying runoff coefficient due to the compositions of the roofs and the intensity of the added block rain. However, it was the roof systems with an extruded clay layer (R1, R8, and R10) that had a significant lower runoff coefficient than all the other types, which indicates the need for a porous flow-based layer for the detention of precipitation.

The field measurements gave a smaller and more varying runoff coefficient for the same roof configurations compared to the laboratory setups. The soil moisture level on the onset of precipitation can explain this variation, which also can be derived as the single most important parameter for the performance of a green roof system.

The results of this study demonstrate the challenge of using a suitable runoff coefficient measured in the laboratory for a given roof in the calculations of stormwater runoff. However, laboratory analysis aids our understanding of how water moves through the layers and is important to understand the underlying flow processes in the different layers in a detention-based roof. In the thicker layers, like the extruded clay aggregate, there will be flow through porous media, which is governed by Darcy's equation, while flow across the drainage boards is governed by Manning's equation. Characterizing the differences in the flow through the different media can aid our understanding of the field observations and by this improve design calculations in urban stormwater management. Further, moving towards a flow characteristic volume-based approach will improve the design of these systems.

ACKNOWLEDGEMENTS

The research is made possible in part by the Centre for Research-based Innovation – Klima 2050 for supporting the work (www.klima2050.no).

REFERENCES

Andenæs, E., Kvande, E., Muthanna, T. N. & Lohne, J. 2018 Performance of blue-green roofs in cold climates: a scoping review. *Buildings* **8** (4), 55. https://doi.org/10.3390/buildings8040055.

Berretta, C., Pöe, S. & Stovin, V. 2014 Moisture content behavior in extensive green roofs during dry periods: the influence of vegetation and substrate characteristics. *Journal of Hydrology* **511**, 374–386.

Busklein, J. O., Thodesen, B. O. & Balmand, E. 2014 *Testing of Leca's® Attributes and Use in Green Roofs Solutions* (SBF 2012 F0254).

FLL 2008 Guideline for the Planning, Construction and Maintenance of Green–Roofing–Green Roofing Guideline. Forschungsgesellschaft Landschaftsentwicklung Landschaftsbau e.V.

Hamouz, V., Lohne, J., Wood, J. R. & Muthanna, T. M. 2018 Hydrological performance of LECA-based roofs in cold climates. *Water* **10**, 263. https://doi.org/10.3390/w10030263.

Hanssen-Bauer, I., Førland, E. J., Haddeland, I., Hisdal, H., Mayer, S., Nesje, A., Nilsen, J. E. Ø., Sandven, S., Sandø, A. B., Sorteberg, A. & Ådlandsvik, B. 2015 *Klima i Norge 2100 (The Climate in Norway 2100)* (NCCS no. 2/201).

IPCC 2013 *Climate Change 2013: The Physical Science Basis.* Contribution of Working Group I to the Fifth Assessment Report of the Intergovernmental Panel on Climate Change (T. F. Stocker, D. Qin, G.-K. Plattner, M. Tignor, S. K. Allen, J. Boschung, A. Nauels, Y. Xia, V. Bex and P. M. Midgley, eds). Cambridge University Press, Cambridge, United Kingdom and New York, NY, USA. http://www.climatechange2013.org/images/report/WG1AR5_TS_FINAL.pdf.

Johannessen, B. G., Hanslin, H. M. & Muthanna, T. M. 2017 Green roof performance in cold and wet regions. *Ecological Engineering* **106**, 436–447.

Johannessen, B. G., Muthanna, T. M. & Braskerud, B. C. 2018 Detention and retention behaviour of four extensive green roofs in three nordic climate zones. *Water* **10** (671). doi:10.3390/w10060671.

Kuichling, E. 1889 The relation between the rainfall and the discharge of sewers in populous areas. *Transaction, American Society of Civil Engineers* **20**, 1–56.

Lindholm, O., Endresen, S., Thorolfsson, S., Sægrov, S., Jakobsen, G. & Aaby, L. 2008 Veiledning i klimatilpasset overvannshåndtering, Norsk Vann Rapport 162. 2008. Rapport 162-2008.

Norwegian Centre for Climate Services (NCCS) 2018. Available from: https://klimaservicesenter.no (accessed 28 March, 14 May 2018).

Norges offentlige utrdening (NOU) 2015: 16. *Overvann i byer og tettsteder: Som problem og ressurs.* Departementenes sikkerhets- og serviceorganisasjon, Informasjonsforvaltning, Oslo.

Peel, M. C., Finlayson, B. L. & McMahon, T. A. 2007 Updated world map of the Köppen-Geiger climate classification. *Hydrology and Earth System Sciences Discussions* **4** (2), 439–473.

Sobczyk, M. & Mrowiec, M. 2016 *Retention Capacity of Extensive Green Roofs.* JWLD.

Stovin, V. R., Vesuviano, G. & Kasmin, H. 2012 The hydrological performance of a green roof test bed under UK climatic conditions. *Journal of Hydrology* **414–415**, 148–161.

Stovin, V., Poë, S., De-Ville, S. & Berretta, C. 2015 The influence of substrate and vegetation configuration on green roof hydrological performance. *Ecological Engineering* **85**, 159–172.

Water Environment Federation, Design of Urban Stormwater Controls Task Force 2012 *Design of Urban Stormwater Controls*, 2nd edn. McGraw-Hill Professional McGraw-Hill Distributor, New York and London.

First received 3 April 2019; accepted in revised form 4 March 2020. Available online 30 March 2020

Detection and attribution of flood responses to precipitation change and urbanization: a case study in Qinhuai River Basin, Southeast China

Guodong Bian, Jinkang Du, Mingming Song, Xueliang Zhang, Xingqi Zhang, Runjie Li, Sengyao Wu, Zheng Duan and Chong-Yu Xu

ABSTRACT

Both flood magnitude and frequency might change under the changing environment. In this study, a procedure combining statistical methods, flood frequency analysis and attribution analysis was proposed to investigate the response of floods to urbanization and precipitation change in the Qinhuai River Basin, an urbanized basin located in Southeast China, over the period from 1986 to 2013. The Mann–Kendall test was employed to detect the gradual trend of the annual maximum streamflow and the peaks over threshold series. The frequency analysis was applied to estimate the changes in the magnitude and frequency of floods between the baseline period (1986–2001) and urbanization period (2002–2013). An attribution analysis was proposed to separate the effects of precipitation change and urbanization on flood sizes between the two periods. Results showed that: (1) there are significant increasing trends in medium and small flood series according to the Mann–Kendall test; (2) the mean and threshold values of flood series in the urbanization period were larger than those in the baseline period, while the standard deviation, coefficient of variation and coefficient of skewness of flood series were both higher during the baseline period than those during the urbanization period; (3) the flood magnitude was higher during the urbanization period than that during the baseline period at the same return period. The relative changes in magnitude were larger for small floods than for big floods from the baseline period to the urbanization period; (4) the contributions of urbanization on floods appeared to amplify with the decreasing return period, while the effects of precipitation diminish. The procedure presented in this study could be useful to detect the changes of floods in the changing environment and conduct the attribution analysis of flood series. The findings of this study are beneficial to further understanding interactions between flood behavior and the drivers, thereby improving flood management in urbanized basins.

Key words | attribution analysis, flood, frequency analysis, Mann–Kendall test, Qinhuai River Basin, urbanization

Guodong Bian
Jinkang Du (corresponding author)
Mingming Song
Xueliang Zhang
Xingqi Zhang
Runjie Li
Sengyao Wu
School of Geography and Ocean Science,
Nanjing University,
Nanjing,
China
E-mail: *njudjk@163.com*

Guodong Bian
Mingming Song
Nanjing Hydraulic Research Institute,
Nanjing 210029,
China

Jinkang Du
Xueliang Zhang
Runjie Li
Sengyao Wu
Jiangsu Center for Collaborative Innovation in Geographical Information Resource Development and Application,
Nanjing,
China

Zheng Duan
Department of Physical Geography and Ecosystem Science,
Lund University,
Sölvegatan 12 SE-223 62 Lund,
Sweden

Chong-Yu Xu
Department of Geosciences,
University of Oslo,
P.O. Box 1047 Blindern N-0316 Oslo,
Norway

INTRODUCTION

Flood disaster is one of the most serious natural hazards, as floods often result in serious property damage and casualties worldwide (Kundzewicz *et al.* 2013; Ragettli *et al.* 2020; Venvik *et al.* 2020). A better understanding of flood characteristics and their potential driving forces is of importance to manage river flows and effectively mitigate flood disasters. Generally, climate change and human activities are identified as the two primary causes for changes in flood

doi: 10.2166/nh.2020.063

regime (Hou *et al.* 2020; Ragettli *et al.* 2020). In recent decades, extreme rainfall events have occurred frequently due to climate change, and the occurrence of flood disasters and the extent of flood-induced damage has an increasing trend (He *et al.* 2017; Ledingham *et al.* 2019). For example, the damage caused by floods in China in 2017 were estimated as US $31 billion (China Floods and Droughts Disasters Bullet in 2017).

Human activities, such as land-use change and urbanization, are important factors influencing the hydrological cycle. Increased urbanization has led to considerable changes in natural catchment characteristics by increasing the amount of impervious surface areas, which reduce infiltration of precipitation during storm events and increase direct runoff (Yao *et al.* 2015; Darabi *et al.* 2019; Li *et al.* 2019; Okoli *et al.* 2019). Additionally, the reduction of forest and wetland coverage due to the urbanization would result in a decrease of the buffering capacity of these ecosystems in flood events. Consequently, runoff increases in proportion to the expansion of impervious areas in a watershed, and the stormflow, peak discharges and flood risk also tend to increase in the urban area (Muis *et al.* 2015; Mahmoud & Gan 2018; Li *et al.* 2020). Moreover, several studies have revealed that a decrease in the infiltration of precipitation due to an increase of impervious areas leads to a higher increase in the volume and flood peak of storm runoff for the medium and small floods than that for the really large and rare events (Braud *et al.* 2013). Kaspersen *et al.* (2015) pointed out that an increase in impervious areas had more effects on the hydrological response for more frequent flood events while only a lesser degree effects for less frequent events. They attributed this difference to the fact that the natural surface was able to reach saturation faster during very extreme events and started to behave like the impervious surface rather quickly after the onset of the events. Inversely, the time to saturation is commonly much longer during less extreme events. However, the effect of land-use change on the flood regime cannot always be straightforwardly investigated. Some studies revealed a specific difficulty in detecting flood trends due to several signals overlapping in the analysis process and found little influence of land-use change predominant urbanization on floods (Blöschl *et al.* 2007; Hannaford *et al.* 2013; Madsen *et al.* 2014). Furthermore, the reality is often further complicated by the impacts of several driving factors. Hence, an intensive study is essential to be conducted in the specific basins before a generalized conclusion can be drawn. This constitutes the motivation for this study.

The question whether the magnitude and frequency of floods have changed in the changing environment is of significance. Amounts of various studies have detected trends in flood changes using a range of statistical tests from long-term flood records (Mediero *et al.* 2014; Yin *et al.* 2015; Stevens *et al.* 2016; Balistrocchi & Bacchi 2017; Mudersbach *et al.* 2017; Kundzewicz *et al.* 2018; Dehghanian *et al.* 2020). Hall *et al.* (2014) stated that in the simplest case of change detection of the flood peak records, the Pettitt test and Mann–Whitney *U* test were frequently applied for step changes, and Mann–Kendall test and Spearman's test were widely performed for gradual changes. Modarres *et al.* (2016) applied the Mann–Kendall test and two pre-whitening trend tests to detect trends for the annual flood peaks in Iran and found a significant increasing trend in flood magnitude in most hydrological basins. Alternatively, the non-stationary flood frequency analysis has also been widely used to detect the variation characteristics of floods in many studies (Viglione *et al.* 2013; Ahmad *et al.* 2015; Šraj *et al.* 2016; Jiang *et al.* 2019). The basic idea of this approach is that it allows the parameters of the flood frequency distribution to change in time as a function of covariates. Delgado *et al.* (2010) mentioned that the non-stationary generalized extreme value function (NSGEV) is advantageous to detect changes in different flood magnitudes. They found a positive trend in the scale parameter of a fitted distribution, the frequencies of both large floods and small floods increased in the Mekong River. However, the above studies merely focused on the detection of river peak flow variation by performing statistical analyses of flood time series, and more concerted efforts are required for attributing trends in floods, as highlighted by Merz *et al.* (2012).

Quantitatively evaluating the contributions of precipitation change and urbanization to the flood response is essential for managing flood risk and is also of practical importance for designing measures to mitigate hazards. Hydrological modeling and statistical analysis are two widely used approaches to differentiate the individual effect of precipitation change and urbanization on floods. Many studies assessed the impact of urbanization on flood peak using hydrological modeling for different land-use scenarios (Jothityangkoon *et al.* 2013; Wolski *et al.* 2014; Aich *et al.* 2015; Zope *et al.* 2017). For example, Du *et al.* (2015) used the Soil Conservation Service model to simulate the hydrological process for three land-use scenarios in the Pearl River Delta to assess the impacts of urbanization on floods. They found a significant increasing trend in floods and attributed this trend to the effect of the expansion of impervious surfaces and the displacement of farmland in forested hills in the urbanization process. Chen *et al.* (2015) analyzed the effects of urbanization on flood characteristics by setting up the HEC-HMS (Hydrologic Engineering Center's Hydrologic Modeling System) model to simulate flood processes for different land-use scenarios, and they concluded that the peak discharge and flood volume increased in the rapid urbanization expansion process. Furthermore, the non-stationary extreme value statistics are frequently adopted for the attribution of floods. The parameters of the flood frequency distribution are allowed to change by various driving factors that are included as covariates, followed by a test if such a model expansion results in a significantly better fit to the extreme values (López & Francés 2013; Machado *et al.* 2015; Viglione *et al.* 2016). For instance, Villarini & Strong (2014) selected precipitation and land cover indicators as external covariates and applied the Generalized Additive Models for Location, Scale and Shape (GAMLSS) to attribute the flood changes to rainfall variations in Iowa. Prosdocimi *et al.* (2015) used a GEV distribution and considered both precipitation and urbanization as covariates. They found that the increasing urbanization level has the dominant effect on flood peak in the UK. To our knowledge, the hydrological models and non-stationary extreme value statistics have been widely applied in the detection and attribution of changes in floods under the context of the changing environment. However, complex structures, parameters identification of the hydrological models and the uncertainty involved in the simulations are important limitations of the application in the attribution analysis. The far more difficult problem of non-stationary extreme value statistics is the inability to quantify the relative importance of the different drivers in flood time series, which is suggested by Merz *et al.* (2012) who think that more effort is needed to quantitatively evaluate the attribution for flood time series.

Therefore, the main research aim of the study is to propose a procedure to investigate how flood regime changes under changing precipitation and land-use condition, and quantify the attribution of flood changes to precipitation and urbanization. The specific objectives of this study are (1) to detect changes in characteristics of flood series, (2) to explore the variations of the flood frequency after the urbanization and (3) to separate the contributions of precipitation change and urbanization to flood regime changes.

STUDY AREA AND DATA

The Qinhuai River, one of the tributaries of the lower Yangtze River, is located in the southwest of Jiangsu province (Figure 1). The basin area is 2,631 km^2. It is located in the subtropical monsoon climate zone, where the local climate is controlled by the East Asia summer monsoon. The annual mean air temperature is 15.4 °C, and average annual precipitation is 1,116 mm during the period 1986–2013. The rainy season is from April to October.

As a typical watershed in Yangtze delta plain, the Qinhuai River Basin has a marked advancement of urbanization since the beginning of the 2000s (Du *et al.* 2012; Chen & Du 2014; Hao *et al.* 2015). The impervious areas increased from 3.8% to 8.6% from 1986 to 2001; however, from 2002 to 2013, the impervious areas increased from 8.8% to 17.5% (Bian *et al.* 2017). Details of land use/land cover at different periods can be found in the literature by Hao *et al.* (2015).

The 28-year (1986–2013) daily rainfall data from seven rain-gauge stations and daily discharge data at the Inner Qinhuai station and Wudingmen station were obtained from the local hydrological bureau. The basin location, elevation, network and the distribution of the two hydrological stations and seven rain-gauge stations are shown in Figure 1, and the two streamflow gauging stations are located at the two outlets of the basin.

METHODS

In this study, the annual maximum streamflow (AMS) series and the peaks over threshold (POT) sampling method were

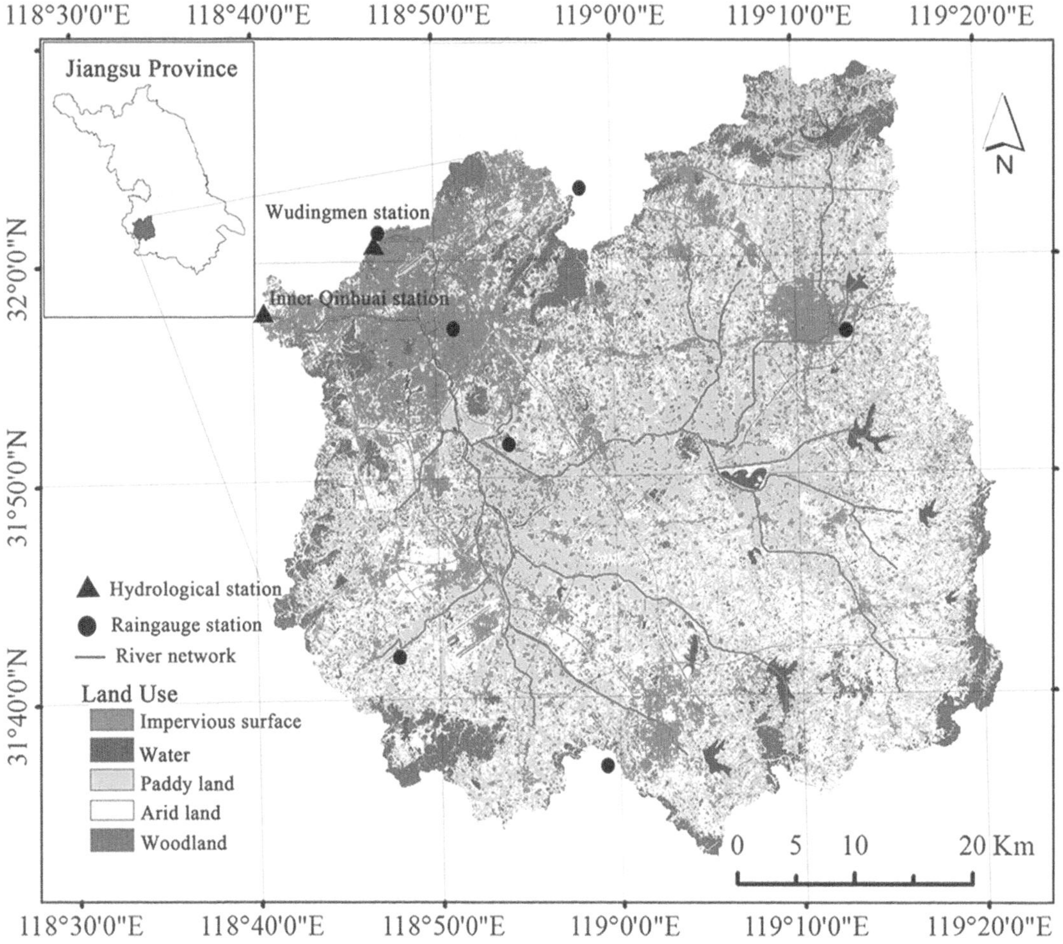

Figure 1 | Locations of Qinhuai River Basin and rain gauge and hydrological stations, and the spatial distributions of the river network and land use (2013) of the basin.

used to obtain the flood series from the daily streamflow data first. Secondly, the nonparametric Mann–Kendall test was applied to detect changes in trends of flood series. Thirdly, the frequency analysis was conducted to assess the frequency variations of flood series. Finally, the attribution analysis proposed in this study was used to quantitatively evaluate the contributions of precipitation change and urbanization to the flood changes.

Selection of flood series

The most common indicator used in flood trend studies is the AMS. In this study, the AMS series were acquired from the daily streamflow data of 1986–2013 for the flood trend analysis. However, this sampling method can result in a loss of information of floods (Bezak *et al.* 2014). For example, some low discharge values in dry years, as the largest flood in a year, could be included in AMS series, and some relatively large floods in wet years may not be considered in AMS series, as they are not the annual largest floods. Therefore, the POT sampling method is also used to overcome those drawbacks by selecting all the floods that exceed a given threshold regardless of the time they occurred (Mediero *et al.* 2014). The first step of the POT method is consideration of the independence of floods. The independence criteria were evaluated following Silva *et al.* (2012):

$$D > 5 + \log(A) \tag{1}$$

$$Q_{min} < \frac{3}{4}\min(Q_1, Q_2) \tag{2}$$

where D denotes the interval time between two flood peaks in days; A is the basin area in km^2; Q_1 and Q_2 denote the magnitudes of two flood peaks in m^3/s, respectively.

It is commonly assumed that a POT series improves an AMS series in the case of a minimum of two or three events per year on average (Mediero *et al.* 2014). In order to identify the changes of large, medium and small floods under precipitation change and urbanization, we selected daily flood series with one, two and three events on average per year for the POT time series (referred to as POT1, POT2 and POT3 hereafter, respectively).

Detect trend of the flood series

The temporal trends in AMS and POT time series of flood can be detected by nonparametric trend tests which are more robust to outliers and do not need any assumption about the distribution. In this study, the gradual trend test was performed using the rank-based nonparametric Mann–Kendall (MK) test recommended by the World Meteorological Organization (Chebana *et al.* 2013). The MK trend statistic S is calculated following Mann and Kendall (Mann 1945; Kendall 1975). However, if time series data show serial autocorrelation, robust results of the MK test cannot be achieved (Fateh *et al.* 2013). Therefore, a test of the autocorrelation of the time series must be conducted before applying the MK test. If significant autocorrelation is detected, the trend-free pre-whitening procedure proposed by Yue & Wang (2002) will be adopted to remove the effect of serial correlation.

Sen's non-parametric method was applied to calculate the change per year for an existing trend by Sen's slope (Sen 1968). The MK test and Sen's estimation together are also called the Sen-MK test.

Frequency analysis

The flood frequency analysis involves the selection of an appropriate probability density function (PDFs) and a parameter estimation method to modeling the distribution of flood data series. Series studies have shown that the three-parameter Pearson type (P-III) distribution was the most appropriate method to quantify the frequency of AMS series in China (Chen *et al.* 2001; Yin *et al.* 2015). The hydraulic design manual of China also suggests the use of P-III distribution for AMS in China. Hence, the P-III distribution was used to fit the AMS series in our study. The generalized Pareto distribution (GP) was used to fit the distribution of POT series which well fits to partial duration series in almost all cases (Guru & Jha 2015; Solari *et al.* 2017). Due to the short length of the series, the parameters of PDFs were determined by the visual evaluation of the goodness-of-fit.

In the study, the computation formula raised by Rosbjerg (1985) was adopted to convert the frequencies of POT and AMS series to the return period for direct comparison. The changes in the return period with same flood size and changes in flood size with the same return period were also analyzed.

Attribution analysis

The changes in flood regime are generally caused by precipitation change and urbanization. Quantifying the contributions of precipitation change and urbanization to the flood changes is important for flood prediction and flood-induced disaster mitigation. An attribution analysis method, illustrated in Figure 2, was proposed and used to separate the influence of precipitation change and urbanization on floods in this study. This attribution method is

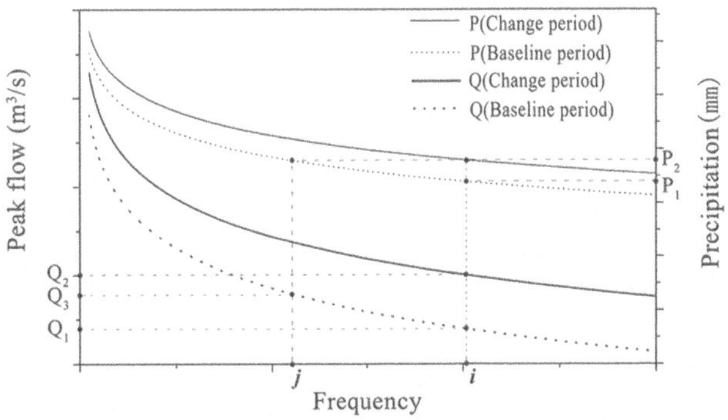

Figure 2 | The separation of precipitation change and urbanization impacts on flood peak. The thick solid and dashed lines are the flood frequency distribution curves of the two periods, respectively. The thin solid and dashed lines are distribution curves of the causative precipitations corresponding to floods for the two periods, respectively. Q_1 denotes the flood size of the baseline period at the frequency i, Q_2 denotes the flood size of the change period at the same frequency i. Q_3 denotes the flood size of the baseline period as the same precipitation condition as Q_2 at the frequency j. P_1 is the precipitation corresponding to the flood size Q_1, and P_2 is the causative precipitation corresponding to the flood size Q_2.

based on the flood frequency curves and the causative precipitation distribution curves in the baseline period and change period. Therefore, the precondition for this method is that the whole study period can be divided into the baseline period and change period. Then, the causative precipitation distribution curves of the two periods were built corresponding to flood series. The ordinate value of any point on the precipitation distribution curve presents the amount of a causative precipitation sample, while the abscissa value presents the corresponding sample's flood frequency.

From Figure 2, we can see that flood size change ΔQ due to precipitation change and urbanization from the baseline period to the change period at the flood frequency i is

$$\Delta Q = Q_2 - Q_1 \tag{3}$$

where Q_1 denotes the flood size of the baseline period at the certain frequency, and Q_2 denotes the flood size of the change period at the same frequency.

The flood size change caused by urbanization ΔQ_{urban} is the difference between Q_2 and Q_3, which can be written as follows:

$$\Delta Q_{urban} = Q_2 - Q_3 \tag{4}$$

where Q_3 denotes the flood size of the baseline period with the same causative precipitation of Q_2.

Then, the precipitation-induced flood size change ΔQ_{pre} can be expressed as

$$\Delta Q_{pre} = \Delta Q - Q_{urban} \tag{5}$$

Detailed notations of Equations (3)–(5) are also defined in the caption of Figure 2.

RESULTS

Previous studies found that the annual runoff coefficient of Qinhuai River abruptly changed around 2002 and 2003, and the change in annual runoff coefficient after 2002 in the area was attributed mainly to land-use/land cover changes predominated by urbanization (Du *et al.* 2012; Hao *et al.* 2015; Bian *et al.* 2017). Therefore, the long-term flood series under the changing environment condition were divided into the baseline period (1986–2001) and urbanization period (2002–2013). The F-test (Mckerchar & Henderson 2003), for the equality of the variances, was performed to test the variance of each of the flood series in the baseline period and urbanization period. In addition, it was found that the results of the F-test are insignificant, which indicated that the variance of each of the flood series in two periods is homogeneous. Then, the statistical methods and flood frequency estimation and attribution analysis were performed for both periods to analyze the flood changes, and the results are discussed in the following sections.

Changes in characteristics of flood series from the baseline period to the urbanization period

Some statistics of all flood series such as mean, standard deviation, coefficient of variation and coefficient of skewness were calculated and shown in Table 1, and threshold values of POT series and lowest values of AMS were also

Table 1 | The mean, standard deviation, coefficient of variation and coefficient of skewness of eight flood series in the baseline period (1986–2001) and the urbanization period (2002–2013)

	AMS		POT1		POT2		POT3	
Flood series	**Baseline period**	**Urbanization period**	**Baseline period**	**Urbanization period**	**Baseline period**	**Urbanization period**	**Baseline period**	**Urbanization period**
Mean (m^3/s)	487.7	744.3	706.0	790.0	491.3	588.4	374.6	464.0
Standard deviation (m^3/s)	337.2	245.4	271.2	185.0	290.5	245.5	289.2	267.9
Coefficient of variation	0.69	0.33	0.38	0.23	0.59	0.42	0.77	0.58
Coefficient of skewness	0.84	0.10	0.90	1.20	1.32	0.89	1.39	0.92

identified and shown in Figure 3. It can be seen from Table 1 that the mean values of flood series in the urbanization period are larger than those in the baseline period, while the standard deviation, coefficient of variation and coefficient of skewness of flood series are both higher during the baseline period than those during the urbanization period. It can also be seen from Figure 3 that the threshold values are larger in the urbanization period than those in the baseline period for all POT flood series. In addition, the lowest flood size is larger in the urbanization period than that in the baseline period for AMS series.

The larger threshold and mean values during the urbanization period indicate that the effects of urbanization and precipitation change led to the increase in the flood magnitudes after the year 2002. The decrease in the standard deviation, coefficient of variation and coefficient of skewness of flood series implies the decrease in the differences in flood magnitudes of each flood series during the urbanization period, which is consistent with the results that magnitude of a small flood increased greater than a big flood in the urbanization period. Another possible explanation will be given in the next section.

Trends of flood series from the baseline period to the urbanization period

Four flood series of AMS, POT1, POT2 and POT3 during the baseline period and additional four series during the urbanization period were constructed, respectively, from the daily streamflow of the two periods. The flood series and trends are shown in Figure 3.

The autocorrelation was checked first for all flood series, and no significant autocorrelation data series were found at the 5% significance level. The results of the Sen-MK test for gradual trends of AMS series showed no significant increasing trends at the significance level of 0.1 ($Z < 1.65$ and $\beta > 0$). Moreover, for the whole period, the POT1 and POT2 series show insignificant trends at the significance level of 0.1 ($Z < 1.65$ and $\beta > 0$), but the POT3 series manifested a significant increasing trend at the

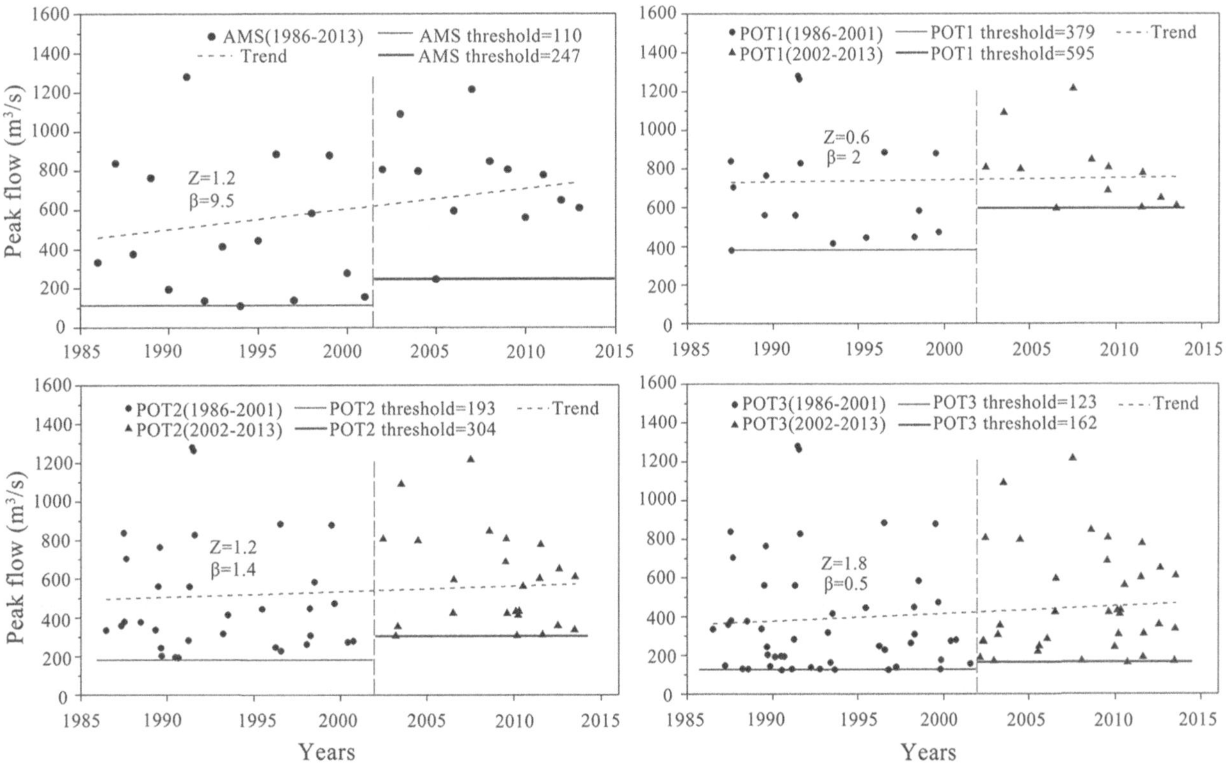

Figure 3 | Flood peaks of POT1, POT2 and POT3 series during the baseline and urbanization periods. Dashed lines show the trends of all series. Thin solid lines show the thresholds in POT1, POT2 and POT3 series during the baseline period. Thick solid lines show the thresholds in POT1, POT2 and POT3 series during the urbanization period.

significance level of 0.1 ($Z > 1.65$ and $\beta > 0$), suggesting that the magnitudes of small floods increased in the urbanization period. To gain more insight into changes in flood frequency, we defined two new times series based on the POT1–3 series. POT1 can be referred as the large flood series, flood sizes between the POT2 and POT1 thresholds can be referred as medium floods, and flood sizes between the POT3 and POT2 thresholds can be referred as small floods. The medium flood series showed a significant increasing trend at the significance level of 0.05 ($Z > 1.96$ and $\beta > 0$), while the small flood series showed a significant increasing trend at the significance level of 0.01 ($Z > 2.56$ and $\beta > 0$) (Figure 4). This indicates that the smaller the magnitude of flood, the larger the increase in the magnitude during the whole period.

Changes in the return period of flood series from the baseline period to the urbanization period

In the previous sections, three different POT series, i.e. POT1, POT2 and POT3 have been selected and analyzed to identify the changes of large, medium and small peak flows under precipitation change and urbanization, respectively. In this section and the following section, in order to quantify and attribute the changes in the return period or the frequency of exceedance of flood series from the baseline period to the urbanization period, two extreme runoff series were investigated, i.e. the AMS series and the standard POT series. The standard POT series are selected based on the optimal threshold values which are determined by the tradeoff of three criteria, i.e. the mean number of over-threshold events, mean exceedance above threshold and dispersion index (Mediero *et al.* 2014). In this study, the optimal threshold values of POT series before and after 2003 were selected to be 139 and 243 m^3/s, respectively, corresponding to an average of 2.3 flood events per year.

The AMS and POT series were fitted to the P-III and GP distributions, respectively, the results are shown in Figure 5. It can be seen from Figure 5 that flood magnitudes for both AMS and POT series are always higher at the same return period during the urbanization period than those during the baseline period.

The root mean square error (RMSE) and probability plot correlation coefficient (PPCC) (Heo *et al.* 2008) were used to evaluate the goodness of fit for AMS and POT series, and the results are shown in Table 2. It can be found that the values of PPCC are more than 0.9 for each flood series in the baseline period and urbanization period, but the values of RMSE for POT series are less than those for AMS series, which indicates that the goodness of fit of GP distribution for POT series is better than P-III distribution for AMS series in both baseline and urbanization periods.

The frequencies of POT and AMS series cannot be directly compared, while they must be converted to the return period by the computational formula raised by Rosbjerg (1985) for the purpose of comparison. Table 3 shows the changes in the return period for the same flood size from the baseline to the urbanization period for both AMS and POT series. The return period decreases for the same flood magnitude from the baseline to the urbanization period in both AMS and POT series. The relative decrease in

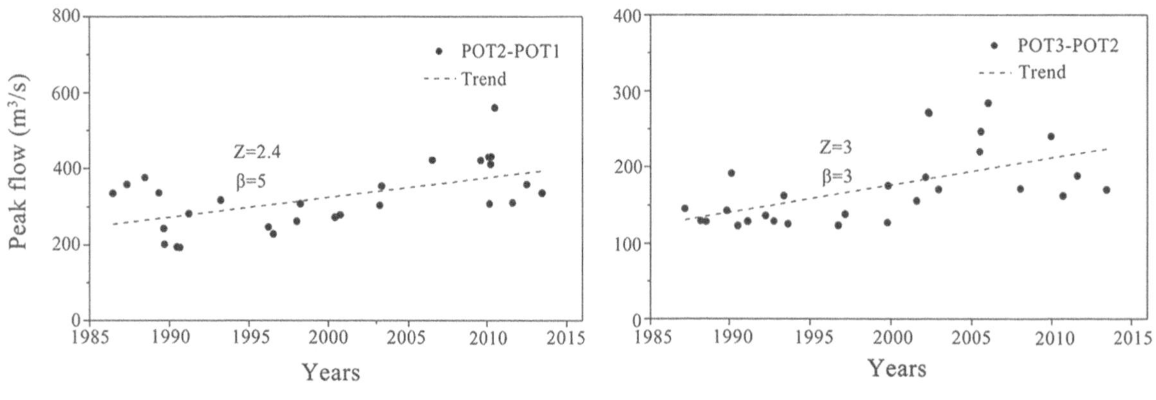

Figure 4 | Flood peaks of medium (the left) and small (the right) flood events derived from two periods, respectively. Dashed lines show the trends of both series.

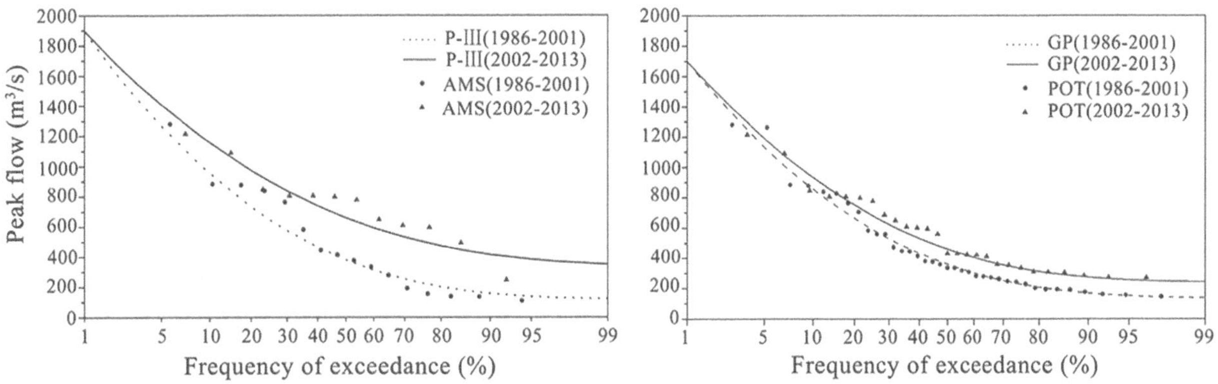

Figure 5 | The frequency distributions of AMS and POT series during the two periods. The dashed lines are fitted frequency distribution curves of POT flood series during the baseline period. The solid lines are fitted frequency distribution curves of POT flood series during the urbanization period.

Table 2 | The test of goodness of fit for AMS and POT series in the baseline period (1986–2001) and the urbanization period (2002–2013)

	AMS		**POT**	
Indicators	**Baseline period**	**Urbanization period**	**Baseline period**	**Urbanization period**
RMSE (m^3/s)	76.7	87.2	40.8	47.4
PPCC (-)	0.98	0.94	0.99	0.98

Table 3 | Changes in return period with same flood size for AMS and POT series

	AMS		**POT**	
Peak flow (m^3/s)	**Baseline period (yr)**	**Urbanization period (yr)**	**Baseline period (yr)**	**Urbanization period (yr)**
400	2.1	1.1	1.0	0.7
600	3.5	1.7	1.8	1.3
800	5.9	3.0	3.3	2.5
1,000	10.0	5.4	5.9	4.7
1,200	16.8	10.2	10.6	8.9
1,400	28.1	19.4	18.9	16.7
1,600	47.2	37.5	33.4	31.1

the return period decreased with increasing flood magnitude except for a peak flow of 400 m^3/s in AMS series, indicating that the same flood event will occur more frequently during the urbanization period. In addition, the smaller the flood event, the larger the return period decreases.

Table 4 shows the changes in magnitude with the same return period from the baseline to the urbanization period in AMS and POT series. It can be seen that the flood magnitudes increase for the same return period from baseline to urbanization periods both in AMS and POT series, suggesting that the flood event with larger peak flow will occur during the urbanization period with the same return period of the baseline period. In addition, the relative increments in magnitude increase with the decreasing return period (i.e. with the decrease of flood magnitude). This is why the coefficient of variation and coefficient of skewness of flood series decrease during the urbanization period.

By comparing the results of the frequency analysis for AMS and POT series, we can obtain the following findings: (1) with the same return period, the flood size in AMS series was underestimated compared with those in POT series in both baseline and urbanization periods, as some small floods in dry years are included and several large floods in wet years are usually overlooked in AMS series; (2) the estimated changes in the return period and magnitude of AMS series are apparently larger than those of POT series; (3) due to some small floods in dry years included in the baseline period of AMS series, the flood size with the short return period is much smaller than those in the urbanization period, leading to a remarkable increase in estimated flood size between the two periods. Hence, it should be noted that the results obtained from AMS series could not be sufficiently reasonable because of the smaller sample size of AMS series than that of POT series and the drawback

Table 4 | Changes in flood size with the same return period for AMS and POT series

	AMS			POT		
Return period (yr)	**Baseline period (m^3/s)**	**Urbanization period (m^3/s)**	**Increased by (%)**	**Baseline period (m^3/s)**	**Urbanization period (m^3/s)**	**Increased by (%)**
30	1,426.1	1,533.4	7.5	1,564.8	1,580.7	1.0
20	1,268.8	1,409.5	11.1	1,422.6	1,453.5	2.2
10	1,000.8	1,194.9	19.4	1,182.1	1,235.5	4.5
5	733.3	973.7	32.8	945.0	1,016.8	7.6
2	380.9	661.0	73.5	636.4	726.5	14.1
1	119.2	338.4	183.9	406.8	506.0	24.4

of the sampling strategy as detailed earlier in Selection of flood series.

Evaluation of causative precipitation and urbanization impacts on changes in flood size

The attribution analysis proposed in this study needs causative precipitation corresponding to each flood event. The correlation analysis method was used to find best relationship between the flood size and the accumulated precipitation of one-day, two-day, until seven-day at and before the date of the flood event. The six-day accumulated precipitation series was found to have highest correlation ($r = 0.86$, significant at level of 0.05) with the flood sizes of the POT series. Therefore, the six-day accumulated precipitation was taken as the causative precipitation corresponding for each flood event.

Based on flood frequency curves of the POT series during the baseline and urbanization periods, the two causative precipitation distribution curves of the two periods were built with PDFs of GP distribution, as the causative precipitation series corresponds to POT flood series, and those curves were fitted by visualization evaluation because of short time series (Figure 6).

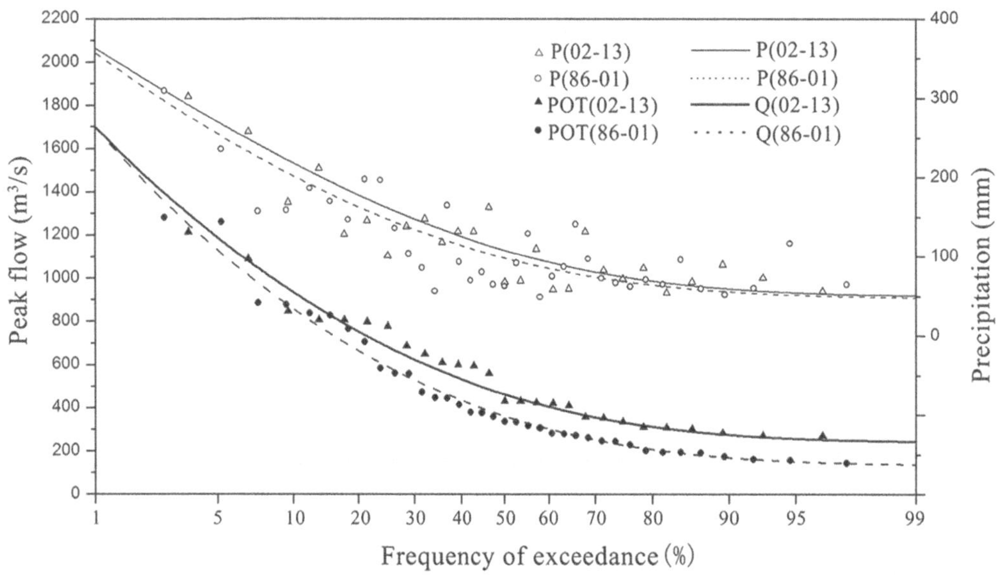

Figure 6 | The separation of the contributions of precipitation change and urbanization to flood sizes. The frequency distributions of flood magnitudes with the thick solid line for the urbanization period (2002–2013) and the thick dashed line for the baseline period (1986–2001). The distribution curves of causative precipitations with the thin solid line for the urbanization period and the thin dashed line for the baseline period.

The individual contribution of urbanization and causative precipitation change was calculated at different return periods based on the four curves by the method described under Attribution analysis, and the results are shown in Table 5. It can be seen that the effects of both causative precipitation change and urbanization are larger on normal floods than those on large floods. However, they exert different degrees of influences on different flood sizes. For example, for the heavy flood with the return period of 30 years, the causative precipitation change and urbanization contributed by 98.8% and 1.2%, to the 1% increase in flood size. However, for small flood with the return period of one year, causative precipitation change and urbanization contributed by 42.4% and 57.6%, to the 24.4% increase of flood size. Therefore, we can conclude that the impacts of urbanization on flood size are larger for small floods and diminish as the flood magnitude increases (i.e. flood frequency decreases).

These results are consistent with findings reported in the literature (Braud *et al.* 2013; Kaspersen *et al.* 2015). They found that the effects of urbanization on floods are proportionally greater for high-frequency events. Sillanpää & Koivusalo (2015) also confirmed this finding by dividing the event data into groups according to the event size and identifying statistically significant differences between the periods for the group of minor rainfall events and no significant differences between the two periods for major events. This difference can be attributed to the fact that the high intensity and large volume of rainfall exceed infiltration capacity of pervious areas in extreme events, causing the natural surface to behave like an impervious surface. Therefore, the relative influence of the urban areas will decrease in large flood events. A reduction in the natural infiltration due to the growth of impervious areas can be expected to lead to an increase in the volume of storm runoff for the medium and small floods.

DISCUSSION

Effects of two different sampling methods on flood change evaluation

The AMS and peak over threshold (POT) sampling are usually used for the flood frequency analysis. The POT sampling has the following advantages over the AMS sampling (Bezak *et al.* 2014; Mediero *et al.* 2014): (1) it provides an approach to control the number of selected floods by choosing an appropriate threshold and thereby use a larger dataset of floods; (2) it ignores useless small floods in dry years and thereby concentrates only on the higher maximum values, which contain most of the information about flood processes. Even though the AMS sampling has many drawbacks compared to the POT sampling, it still remains the most popular approach to analyze flood changes (Bai *et al.* 2015; Yan *et al.* 2017). The AMS sampling is useful and effective in flood trend detection and frequency analysis as long as the time period is long enough.

In this study, the frequency curves derived by P-III distribution based on AMS series are not appropriate to analyze

Table 5 | The attributions of causative precipitation change and urbanization to flood sizes

Return period (yr)	Q_1 (m^3/s)	Q_2 (m^3/s)	P_1 (mm)	P_2 (mm)	Q_3 (m^3/s)	Total (%)	Contribution from urbanization (%)	Contribution from causative precipitation (%)
30	1,564.8	1,580.7	341.8	344.0	1,580.5	1.0	1.2	98.8
20	1,422.6	1,453.5	312.3	321.4	1,453.0	2.2	1.8	98.2
10	1,182.1	1,235.5	267.4	280.3	1,233.7	4.5	3.5	96.5
5	945.0	1,016.8	221.3	235.9	1,006.4	7.6	14.5	85.5
2	636.4	726.5	158.1	171.8	694.0	14.1	36.1	63.9
1	406.8	506.0	108.6	118.8	448.8	24.4	57.6	42.4
0.5	180.3	284.7	57.7	61.6	196.5	57.9	84.5	15.5

Q_1 and Q_2 are the flood sizes of the baseline and urbanization periods with the same frequency, respectively. P_1 and P_2 are the corresponding causative precipitation amount to Q_1 and Q_2, respectively. Q_3 is the flood size during the baseline period which would be contributed by P_2.

the flood changes for a number of reasons (more details have already been given in the above results): the high flood quantiles are underestimated (Table 4), The changes in the return period and magnitudes are overestimated (Tables 3 and 4), and an incredible increase in estimated flood sizes for the short return period (Table 4). All these issues can be attributed to the small datasets (16 and 12 years for the baseline and urbanization periods, respectively) and the sampling strategy (some small floods are included and other large floods are missing) for the AMS. On the contrary, the POT sampling concentrates mainly on the higher maximum values, and it obtains larger number of flood events and selects all large floods. The results based on POT series are reasonable even though the time period is short. Our study shows that the POT sampling is preferred over the AMS sampling when applied to a short time period.

Attribution of trends in flood time series

Studies on flood trend attribution have been of considerable interest because of flood risk in the urbanized area has an increasing trend due to the effects of precipitation change and urbanization. Theoretically, the drivers that may have impacted the flood behavior should be quantitatively investigated. However, the fact that many factors affect flood behavior complicates attribution analysis, and the current state of flood trend attribution is poor as pointed out by Merz *et al.* (2012).

In this study, the attribution analysis method based on the frequency distribution of flood size and causative precipitation distribution curve was proposed and used to separate the contributions of precipitation change and urbanization to flood changes between two periods. Theoretically, the method can quantify the relative contributions of the two factors with a certain degree of accuracy. However, when the method was applied to the case study in this paper, it was found that the causative precipitation distribution curve was more scatter than the flood frequency distribution, which indicates that several other factors might influence the flood sizes, including the duration and variations in intensity during one event as well as antecedent soil moisture condition, changes in land surface patterns and flood control measures. Therefore, the attribution results by this method can provide approximate ratios of the causative precipitation change and urbanization effects on the flood sizes based on the fitted causative precipitation curves. Anyway, the results of the attribution analysis in this study provide evidence that the urbanization had a significant effect on flood peaks for smaller floods, while larger floods are mainly affected by rainfall amounts, which suggests the usefulness and applicability of the proposed method.

It should be noted that one frequency distribution was assigned to each of the flood and causative precipitation series in this paper. We believe that applying different distributions might provide different results and is a topic for further study.

CONCLUSIONS

This study presented a procedure combining statistical methods, flood frequency analysis and attribution analysis to examine the response of floods to urbanization and precipitation change in the Qinhuai River Basin, an urbanized basin located in southeast China, over the period from 1986 to 2013. We analyzed AMS, POT1, POT2 and POT3 series, where the three latter series were created by selecting independent peaks over three different thresholds resulting in 1, 2 and 3, flood events per year, respectively. In addition, we considered floods above the POT1 threshold as large floods, floods between the POT2 and POT1 thresholds as medium floods and flood sizes between the POT3 and POT2 thresholds as small floods. All flood series were constructed from daily streamflow of the baseline period and urbanization period.

The following conclusions can be drawn from this study:

1. The AMS, POT1 and POT2 series showed no significant increasing trends at the significance level of 0.1, and the POT3, medium and small flood series showed significant positive trends at the significance level of 0.1, 0.05 and 0.01, respectively.
2. The mean and threshold values of AMS and different POT series in the urbanization period (1986–2001) were larger than those in the baseline period (2002–2013), while standard deviation, coefficient of variation and coefficient of skewness of AMS and different POT

series were higher in the baseline period than those in the urbanization period.

3. The flood magnitude was higher during the urbanization period than that during the baseline period at the same flood frequency (or return period) of exceedance. The changes in magnitudes of small floods were relatively larger than those of large floods from the baseline period to the urbanization period.
4. The precipitation changes and urbanization are the main driving factors leading floods change in the Qinhuai River Basin. The contributions of urbanization on floods appeared to amplify with decreasing flood size, while the effects of precipitation diminish.

The procedure proposed in this study has been demonstrated to be useful for the trend and attribution analysis of flood series. The findings of this study can advance our understanding of interactions between flood behavior and the drivers, thereby improving flood management in urbanized basins.

ACKNOWLEDGEMENTS

This work was supported by the National Natural Science Foundation of China (Grant No. 41771029 and N. 41371044), the Belt and Road Special Foundation of the State Key Laboratory of Hydrology-Water Resources and Hydraulic Engineering (2018490311) and the Research Council of Norway (FRINATEK Project 274310), which are greatly appreciated.

REFERENCES

Ahmad, I., Fawad, M. & Mahmood, I. 2015 At-site flood frequency analysis of annual maximum stream flows in Pakistan using robust estimation methods. *Polish Journal Environmental Studies* **24** (6), 2345–2353.

Aich, V., Liersch, S., Vetter, T., Andersson, J. C. M., Müller, E. N. & Hattermann, F. F. 2015 Climate or land use? – Attribution of changes in river flooding in the Sahel Zone. *Water* **7** (6), 2796–2820.

Bai, P., Liu, X., Liang, K. & Liu, C. 2015 Investigation of changes in the annual maximum flood in the Yellow River basin, China. *Quaternary International* **392**, 168–177.

Balistrocchi, M. & Bacchi, B. 2017 Derivation of flood frequency curves through a bivariate rainfall distribution based on copula functions: application to an urban catchment in northern Italy's climate. *Hydrology Research* **48** (3), 749–762.

Bezak, N., Brilly, M. & Šraj, M. 2014 Comparison between the peaks-over-threshold method and the annual maximum method for flood frequency analysis. *Hydrological Science Journal* **59** (5), 959–977.

Bian, G. D., Du, J. K., Song, M. M., Xu, Y. P., Xie, S. P., Zheng, W. L. & Xu, C.-Y. 2017 A procedure for quantifying runoff response to spatial and temporal changes of impervious surface in Qinhuai River basin of southeastern China. *Catena* **157**, 268–278.

Blöschl, G., Ardoin-Bardin, S., Bonell, M., Dorninger, M., Goodrich, D., Gutknecht, D., Merz, B., Shand, P. & Szolgay, J. 2007 At what scales do climate variability and land cover change impact on flooding and low flows? *Hydrological Process* **21**, 1241–1247.

Braud, I., Breil, P., Thollet, F., Lagouy, M., Branger, F., Jacqueminet, C., Kermadi, S. & Michel, K. 2013 Evidence of the impact of urbanization on the hydrological regime of a medium-sized periurban catchment in France. *Journal of Hydrology* **485** (4), 5–23.

Chebana, F., Ouarda, T. B. M. J. & Duong, T. C. 2013 Testing for multivariate trends in hydrological frequency analysis. *Journal of Hydrology* **486**, 519–530.

Chen, A. L. & Du, J. K. 2014 Simulation and forecast of land cover pattern in Qinhuai River Basin based on the CA-Markov model. *Remote Sensing for Land & Resources* **26**, 184–189. (In Chinese).

Chen, Y. F., Sha, Z. G., Chen, J. C. & Chen, M. 2001 Study on L-moment estimation method for P-III distribution with historical flood. *Journal of Hohai University* **29** (4), 76–80 (in Chinese).

Chen, X., Tian, C., Meng, X., Xu, Q., Gui, G., Zhang, Q. & Xiang, L. 2015 Analyzing the effect of urbanization on flood characteristics at catchment levels. *Proceedings of the International Association of Hydrological Sciences* **370**, 33–38.

Darabi, H., Haghighi, A. T., Mohamadi, M. A., Rashidpour, M., Ziegler, A. D., Hekmatzadeh, A. A. & Kløve, B. 2019 Urban flood risk mapping using data-driven geospatial techniques for a flood-prone case area in Iran. *Hydrology Research*, in press. https://doi.org/10.2166/nh.2019.090.

Dehghanian, N., Nadoushani, S. S. M., Saghafian, B. & Damavandi, M. R. 2020 Evaluation of coupled ANN-GA model to prioritize flood source areas in ungauged watersheds. *Hydrology Research* in press. https://doi.org/10.2166/nh.2020.141.

Delgado, J. M., Apel, H. & Merz, B. 2010 Flood trends and variability in the Mekong River. *Hydrology and Earth System Science* **14** (3), 407–418.

Du, J. K., Li, Q., Rui, H. Y., Zuo, T., Zheng, D., Xu, Y. P. & Xu, C.-Y. 2012 Assessing the effects of urbanization on annual runoff and flood events using an integrated hydrological modeling system for Qinhuai River basin, China. *Journal of Hydrology* **464** (5), 127–139.

Du, S. Q., Rompaey, A. V., Shi, P. J. & Wang, J. 2015 A dual effect of urban expansion on flood risk in the Pearl River Delta (China) revealed by land-use scenarios and direct runoff simulation. *Natural Hazards* **77** (1), 111–128.

Fateh, C., Taha, B. M. J. O. & Thuy, C. D. 2013 Testing for multivariate trends in hydrologic frequency analysis. *Journal of Hydrology* **486** (8), 519–530.

Guru, N. & Jha, R. 2015 Flood frequency analysis of Tel basin of Mahanadi river system, India using annual maximum and POT flood data. *Aquatic Procedia* **4**, 427–434.

Hall, J., Arheimer, B., Borga, M., Brázdil, R., Claps, P., Kiss, A., Kjeldsen, T. R., Kriauciuniene, J., Kundzewicz, Z. W., Lang, M., Llasat, M. C., Macdonald, N., Mcintyre, N., Mediero, L., Merz, B., Merz, R., Molnar, P., Montanari, A., Neuhold, C., Parajka, J., Perdigao, R. A. P., Plavcova, L., Rogger, M., Salinas, J. L., Sauquet, E., Schär, C., Szolgay, J., Viglione, A. & Blöschl, G. 2014 Understanding flood regime changes in Europe: a state-of-the-art assessment. *Hydrology and Earth System Science* **18** (7), 2735–2772.

Hannaford, J., Buys, G., Stahl, K. & Talaksen, L. M. 2013 The influence of decadal-scale variability on trends in European streamflow records. *Hydrology and Earth System Science* **17** (7), 2717–2733.

Hao, L., Sun, G., Liu, Y., Wan, J., Qin, M., Qian, H., Liu, C., Zheng, J., John, R., Fan, P. & Chen, J. 2015 Urbanization dramatically altered the water balances of a paddy field-dominated basin in southern China. *Hydrology and Earth System Science* **19**, 3319–3331.

He, B., Huang, X., Ma, M., Chang, Q., Tu, Y., Li, Q., Zhang, K. & Hong, Y. 2017 Analysis of flash flood disaster characteristics in China from 2011 to 2015. *Natural Hazards* **90** (3), 1–14.

Heo, J. H., Kho, Y. W., Shin, H., Kim, S. & Kim, T. 2008 Regression equations of probability plot correlation coefficient test statistics from several probability distributions. *Journal of Hydrology* **355**, 1–15.

Hou, J. M., Li, B. Y., Tong, Y., Ma, L. P., Ball, J., Luo, H., Liang, Q. H. & Xia, J. Q. 2020 Cause analysis for a new type of devastating flash flood. *Hydrology Research* in press. https://doi.org/10.2166/nh.2019.091.

Jiang, C., Xiong, L. H., Yan, L., Dong, J. F. & Xu, C.-Y. 2019 Multivariate hydrologic design methods under nonstationary conditions and application to engineering practice. *Hydrology and Earth System Sciences* **23**, 1683–1704.

Jothityangkoon, C., Hirunteeyakul, C., Boonrawd, K. & Sivapalan, M. 2013 Assessing the impact of climate and land use changes on extreme floods in a large tropical catchment. *Journal of Hydrology* **490** (20), 88–105.

Kaspersen, P. S., Ravn, N. H., Arnbjerg-Nielsen, K., Madsen, H. & Drews, M. 2015 Influence of urban land cover changes and climate change for the exposure of European cities to flooding during high-intensity precipitation. *Proceedings of the IAHS* **370**, 21–27.

Kendall, M. G. 1975 *Rank Correlation Methods*. Griffin, London, UK.

Kundzewicz, Z. W., Kanae, S., Seneviratne, S. I., Handmer, J., Nicholls, N., Peduzzi, P., Mechler, R., Bouwer, L. M., Arnell, N. W., Mach, K. J., Muir-Wood, R., Brakenridge, G. R., Kron, W., Benito, G., Honda, Y., Takahashi, K. & Sherstyukov, B. 2013 Flood risk and climate change: global and regional perspectives. *Hydrological Sciences Journal* **59** (1), 1–28.

Kundzewicz, Z. W., Pińskwar, I. & Brakenridge, G. R. 2018 Changes in river flood hazard in Europe: a review. *Hydrology Research* **49** (2), 294–302.

Ledingham, J., Archer, D., Lewis, E., Fowler, H. & Kilsby, C. 2019 Contrasting seasonality of storm rainfall and flood runoff in the UK and some implications for rainfall-runoff methods of flood estimation. *Hydrology Research* **50** (5), 1309–1323.

Li, Y. Y., Luo, L. F., Wang, Y. M., Guo, A. J. & Ma, F. 2019 Spatiotemporal impacts of land use land cover changes on hydrology from the mechanism perspective using SWAT model with time-varying parameters. *Hydrology Research* **50** (1), 244–261. doi: 10.2166/nh.2018.006.

Li, H., Gao, H. K., Zhou, Y. L., Xu, C.-Y., Ortega, M. R. Z. & Sælthun, N. R. 2020 Usage of SIMWE model to Model Urban Overland Flood: a case study in Oslo. *Hydrology Research* (in press). doi: 10.2166/nh.2020.068.

López, J. & Francés, F. 2013 Non-stationary flood frequency analysis in continental Spanish rivers, using climate and reservoir indices as external covariates. *Hydrology and Earth System Science* **17**, 3189–3203.

Machado, M. J., Botero, B. A., López, J., Francés, F., Díez-Herrero, A. & Benito, G. 2015 Flood frequency analysis of historical flood data under stationary and non-stationary modelling. *Hydrology and Earth System Science* **19** (6), 2561–2576.

Madsen, H., Lawrence, D., Lang, M., Martinkova, M. & Kjeldsen, T. R. 2014 Review of trend analysis and climate change projections of extreme precipitation and floods in Europe. *Journal of Hydrology* **519**, 3634–3650.

Mahmoud, S. H. & Gan, T. Y. 2018 Urbanization and climate change implications in flood risk management: developing an efficient decision support system for flood susceptibility mapping. *Science of the Total Environment* **636**, 152–167.

Mann, H. B. 1945 Nonparametric tests against trend. *Econometrica* **13** (3), 245–259.

Mckerchar, A. I. & Henderson, R. D. 2003 Shifts in flood and low-flow regimes in New Zealand due to interdecadal climate variations. *Hydrological Sciences Journal* **48** (4), 637–654.

Mediero, L., Santillán, D., Garrote, L. & Granados, A. 2014 Detection and attribution of trends in magnitude, frequency and timing of floods in Spain. *Journal of Hydrology* **517**, 1072–1088.

Merz, B., Vorogushyn, S., Uhlemann, S., Delgado, J. & Hundecha, Y. 2012 HESS Opinions "More efforts and scientific rigour are needed to attribute trends in flood time series". *Hydrology and Earth System Science* **16** (5), 1379–1387.

Modarres, R., Sarhadi, A. & Burn, D. H. 2016 Changes of extreme drought and flood events in Iran. *Global and Planetry Change* **144**, 67–81.

Mudersbach, C., Bender, J. & Netzel, F. 2017 An analysis of changes in flood quantiles at the gauge Neu Darchau (Elbe

River) from 1875 to 2013. *Stochastic Environmental Research and Risk Assessment* **31** (1), 145–157.

Muis, S., Güneralp, B., Jongman, B., Aerts, J. & Ward, P. 2015 Flood risk and adaptation strategies in Indonesia: a probabilistic analysis using globally available data. *Science of the Total Environment* **538** (4), 445–457.

Okoli, K., Mazzoleni, M., Breinl, K. & Di Baldassarre, G. 2019 A systematic comparison of statistical and hydrological methods for design flood estimation. https://doi.org/10.2166/nh.2019.188

Prosdocimi, I., Kjeldsen, T. R. & Miller, J. D. 2015 Detection and attribution of urbanization effect on flood extremes using nonstationary flood-frequency models. *Water Resources Research* **51** (6), 4244–4262.

Ragettli, S., Tong, X., Zhang, G., Wang, H., Zhang, P. & Stähli, M. 2020 Climate change impacts on summer flood frequencies in two mountainous catchments in China and Switzerland. *Hydrology Research* in press. https://doi.org/10.2166/nh.2019.118.

Rosbjerg, D. 1985 Estimation in partial duration series with independent and dependent peak values series. *Journal of Hydrology* **76**, 183–195.

Sen, P. K. 1968 Estimates of the regression coefficient based on Kendall's tau. *Journal of American Statistical Association* **63**, 1379–1389.

Sillanpää, N. & Koivusalo, H. 2015 Impacts of urban development on runoff event characteristics and unit hydrographs across warm and cold seasons in high latitudes. *Journal of Hydrology* **521**, 328–340.

Silva, A. T., Portela, M. M. & Naghettini, M. 2012 Nonstationarities in the occurrence rates of flood events in Portuguese watersheds. *Hydrology and Earth System Sciences* **16** (1), 241–254.

Solari, S., Egüen, M., Polo, M. & Losada, M. A. 2017 Peaks over threshold (POT): a methodology for automatic threshold estimation using goodness of fit *p*-value. *Water Resources Research* **53**, 2833–2849.

Šraj, M., Viglione, A., Parajka, J. & Blöschl, G. 2016 The influence of non-stationarity in extreme hydrological events on flood frequency estimation. *Journal of Hydrology and Hydromechanics* **64**, 426–437.

Stevens, A. J., Clarke, D. & Nicholls, R. J. 2016 Trends in reported flooding in the UK: 1884–2013. *Hydrological Sciences Journal* **61**, 50–63.

Venvik, G., Bang-Kittilsen, A. & Boogaard, F. C. 2020 Risk assessment for areas prone to flooding and subsidence – a concept model with case study from Bergen, Western Norway. *Hydrology Research* in press. doi: 10.2166/nh.2019.030.

Viglione, A., Merz, R., Salinas, J. L. & Blöschl, G. 2013 Flood frequency hydrology: 3. A Bayesian analysis. *Water Resources Research* **49** (2), 675–692.

Viglione, A., Merz, B., Dung, N. V., Parajka, J., Nester, T. & Blöschl, G. 2016 Attribution of regional flood changes based on scaling fingerprints. *Water Resources Research* **52**, 5322–5340.

Villarini, G. & Strong, A. 2014 Roles of climate and agricultural practices in discharge changes in an agricultural watershed in Iowa. *Agriculture, Ecosystems & Environment* **188**, 204–211.

Wolski, P., Stone, D., Tadross, M., Wehner, M. & Hewitson, B. 2014 Attribution of floods in the Okavango basin, Southern Africa. *Journal of Hydrology* **511** (7), 350–358.

Yan, L., Xiong, L., Liu, D., Hu, T. & Xu, C.-Y. 2017 Frequency analysis of nonstationary annual maximum flood series using the time-varying two-component mixture distributions. *Hydrological Processes* **31**, 69–89.

Yao, L., Chen, L. & Wei, W. 2015 Assessing the effectiveness of imperviousness on storm-water runoff in micro urban catchments by model simulation. *Hydrological Processes* **30** (12), 1836–1848.

Yin, J., Ye, M., Yin, Z. & Xu, S. 2015 A review of advances in urban flood risk analysis over China. *Stochastic Environmental Research and Risk Assessment* **29** (3), 1063–1070.

Yue, S. & Wang, C. Y. 2002 Regional streamflow trend detection with consideration of both temporal and spatial correlation. *International Journal of Climatology* **22** (8), 933–946.

Zope, P. E., Eldho, T. I. & Jothiprakash, V. 2017 Hydrological impacts of land use–land cover change and detention basins on urban flood hazard: a case study of Poisar River basin, Mumbai, India. *Natural Hazards* **87** (3), 1267–1283.

First received 10 May 2019; accepted in revised form 17 February 2020. Available online 26 March 2020

Usage of SIMWE model to model urban overland flood: a case study in Oslo

Hong Li, Hongkai Gao, Yanlai Zhou, Chong-Yu Xu IWA, Rengifo Z. Ortega M. and Nils Roar Sælthun

ABSTRACT

There has been a surge of interest in the field of urban flooding in recent years. However, current stormwater management models are often too complex to apply on a large scale. To fill this gap, we use a physically based and spatially distributed overland flow model, SIMulated Water Erosion (SIMWE). The SIMWE model requires only rainfall intensity, terrain, infiltration, and surface roughness as input. The SIMWE model has great potential for application in real-time flood forecasting. In this study, we use the SIMWE model at two resolutions (20 m and 500 m) for Oslo, and at a high resolution (1 m) at the Grefsen area, which is approximately 1.5 km^2 in Oslo. The results show that the SIMWE model can generate water depth maps at both coarse and high resolutions. The spatial resolution has strong impacts on the absolute values of water depth and subsequently on the classification of flood risks. The SIMWE model at a higher spatial resolution produces more overland flow and higher estimation of flood risk with low rainfall input, but larger areas of risk with high rainfall input. The Grefsen case study shows that roads act as floodways, where overland flow accumulates and moves fast.

Key words | Nordic cities, Oslo, overland flow, SIMWE, urban flood

Hong Li
Yanlai Zhou
Chong-Yu Xu IWA (corresponding author)
Nils Roar Sælthun
Department of Geosciences,
University of Oslo,
Blindern, 0316 Oslo,
Norway
E-mail: *c.y.xu@geo.uio.no*

Hongkai Gao
Key Laboratory of Geographic Information Science (Ministry of Education of China),
East China Normal University,
Shanghai,
China
and
School of Geographical Sciences,
East China Normal University,
Shanghai,
China

Rengifo Z. Ortega M.
Norwegian Water Resources and Energy Directorate,
Majorstua, 0301 Oslo,
Norway

INTRODUCTION

Humans are 'urban species' nowadays and the urban population is still increasing. Two centuries ago, only 3% of the world's population lived in cities (Florida 2011). However, the number grew to 34% in 1961 and 55% in 2017 (United Nations Population Division 2018). The global urban population is expected to grow approximately 1.84% per year until 2020 (WHO 2015). By the year 2030, more than five billion people (60% of the global population) will live in urban areas (Florida 2011). In Norway, 82% of the population is living in cities and towns, and more than one million are living in the Oslo urban settlement (SSB 2018a). Meanwhile, the population in Oslo is continuously growing and has increased by 1.2% from 2017 to 2018.

Urbanization has strong impacts on the quantity and quality of local runoff as well as erosion and sediment transport. Urbanization is a process of building more roads, houses, and commercial and industrial buildings (USGS 2018). Subsequently, more wastewater is discharged or leaked into local streams. Roads, buildings, and pavements make the surface impermeable and therefore reduce infiltration. Eventually, urban floods have larger peaks and shorter lag time. Additionally, less infiltration caused by impervious areas leads to declining groundwater recharge, which

doi: 10.2166/nh.2020.068

reduces baseflow during dry seasons and increases the risk of geological hazards, e.g., land subsidence (Donaldson 1968; Yin *et al.* 2015).

Urban floods pose a great threat to the security of residents and their properties in inundation areas. In the UK, the estimated expected annual damage caused by flooding in 2002 was approximately one billion British pounds (Hall *et al.* 2005). In the municipality of Oslo, 2,396 incidents of urban flood-related damage were reported for the period from 2008 to 2014, and the insurance claims reached more than 97 million Norwegian kroner (VAV 2016), which only accounts for approximately 70% of damage events (Finsland, W. 2019, personal communication). In addition to the direct and tangible loss, there are numerous types of indirect and intangible loss. This includes, but is not limited to, the losses from the loss of utilities and supply chain disruption, which is significantly more than the direct cost (NOU 2015).

There are three types of urban floods categorized by cause (Maddox 2014), i.e., coastal (surge flood), fluvial (river flood), and pluvial (surface flood). Coastal and fluvial floods occur in the areas near a coast or along rivers, and the floods take place when the water overflows the barriers. However, in most places, urban floods occur as pluvial flooding, which is due to intensive rainfall that exceeds infiltration rates and drainage capacity of sewage networks. Water may even enter the sewage system in one place and then run out somewhere else and result in flooding.

Climate change, mainly changes in magnitude and frequency of rainfall and snowmelt, makes the issue of urban flooding more complicated. For Norway, annual precipitation has increased by 18% since 1900 and the increasing trend will likely continue (Hanssen-Bauer *et al.* 2015). The magnitude and frequency of extreme rainfall will also likely increase in the future (Hanssen-Bauer *et al.* 2015).

Although there is an urgent need to tackle the issue of urban floods, flood risk measurement, modeling, and prediction are still largely inadequate and insufficient, even in developed countries. For the modeling techniques, the MIKE series and Storm Water Management Model (SWMM) are widely used. Both models allow us to estimate water balance components in urban regions and to route flow in pipeline networks. The MIKE series have various models for different purposes, for example, MIKE Hydro for runoff simulation, MIKE Flood for inundation, and MIKE Urban for cities (DHI 2018). Both MIKE Urban and SWMM use a semi-distributed model structure, which means subcatchment is the basic unit for water balance calculation. The drainage network through connecting point to the pipes collects the surface runoff of each subcatchment. However, due to the model limitations, neither model can produce inundation maps nor estimate water depth over a large area. Finally, yet importantly, to set up these two models requires a great deal of manual work.

In this study, we utilize an open-source and physically based spatially distributed overland flow model called SIMulated Water Erosion (SIMWE). The SIMWE model is integrated into a free and open-source GIS platform (GRASS GIS https://grass.osgeo.org/). The SIMWE model can also run from the bash script, which allows us to work on many projects efficiently. The implementation in GRASS GIS also allows us to update input data or model parameters easily. Finally, yet importantly, the SIMWE model inherits high visualizing skills from GRASS GIS, which is important to identify risk areas and design flow paths. The SIMWE model has been used to assess flash floods in the Malá Svinka Basin, Slovakia (Hofierka & Knutová 2015). The results show the gradual changes in water depth across the basin and confirm the excellent robustness and flexibility of the SIMWE model.

The main purpose of this paper is to examine the applicability of the SIMWE model in simulating urban overland flow. We use this model to produce maps of the inundation area and estimate water depth for the whole of Oslo and for a small catchment at a high resolution. Additionally, we test the model sensitivity to spatial resolution and precipitation input. To our knowledge, this is the first time that the SIMWE model has been used for urban flood simulation, and the first time the urban flood simulation for the whole of Oslo municipality has been undertaken. The results are useful for urban flood mitigation and city planning.

METHODS

SIMWE model

The SIMWE model is a physically based spatially distributed model. The input is net precipitation (rainfall–filtration) and terrain and surface roughness. The output is water depth, flow velocity, and discharge. In our simulation for urban overland flow, we neglect the sewage drainage system. The stormwater drainage system can reduce the overland flow to some extent. However, drainage pipes become full for extreme rainfall events, for example, 50-year rainfall.

The fundamental theory is the Saint Venant equation for continuity of flow. Urban flood usually occurs as shallow overland flow. For this type of flow, spatial variation in velocity with respect to depth can be neglected (Mitas & Mitasova 1998; Hofierka & Knutová 2015). The flow process can be approximated by the bivariate form of the Saint Venant equation for continuity of flow (Equation (1)):

$$\frac{\partial h(\boldsymbol{r}, t)}{\partial t} = i_e(\boldsymbol{r}, t) - \nabla \cdot \boldsymbol{q}(\boldsymbol{r}, t) \tag{1}$$

where, $\boldsymbol{r} = (x, y)$ is the position, t is the time, $h(\boldsymbol{r}, t)$ is the depth of overland flow, $i_e(\boldsymbol{r}, t)$ is the rainfall excess = (rainfall − infiltration). $\boldsymbol{q}(\boldsymbol{r}, t)$ is the water flow per unit width.

For a shallow water flow, with the hydraulic radius approximated by the normal flow depth $h(\boldsymbol{r}, t)$, the unit discharge is given by:

$$\boldsymbol{q}(\boldsymbol{r}, t) = \boldsymbol{v}(\boldsymbol{r}, t)h(\boldsymbol{r}, t) \tag{2}$$

where $\boldsymbol{v}(\boldsymbol{r}, t)$ is the flow velocity. Then, $\boldsymbol{v}(\boldsymbol{r}, t)$ can be derived from Manning's formula:

$$v(\boldsymbol{r}, t) = \frac{h(\boldsymbol{r}, t)^{\frac{2}{3}} |\boldsymbol{s}_f(\boldsymbol{r}, t)|^{\frac{1}{2}}}{n(\boldsymbol{r})} \tag{3}$$

where $n(\boldsymbol{r})$ is the dimensionless Manning's coefficient. $\boldsymbol{s}_f(\boldsymbol{r}, \boldsymbol{t})$ is the negative gradient of the overland flow surface (hydraulic slope direction):

$$\boldsymbol{s}_f(\boldsymbol{r}, t) = \boldsymbol{s}(\boldsymbol{r}) - \nabla h(\boldsymbol{r}, t) \tag{4}$$

where, $\boldsymbol{s}(\boldsymbol{r}) = -\nabla z(\boldsymbol{r})$ is the negative elevation gradient, $z(\boldsymbol{r})$ is the elevation.

Classification flood risk levels of urban flood

At present, there is neither an urban flood warning system in Norway nor the criteria for critical urban flood sizes. In this study, we use the height of four different rainboots to classify the urban flood risk levels (Figure 1 and Table 1).

STUDY AREA AND DATA

Study area

Oslo is the capital city of Norway as well as the economic and governmental center. It is one of the northernmost capitals in the world. The city constitutes both a city and a municipality. The urban area extends beyond the boundaries of the municipality into the surrounding municipality, Akershus.

As of 1 January 2018, the municipality of Oslo had a population of 673,469 and the whole population for the urban area reached 1,099,346 (SSB 2018b). The population was increasing at record rates during the early 2000s, making it the fastest-growing major city in Europe at the time. This growth stems from international immigration,

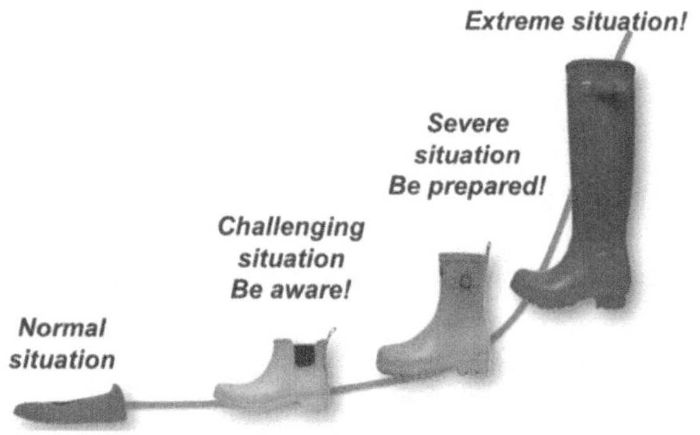

Figure 1 | Popular representation of the flood risk levels, symbolized by boots (Krøgli *et al.* 2018).

Table 1 | Classification of urban floods based on water depth

Risk level	Low/Green	Challenging/ Yellow	Severe/Orange	Extreme/ Red
Water depth	9.5–15.5 cm	15.5–23.5 cm	23.5–43.4 cm	>43.5 cm

high birth rates, and intra-national migration (Wikipedia 2018).

The city is surrounded by a hilly forest and the Oslo Fjord, and they are connected by a number of waterways (Figure 2). The waterways have been subject to a new radical strategy, which has completely reversed the previous approach of enclosing these channels. They are now being actively re-opened in order to make them accessible for people, to efficiently manage stormwater, and to facilitate development and restorations of habitat.

Data

To run the SIMWE model, rainfall, terrain, infiltration, and Manning's n are required inputs. These data can be spatially distributed or uniform. In this study, we use a spatially uniform rainfall due to the unavailability of spatially distributed rainfall. Other data are spatially distributed.

The terrain data are a hybrid product of two terrain data sets. For the Oslo city area, the terrain map is extracted from LiDAR data operated in the summer of 2014 (BLOM 2014). For the forest and hilly area surrounding the city area, the terrain is interpolated from elevation contours from the Norwegian Mapping Authority. The final terrain product with a spatial resolution of 0.5 m includes the surface area outside the city and the surface area in the city as buildings and roads.

The infiltration map is also a hybrid product of descriptive infiltration capability and impermeable surface (Figure 3). The infiltration capacity is produced from a soil product from the Geological Survey of Norway (NGU 2018). The infiltration capacity is not quantitatively described, but classified into four categories, i.e., none, little, middle, and good. Subsequently, we transform the infiltrate capacity to infiltrate rate according to Table 2. The impermeable surface is from the Sentinel satellite image (Stange 2017) and the Norwegian common (fkb) map database (Norwegian Mapping

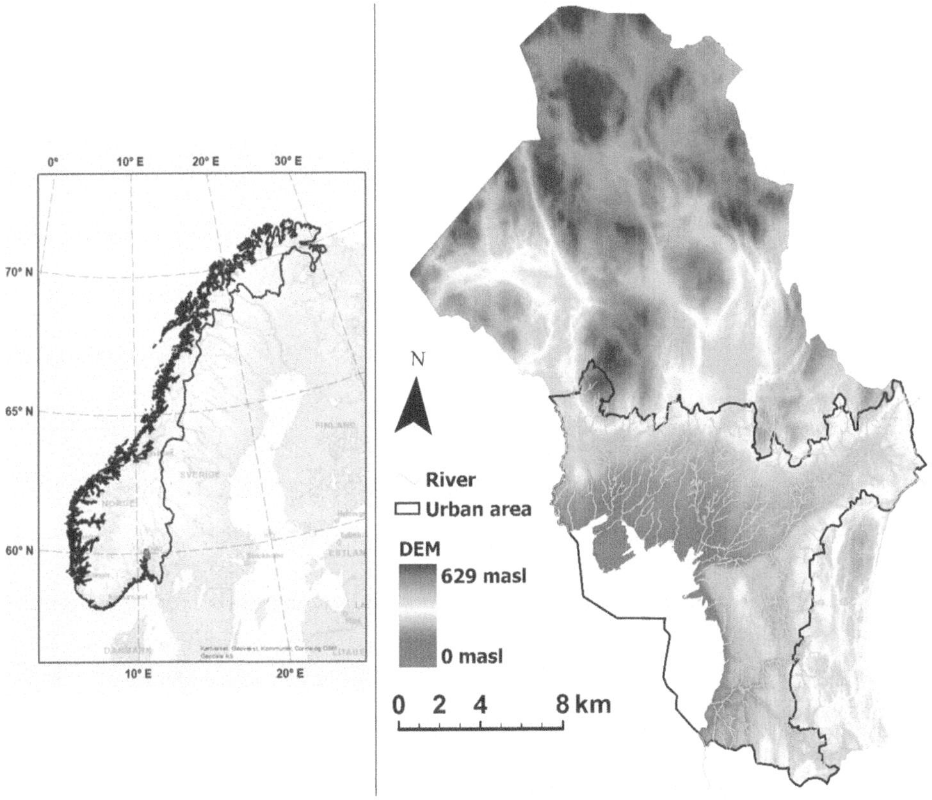

Figure 2 | Location of the Oslo municipality, the terrain, the urban area, and the rivers in the urban area.

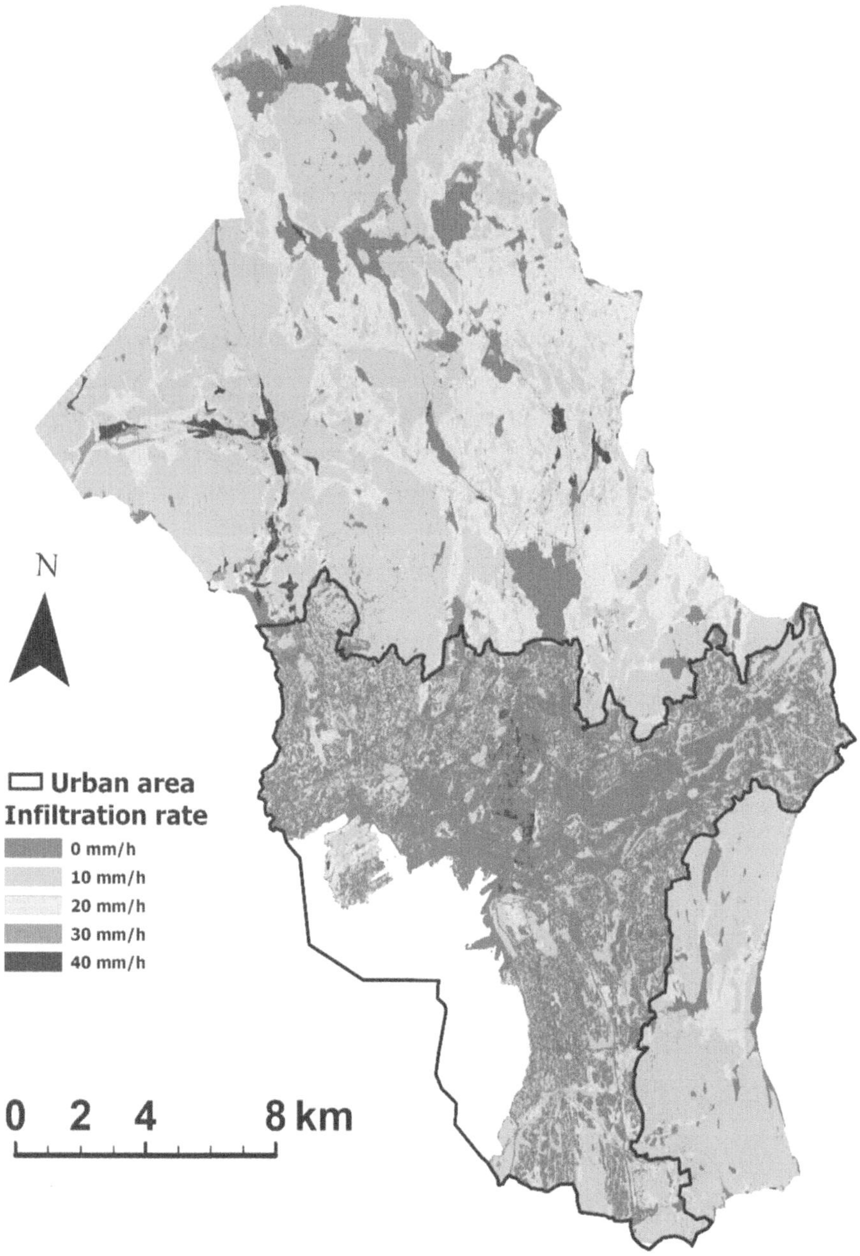

Figure 3 | Infiltration rate used by the SIMWE model. The map is generated based on a descriptive infiltration capacity and an impermeable surface.

Authority 2005). First, the polygons of the infiltrate rate based on NGU and fkb impermeable surface are rasterized into a spatial resolution of 0.5 m. Second, the raster of the Sentinel impermeable surface at a 10-meter grid is resampled by the nearest neighbor to 0.5 m. Lastly, we set the infiltrate rate as 0 mm/h where the surface is impermeable.

Manning's n value is derived from a landuse map from the Norwegian Institute of Bioeconomy Research (NIBIO 2018). There are six types of land surface identified in Oslo, and a large area of the Oslo city area is classified as 'built-up' (Figure 4). The SIMWE model uses Manning's n to calculate the discharge rate and Manning's n value for each landuse type is shown in Table 3.

Table 2 | Infiltration rate based on a descriptive infiltration capacity and impermeable surface

Name	Good	Middle	Little	None/Not classified	Impermeable surface
Infiltration rate (mm/h)	40	30	20	10	0

There are two sources of impermeable surface, i.e., landuse based on Sentinel satellite and the Norwegian common map database.

Experiment design

Due to the limitation in computer memory and time, for the whole of Oslo, we run the SIMWE model at two selected spatial resolutions and three rainfall events. In total, there are six model runs, as summarized in Table 4.

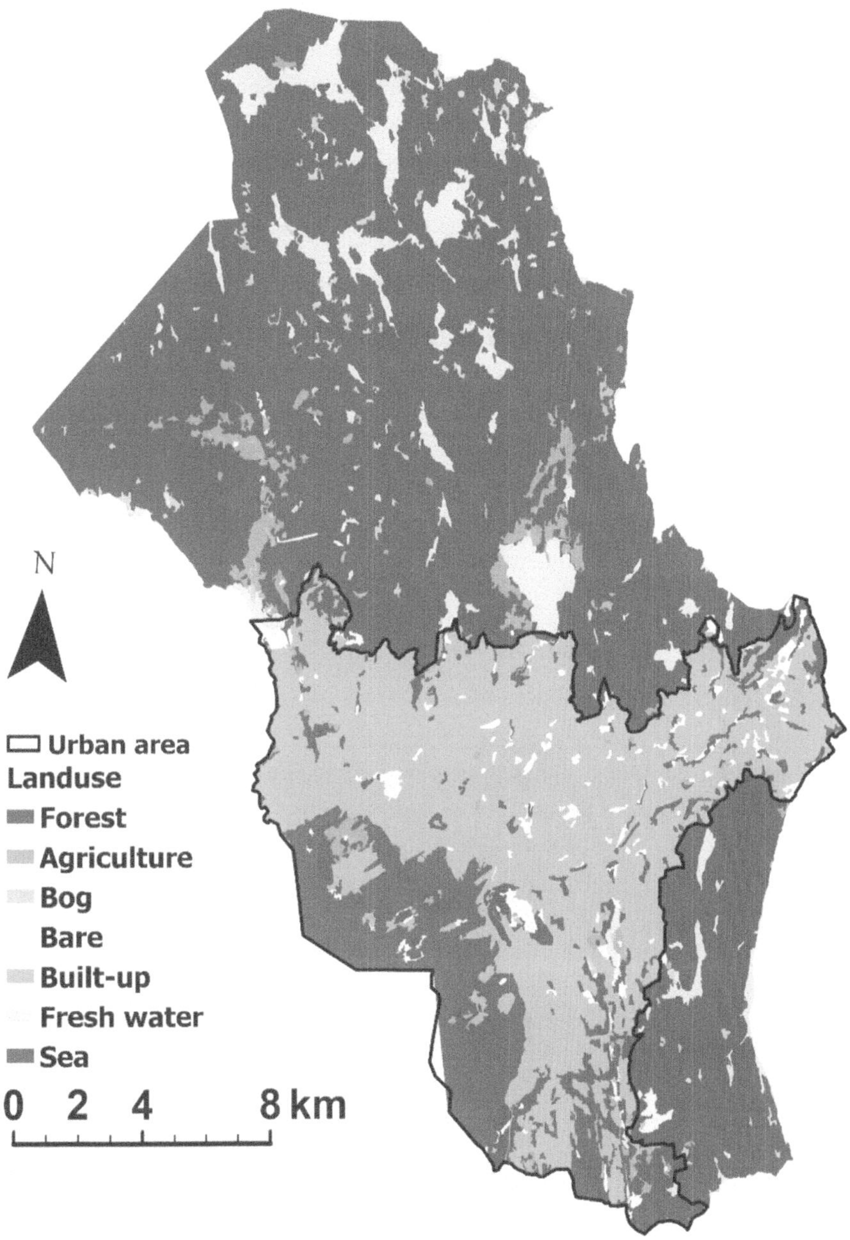

Figure 4 | Map of landuse from NIBIO.

Table 3 | Manning's *n* value for each landuse type in the landuse map (Sigstad 2018)

Class	Description	Manning's *n*
Built area	Rough asphalt	0.016
Agriculture	Mature field crops	0.040
Forest	Heavy stand of timber, few down trees, little undergrowth, flow below branches	0.100
Open area	Cleared land with tree stumps, no sprouts	0.040
Bog	Very weedy reaches, deep pools or floodways with a heavy stand of timber and underbrush	0.100
Water	Clean, winding, some pools, and shoals	0.040

Table 4 | Summary of the model runs for the whole of Oslo

Resolution \ Rain	30 mm/h	50 mm/h	70 mm/h
500 m	run500_30	run500_50	run500_70
20 m	run20_30	run20_50	run20_70

The precipitation intensities of 30 mm/h and 50 mm/h are, respectively, at a return period of 10 years and 200 years at the Blindern station. The Blindern station has the longest rainfall measurements in Oslo and is usually used as a reference station in climatology studies. The precipitation intensity of 70 mm/h is the 200 years' precipitation for the future with climate change. The working flow is summarized in Figure 5.

The model running for the whole of Oslo is used to identify the vulnerable area to flood risk. In addition, we run the SIMWE model at a fine scale, 1 m, at the Grefsen area to examine model behaviors in detail. We select this catchment due to four reasons. First, the sewerage system is relatively simple and there are no pipes to transfer sewage into the catchment. Second, the catchment boundary is well defined. Third, the catchment is relatively steep and flow directions based on terrain are robust. Fourth, the size is good for the model running at a high resolution.

A case study at Grefsen

Grefsen is a residential area in the northern part of Oslo (see Figure 6). The case study area is approximately 1.5 km^2. Most residents live in the eastern part that has a flat terrain. The western part is steep and covered by vegetation. The sewer network is both separated and combined, and

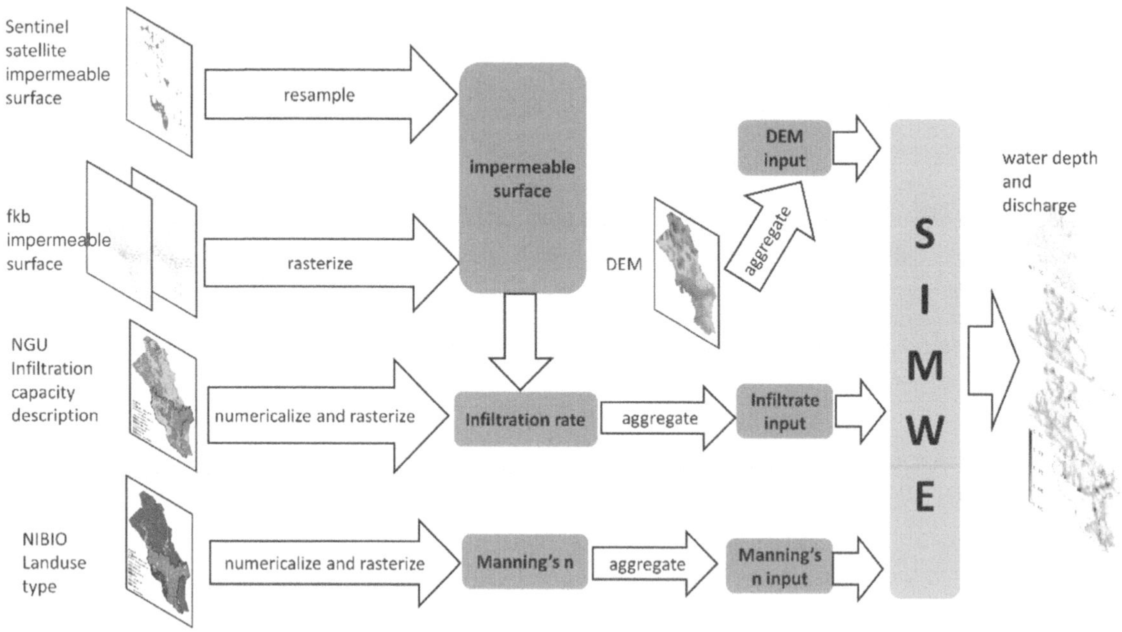

Figure 5 | Workflow of this study.

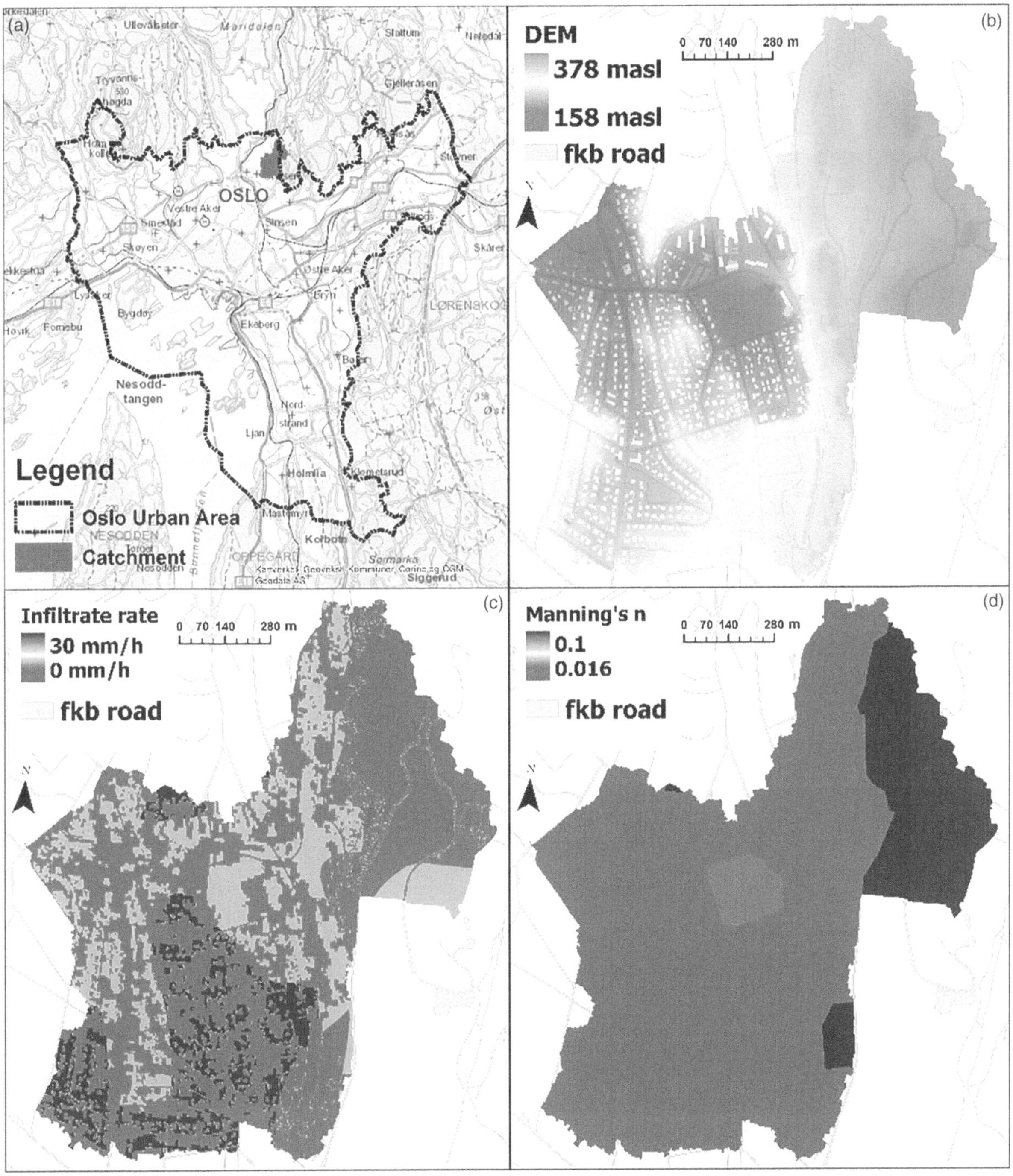

Figure 6 | Grefsen as a case study: (a) location of Grefsen in Oslo, (b) DEM, (c) infiltrate rate, (d) Manning's *n*.

combined sewer overflows occur at the catchment outlet when intensive rainfall events occur. In 2017, combined sewer overflow lasted 5 hours and 23 minutes, and pollution poses a threat to the aquatic life, human health, and groundwater quality in the Akerselva River. Urban hydrological modeling in this catchment is important and meaningful to the water security and environment protection in Oslo.

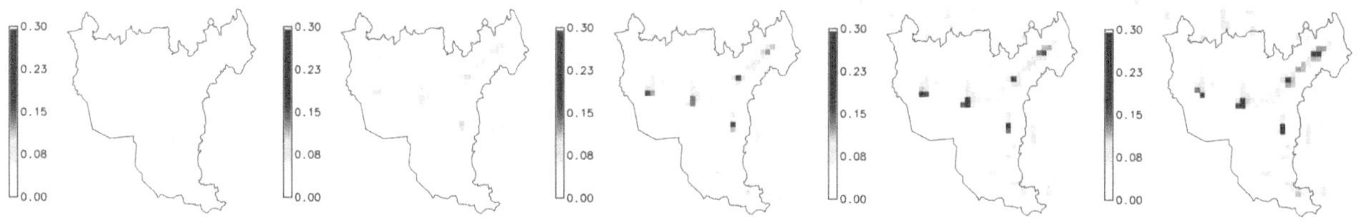

Figure 7 | Maps of inundation area and water depth (m) by the model at 500 m with 50 mm/h rainfall (run500_50) at an elapsed time of 12, 24, 37, 45, and 57 minutes.

RESULTS

Inundation area and water depth

The simulated inundation area and water depth are shown in Figure 7. The rainfall intensity is 50 mm/h at the resolution of 500 m. The figure shows that spatially there is much more water accumulated in the Oslo downtown area than the suburban part. The maximum water depth is 0.3 m after a 1-hour rainfall of 50 mm/h. There are four areas experiencing deep inundation, i.e., Skøyen, Oslo center, Alna, and Østensjø. Among these four areas, Østensjø is a reserved wetland, Alna is along the Fossumbekken/Alna River and Loelva, Skøyen is at the outlet area of three rivers into the fjord. This spatial pattern is the same by the model running at 20 m as shown in Figure 8. The overland flow often appears in low and flat areas with high roughness as well as along the rivers. Therefore, we conclude that the SIMWE model can produce the spatial pattern of overland flow and the inundation area is not sensitive to the spatial resolution.

Unlike the spatial pattern of the inundation area, the absolute value of water depth varies due to the spatial resolution. With the same model inputs, the maximum water depth increases from 0.3 m at the 500 m resolution to more than 3 m at the 20 m resolution, although the number of cells with water deeper than 3 m are very few and sparsely distributed. Figure 9 shows the maximum and mean water depth at the two spatial resolutions. With the higher spatial resolution, the mean of water depth increases whereas the maximum of water depth decreases. The effects of spatial resolution on the water depth come from two aspects. The first is that the SIMWE model uses the first-order partial derivatives of the elevation field to calculate water velocity. At a higher resolution, terrain data have higher derivatives and water flows faster than at coarser resolution. Therefore, the mean water depth will be shallower. The second is that higher resolution can preserve local water ponds better than coarser resolution. The maximum water depth at the local sinks is much larger at the spatial resolution of 20 m than at the spatial resolution of 50 m. The effects of spatial resolution exist for all rainfall events in this study and become more noticeable with time.

Classification of urban flood risks

The classification of flood risk is sensitive to the spatial resolution, as shown in Figure 10. With a small rainfall input (30 mm/h), the high resolution model estimates larger areas in risk than by the coarser resolution model. However, with a high rainfall input (70 mm/h), there are smaller areas in risk estimated by the high resolution model than by the coarser resolution model.

There are larger areas classified as low risk flood and smaller areas classified as high risk by the model results at 500 m than the model results at 20 m. In line with flood warning categories, the red warning level (i.e., water depth more than 43.5 cm) only appears in small

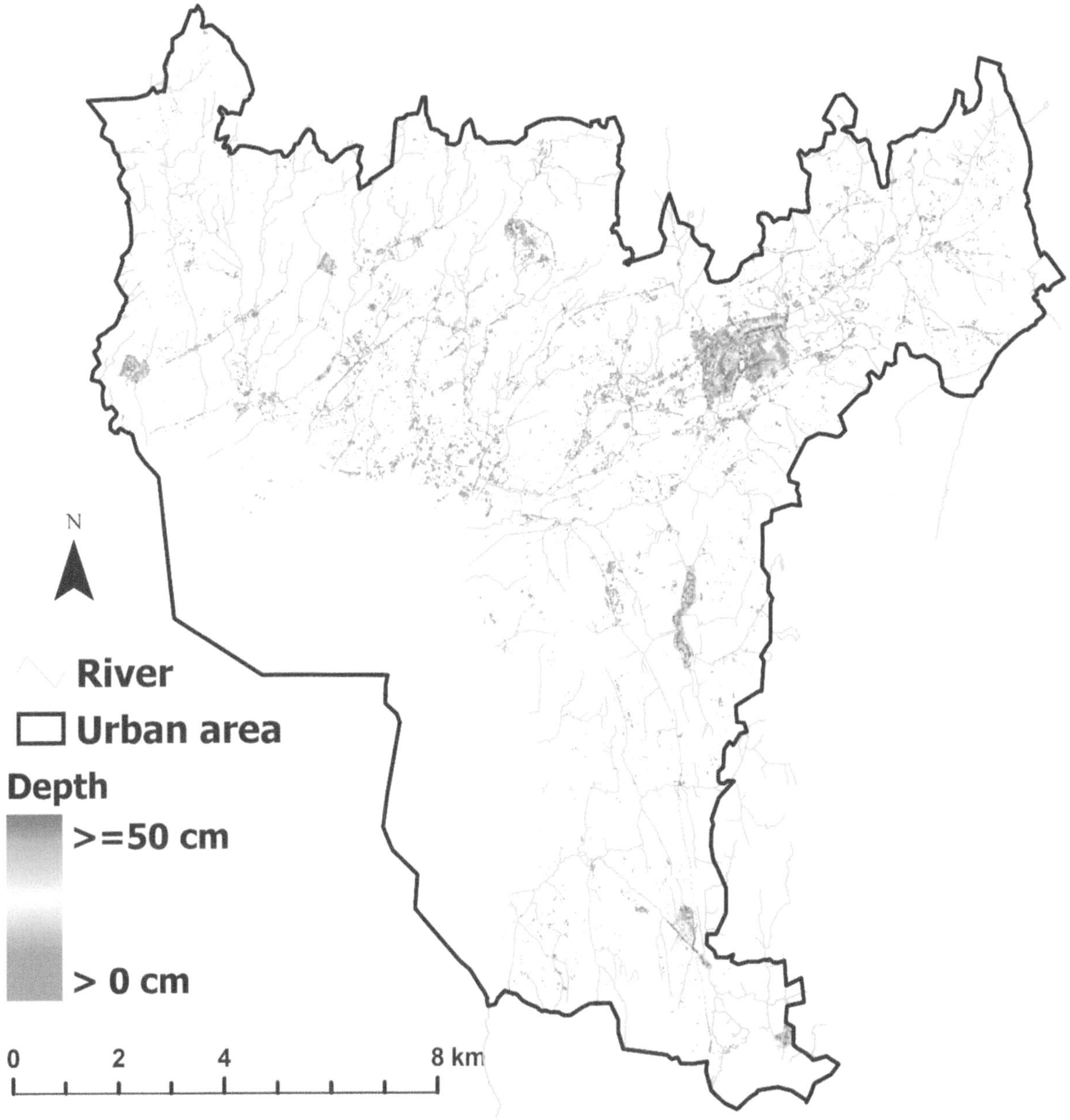

Figure 8 | Map of water depth in the city of Oslo by the model at 20 m with 70 mm/h rainfall (run20_70) at the elapsing time of 1 hour.

areas at 20 m resolution when the rain rate is 70 mm/h, and does not appear at 500 m resolution. The low situation (green warning, water depth between 9.5 and 15.5 cm) appears in all model runs but only appears after 20 minutes' rain at 500 m.

At 20 m resolution for 70 mm/h rain, the area of the low situation reaches its largest extent after 30 minutes' rain and decreases afterwards. At the same time, the area of the challenging situation (yellow warning, water depth between 15.5 and 23.5 cm) and the severe situation (orange warning, water depth between 23.5 and 43.5 cm) increases. This means that the water accumulates in some areas and the low situation changes to a challenging situation.

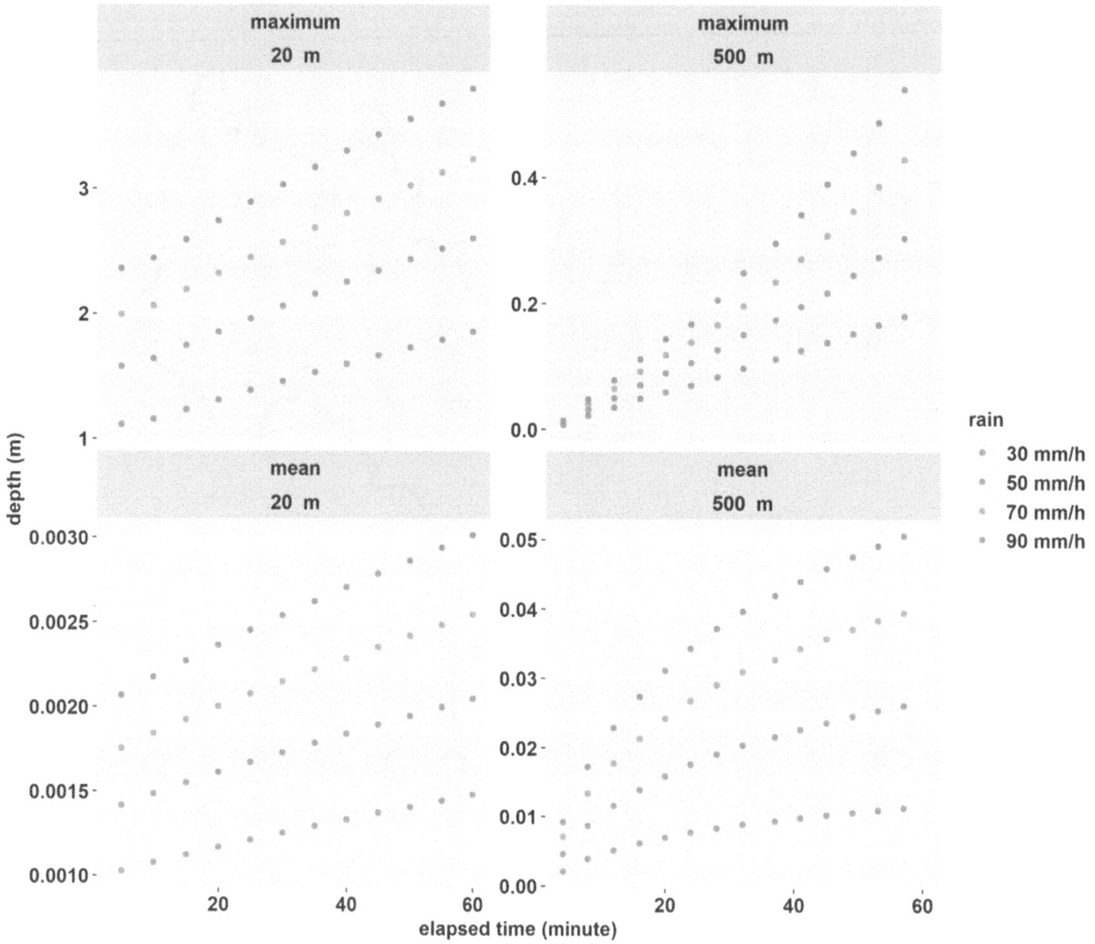

Figure 9 | Mean and maximum of water depth with different rainfall inputs.

A case study at Grefsen

The model results show that roads act as floodways. As shown in Figure 11, both water depth and discharge are high at the side ditches. Additionally, we can find three hotspots in the Grefsen catchment, where there is a relatively large amount of water on the surface. Two of the hotspots are parking lots and one is a football pitch. These three places are relatively low in the surrounding areas and have a low infiltrate rate. To reduce flood risk, first, it is important to keep the floodways open and therefore overland flow can drain quickly. Second, it is possible to make new, or modify, the road networks and other infrastructures to change the flow direction, and consequently to remove the hotspots. Third, it is wise to implement infiltration enhanced nature-based solutions, for example, infiltration trench and bio-retention cells at the location of hotspots.

DISCUSSION

Urban overland flow is usually simulated at a high spatial resolution, from 1 to 5 meters (Hunter *et al.* 2008; Kulkarni *et al.* 2014; Chen *et al.* 2017; Meng *et al.* 2019) due to the complex urban topography. The high resolution model must be able to present the micro-scale topography and blockage effects. However, such a fine scale cannot be applied to a large area due to the limited computation capacity and time. In this study, we utilize the SIMWE model at two

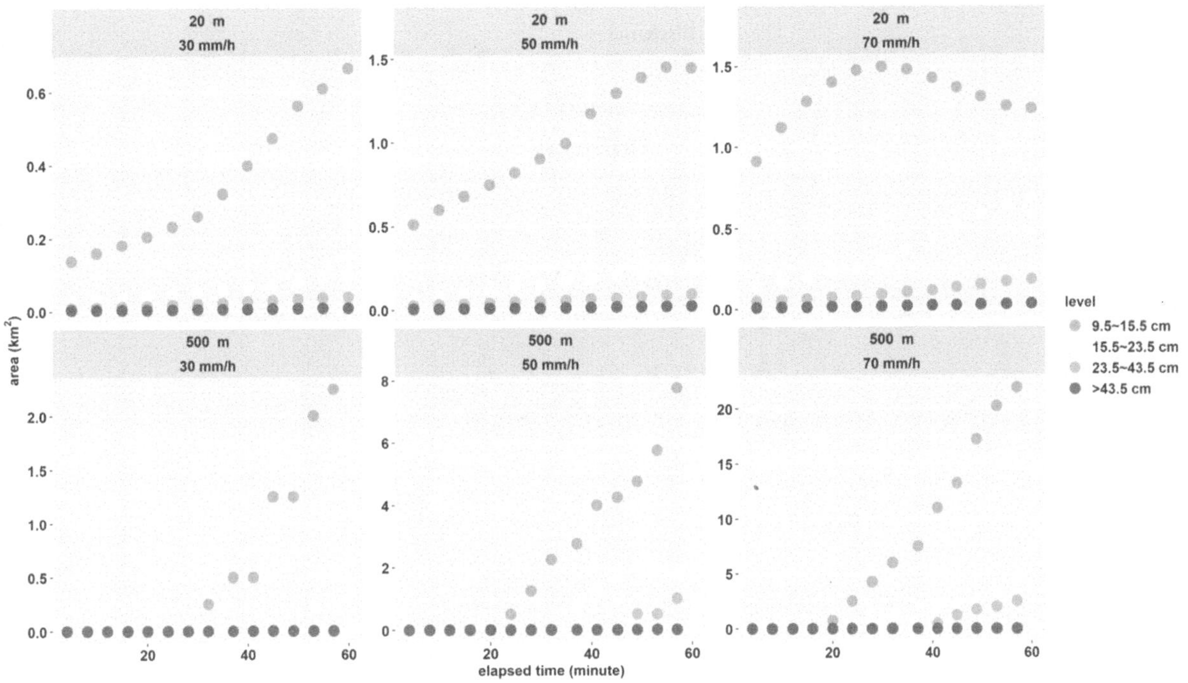

Figure 10 | Area for the flood risk levels.

spatial resolutions, 20 m and 500 m, for the whole of Oslo. The results show that spatial resolution has an impact on the absolute values of water depth and subsequently on the classification of flood risks. However, both spatial resolutions are able to identify the flooding hotspots. This demonstrates that the model results at a coarse resolution can be used to identify flooding hotspots and overflow models with greater detail should be applied in the hotspot areas.

The input data are the main source of uncertainty in urban flood modeling. The input parameters of the SIMWE model are infiltration and surface roughness. They are generated based on maps of impermeable surface and landuse rather than *in situ* measurements. It is worthwhile to note that the SIMWE model is based on overland flow runoff simulation, which is a simplification of complex hydrological processes in urban areas. For example, the infiltration rate changes with soil moisture content, whereas the SIMWE model uses a constant infiltration rate. In such a case, the SIMWE model is more suitable for event design of extreme precipitation when soil and sewage are already saturated.

CONCLUSIONS

Urban flooding is becoming a hot research topic due to the growth of cities and the increase in frequency and magnitude of extreme rainfall events. However, current stormwater management models are too complex to set up and apply on a large scale. In this study, we test a physically based and spatially distributed overland flow model, SIMWE, which is easy to set up and to implement in practice. We drive the SIMWE model at two spatial resolutions with three design rainfall events for the whole of Oslo and at a high spatial resolution at the Grefsen area. The results show that the SIMWE model has high skills in simulating urban overland flood for rainfall events at both coarse and high resolutions. The model at a 20 m spatial resolution estimates a deeper water depth than at a 500 m spatial resolution and this has a strong impact on the classification of flood risks with different rainfall inputs. With a small rainfall input (30 mm/h), there are larger areas in risk estimated by the high resolution model than by the coarse resolution model. However, with a high rainfall

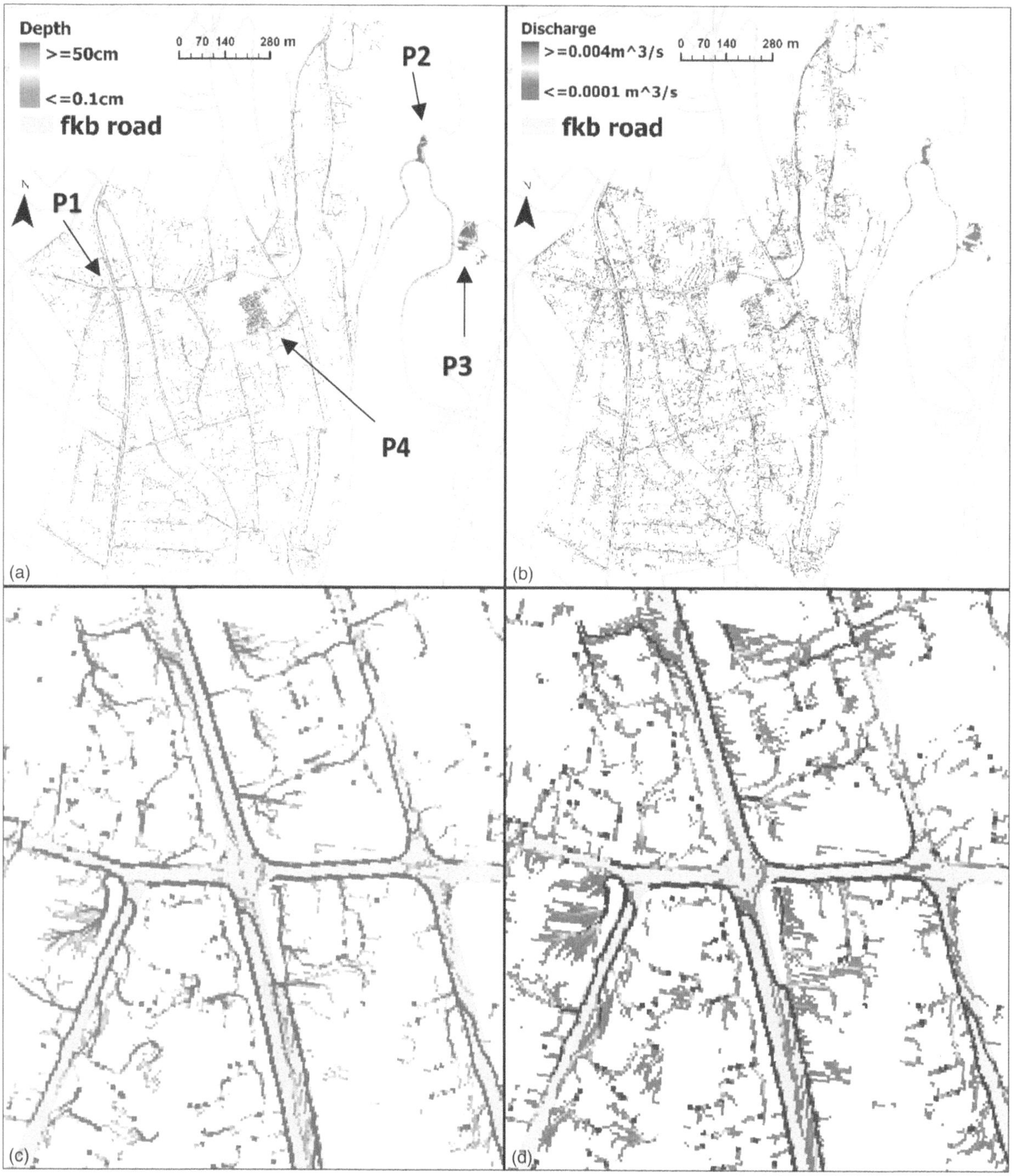

Figure 11 | Depth and discharge at Grefsen (resolution = 1 m, rain = 50 mm/h, time = 60 min). (a) Water depth, (b) discharge, (c) depth at P1 in (a), and (d) discharge at P1. P2, P3, and P4 are hotspots. P2 and P3 are parking lots and P4 is a football pitch.

input (70 mm/h), the high resolution model estimates smaller areas at risk than the coarse resolution model. The case study in the Grefsen area demonstrates that roads are natural floodways. Identification of hotspots provides guidance for implementing flood risk mitigation infrastructure.

ACKNOWLEDGEMENTS

We thank the Norwegian research council for funding through the project New Water Ways: 270742 and data providers. We thank the reviewers and the editor, and their comments and suggestions improve the quality of this study.

REFERENCES

BLOM 2014 *LiDAR-repport*. Oslo, Norway.

Chen, W., Huang, G. & Zhang, H. 2017 Urban stormwater inundation simulation based on SWMM and diffusive overland-flow model. *Water Science and Technology* **76** (12), 3392–3403.

DHI 2018 *Our Products – Materials*. Available from: http://www.polycor.com/eng/products/ (accessed 7 December 2018).

Donaldson, S. 1968 *Effects of Urbanization*. Available from: www.unce.unr.edu (accessed 6 December 2018).

Florida, R. 2011 *Why Cities Matter*. Available from: https://www.citylab.com/design/2011/09/why-cities-matter/123/ (accessed 6 December 2018).

Hall, J. W., Sayers, P. B. & Dawson, R. J. 2005 National-scale assessment of current and future flood risk in England and Wales. *Natural Hazards* **36** (1–2), 147–164. Available from: http://link.springer.com/10.1007/s11069-004-4546-7 (accessed December 7, 2018).

Hanssen-Bauer, I., Førland, E. J., Haddeland, I., Hisdal, H., Mayer, S., Nesje, A., Nilsen, J. E. Ø., Sandven, S., Sandø, A. B., Sorteberg, A. & Ådlandsvik, B. 2015 *Climate in Norway 2100 – a knowledge base for climate adaption*. Available from: www.miljodirektoratet.no/20804 (accessed 7 December 2018).

Hofierka, J. & Knutová, M. 2015 Simulating spatial aspects of a flash flood using the Monte Carlo method and GRASS GIS: a case study of the Malá Svinka Basin (Slovakia). *Open Geosciences* **7**, 118–125.

Hunter, N. M., Bates, P. D., Neelz, S., Pender, G., Villanueva, I., Wright, N. G., Liang, D., Falconer, R. A., Lin, B., Waller, S., Crossley, A. J. & Mason, D. C. 2008 Benchmarking 2D hydraulic models for urban flooding. *Proceedings of the Institution of Civil Engineers: Water Management* **161** (1), 13–30.

Krøgli, I. K., Devoli, G., Colleuille, H., Boje, S., Sund, M. & Engen, I. K. 2018 The Norwegian forecasting and warning service for rainfall-and snowmelt-induced landslides. *Natural Hazards and Earth System Sciences* **18** (5), 1427–1450.

Kulkarni, A. T., Mohanty, J., Eldho, T. I., Rao, E. P. & Mohan, B. K. 2014 A web GIS-based integrated flood assessment modeling tool for coastal urban watersheds. *Computers and Geosciences* **64**, 7–14.

Maddox, I. 2014 *Three Common Types of Flood Explained*. Intermap. Available from: http://www.intermap.com/risks-of-hazard-blog/three-common-types-of-flood-explained (accessed 7 December 2018).

Meng, X., Zhang, M., Wen, J., Du, S., Xu, H., Wang, L. & Yang, Y. 2019 A simple GIS-based model for urban rainstorm inundation simulation. *Sustainability* **11** (10), 1–19.

Mitas, L. & Mitasova, H. 1998 Distributed soil erosion simulation for effective erosion prevention. *Water Resources Research* **34** (3), 505–516.

NGU 2018 *Datasets og Download | Geological Survey of Norway*. Available from: https://www.ngu.no/emne/datasett-og-nedlasting (accessed 10 December 2018).

NIBIO 2018 *Documention av AR50 - Nibio*. Available from: https://www.nibio.no/tjenester/nedlasting-av-kartdata/dokumentasjon/ar50 (accessed 10 December 2018).

Norwegian Mapping Authority 2005 *Common Maps Database (FKB)*. Available from: https://kartkatalog.geonorge.no/metadata/geovekst/felles-kartdatabase-fkb/0e90ca71-6a02-4036-bd94-f219fe64645f (accessed 10 December 2018).

NOU 2015 *Stormwater in cities and towns as a problem and resource*. Available from: https://.fagbokforlaget.no/offpub (accessed 7 December 2018).

Sigstad, H. 2018 *Flood Modelling by HEC-RAS: An Examinenation of Accuracy to Steady Flood Modelling by HEC-RAS With Two Different GIS Representations*. Available from: https://brage.bibsys.no/xmlui/handle/11250/2507021.

SSB 2018a *Population and land area in urban settlements – SSB*. Available from: https://www.ssb.no/en/befolkning/statistikker/beftett/aar (accessed 6 April 2019).

SSB 2018b Population projections. 24(2), 346–360. Available from: http://epp.eurostat.ec.europa.eu/statistics_explained/index.php/Population_projections (accessed 8 December 2018).

Stange, E. 2017 Land cover from Sentinel 2 - Bærum, Lørenskog, Nittedal, Oppegård, Oslo, Sked … Available from: http://urban.nina.no/layers/geonode%3As2_oslo_metro

United Nations Population Division 2018 *Urban Population (% of Total) | Data*. Available from: https://data.worldbank.org/indicator/SP.URB.TOTL.IN.ZS (accessed 6 December 2018).

USGS 2018 *Urbanization and the Hydrologic System*. Available from: https://water.usgs.gov/edu/urbaneffects.html (accessed 6 December 2018).

VAV 2016 *Main report – Strategy for Stormwater Management in the Oslo Municipality*. Available from: https://www.oslo.kommune.no/getfile.php/13174529/Innhold/Politikk og administrasjon/Slik bygger vi Oslo/Vannområde Oslo/Rapporter og planer/2016 Handlingsplan for overvannshåndtering i Oslo kommune – Hovedrapport.pdf (accessed 7 December 2018).

WHO 2015 *Urban Population Growth*. World Health Organization, Geneva. Available from: https://www.who.int/

gho/urban_health/situation_trends/urban_population_growth_text/en/ (accessed 6 December 2018).

Wikipedia. 2018 Oslo. Available from: https://en.wikipedia.org/wiki/Oslo (accessed 7 December 2018).

Yin, J., Yu, D., Yin, Z., Wang, J. & Xu, S. 2015 Modeling the anthropogenic impacts on fluvial flood risks in a coastal mega-city: a scenario-based case study in Shanghai, China. *Landscape and Urban Planning* **136**, 144–155.

First received 17 May 2019; accepted in revised form 27 December 2019. Available online 20 March 2020

IWA Publishing's authorised EU representative for General Product Safety Regulations is Diane D'Arras, 15 rue Duret, 75116 Paris, France, e-mail: safety@iwap.co.uk.

Printed and bound by CPI Group (UK) Ltd, Croydon, CR0 4YY
12/05/2026
02108868-0002